EMBRYOLOGY OF REPTILES

EMBRYOLOGY OF REPTILES

By

A.P. DIWAN

N.K. DHAKAD

ANMOL PUBLICATIONS PVT. LTD.

NEW DELHI-110 002 (INDIA)

ANMOL PUBLICATIONS PVT. LTD.
H.O.: 4374/4B, Ansari Road, Darya Ganj,
New Delhi-110 002 (India)
Ph.: 23278000, 23261597
B.O.: No. 1015, Ist Main Road, BSK IIIrd Stage
IIIrd Phase, IIIrd Block
Bangalore - 560 085 (India)
Visit us at: www.anmolpublications.com

Embryology of Reptiles

First Published, 1995
ISBN 81-7488-102-6

PRINTED IN INDIA

Printed at Mehra Offset Press, Delhi.

Preface

Embryology is a branch of biology which has a most immediate bearing on the problem of life. It has become an immense field whose confines are difficult to delineate. The focus has been and remain on the embryos, on the gradual emergence of form and structure from what appears to be a very modest beginning. The essence of embryonic development is change transition from one stage to another. Embryo is a fleeting stage, a continum along the axis of time. Embryology is a bank division of biology. It stems the anatomy, the histology and the physiology of adult. To understand it well is to aid in the comprehension of the other biological disciplines.

Though, the authors have taken special care to present a current account, yet they are fully aware about their limitations, and the readers may come across the mistakes of various types. For all types of mistakes they extend their due apology.

The authors have freely consulted other standard books and research papers while preparing the manuscript, so the authors claim no originality of the work.

The authors express their thanks to their friends and colleagues whose continuous inspirations have invited them to bring out this text.

The authors are also thankful to Shri J.L Kumar of M/s Anmol Publications Pvt. Ltd. for bringing out a handsome print of this set in a comparatively short time.

Constructive criticisms and suggestions for improvement of the book will thankfully acknowledged.

Authors

Contents

1

The Urinogenital Organs

In vertebrates both excretory and reproductive systems are intimately connected that is why collectively called *urinogenital system*. The excretory components evolved from segmentally or repetitively arranged and uniform tubular structures, which are recognizable in embryogeny. Others such as genital elements, certainly in higher forms, do not show obvious signs of segmental derivation. In anamniote groups it is likely that certain components have not yet developed, i.e., a metanephros and its duct or a corpus luteum. In amniotes the pronephros in most cases is vetigial and physiologically non-functional, even in the embryo and it soons disappears early in development. Some structures show profound changes in function, as well as in structure, during ontogeny and phylogeny. Note for instance the mesonephors and its duct of the male amniote and the specific regions of the oviduct; the later, once a single uniform duct, becomes regionally specialized in structure and function for different reproductive processes. New components have developed during the course of phylogeny, like the urinary bladder or the oviducal placenta, though this organ, like the corpus luteum, may well be considered a variable in the urinogenital system for neither develops in non-reproductive females.

In general the urinogenital system is more complex and morphologically and functionally integrated in higher vertebrates, especially in males. To some extent the reptilian urinogenital system provides an interesting and important link between anamniote fishes and amphibians on the one hand, and amniote birds and mammals on the other. It is indeed noteworthy that many of our fundamental concepts of the origin and development of the vertebrate urinogenital system are based upon research on reptiles.

The Pronephros

As stated already that the excretory tubules are derived from enbryonic mesoderm. The origin and development of various aspects of the reptilian kidney system have been studied to a lesser or greater degree in lizards. The literature relating to the structure of the adult kidney, or derivatives of the different components of the embryonic structures, or of those specialized regions of the metanephros such as the sex segment in males, will be given in the appropriate sections. Many pioneers of reptilian renal embryology either did not mention a pronephros, or could not with certainty distinguish it as a separate regions from the mesonephros. Nevertheless, von Mihalkovics (1885) described three to four pairs of anterior segmental vesicles in *Lacerta agilis* embryos originating from somites. These vesicles at first communicated with the coelom and the protovertebral cavity (Myocoele), while the more posterior vesicles did not. Probably these anterior vesicles were transient pronephric elements. Such elements leading into the open pronephric duct were later on likewise recognized in embryos of *Lacerta Phrynocephalus* crocodiles and tortles and *Sphenodon*. A vestigial pronephros was described in an embryo alligator and *Hoffmann* (1889) and *Wiedersheim* (1890a, b) also reported the early degeneration of the pronephros in lizard, snake crocodilian, and turtle embryos.

The pronephros of *Trionyx* embryos arise behind the auditory region as segmental outgrowths of somites 4 to 10 inclusive. The mesonephros encroached on the posterior pronephric region so that a common area in a single segment supposedly included pronephric and mesonephric tubules, opening by a common funnel into the coelom at one end and into the collecting duct at the other. *Kerens* (1907) denied the coexistence of pronephric and mesonephric tubules in the

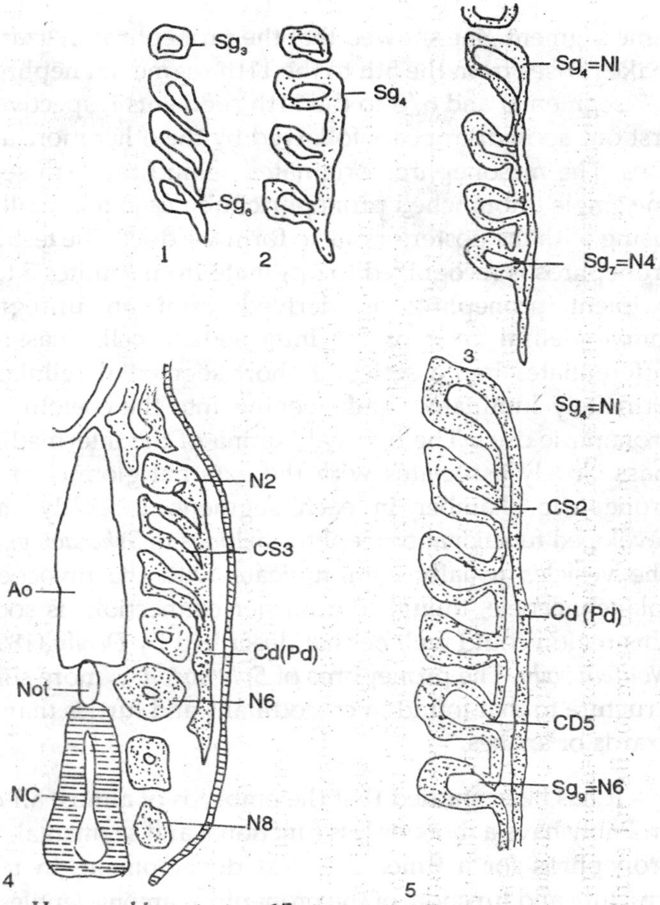

Fig. 1.1. *Hypogeophis rostratus,* 15 segment stage.
Fig. 1.2. *Grandisonia alternans.* 12 segment stage.
Fig. 1.3. *Grandisonia alternans.* 16 segment stage.
Fig. 1.4. Lacerta muralis 15 somite stage.
Fig. 1.5. *Hypogeophis rostratus.* 27 segment stage.

These diagrammatic illustrations show the similarity between the Gymnophiona (Amphibia) and the Reptilia in the primary development of the segmental nephrotomial units and the pronephric duct. The origin of the latter by fusion of the anterior pronephrotomal extensions and thence its independent posterior growth, may well be an ancestral primitive process, possibly discarded in the higher amniotes. Ao, Aorta; Cd, collecting duct; CS 1-5, segmental canals; N 1-8, nephrotomes; Nc, nerve cord, not, notochord; Pd, pronephric duct ; Sg 3-9, body segments 3 to 9.

same segment. She showed that the pronephros in lizards and snakes arises from the 5th to the 11th segmental nephrotomes (6-7 segments) and 6/8 to the 11th segments respectively; the first one soon disappears followed by the other more anterior ones. The mesonephros originates behind the 11th segment. The single unbranched pronephric tubules join laterally, each fusing with the posterior one to form the duct. The testudinian pronephros is recognized to originate from somites 5 to 9; the incipient pronephros is derived from an unsegmented somatopleural crest of the intermediate cells mass, which differentiates into a series of short segmental cellular cords ultimately luminated, and opening into the coelom and the pronephric duct. The internal lamina of the intermediate cell mass closely associates with the external glomus to form a pronephric chamber in each segment. Probably the best developed functional pronephros belongs to *Sphendon punctatus*. The vesicles usually communicate with the myocoele and splanchnocoele, though the former connection is soon lost. This region could well be that described by *Dendy* (1899) as a *Wolffian body*. The pronephros of *Sphenodon* is more similar in structure to that found in crocodilians and turtles than that of lizards or snakes.

It has been studied that the embryos of almost all reptiles probably have a more or less functional and segmental, tubular pronephros for a time. The best development, in terms of structure and function, of the pronephros among reptiles would seem to occur in *Sphenodon*, crocodilians and to a lesser extent, testudines and the least development in lizards and snakes, which have no peritoneal funnels or glomerular tissue and degenerate precociously. In all cases the reptilian pronephros is evanescent, less well-developed in structure and longevity than in anamniotes, but superior in structure to the non-functional vestigial pronephros of birds and mammals.

In vertebrates an external *glomus* (probably fused glomeruli) arises from the dorsal aorta, as an outpushing of coelomic epithelium, and .hangs freely in the splanchnocoele mesial to

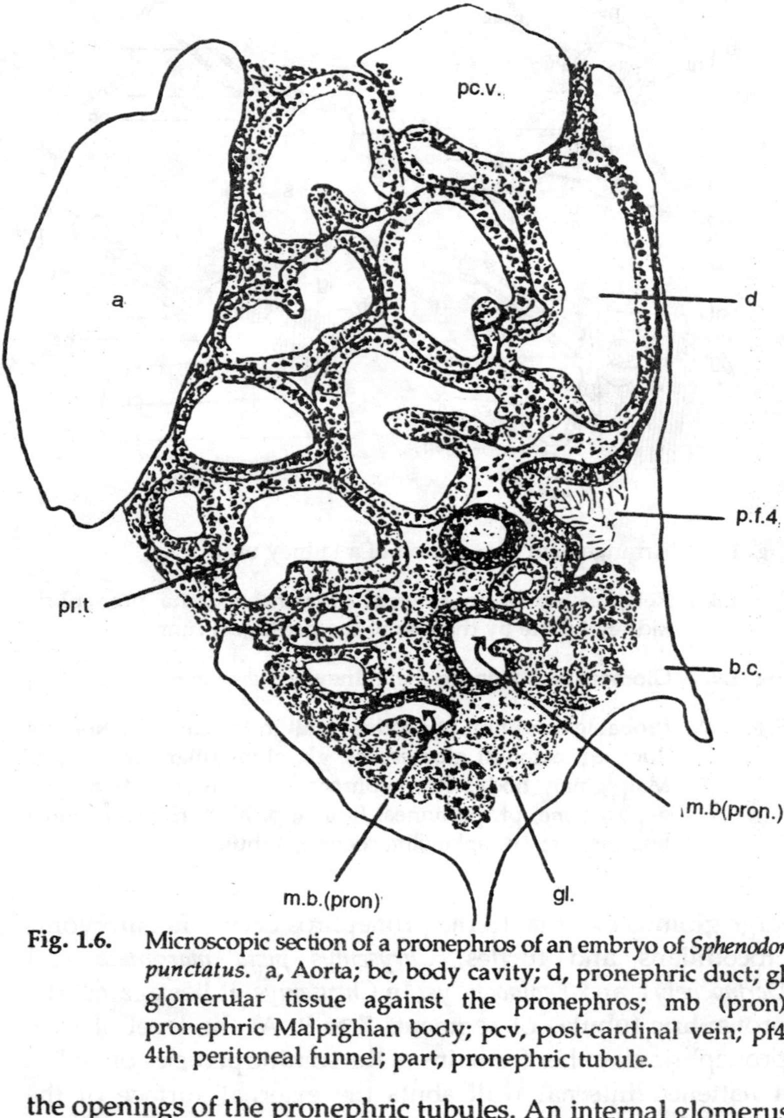

Fig. 1.6. Microscopic section of a pronephros of an embryo of *Sphenodon punctatus*. a, Aorta; bc, body cavity; d, pronephric duct; gl, glomerular tissue against the pronephros; mb (pron), pronephric Malpighian body; pcv, post-cardinal vein; pf4, 4th. peritoneal funnel; part, pronephric tubule.

the openings of the pronephric tubules. An internal glomerus is enclosed within a pronephric (or mesonephric) chamber derived from a nephrotome and its nephrocoele. A large A

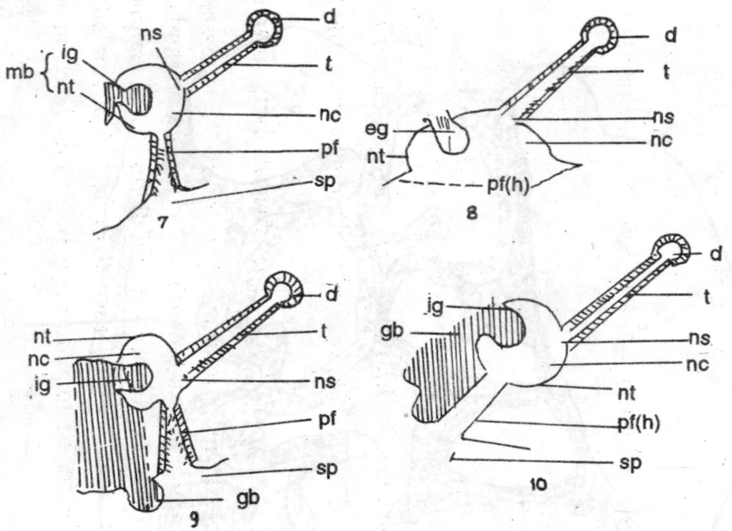

Fig. 1.7. Arrangement of a section of a kidney unit.

Fig. 1.8. Relationship of an external glomerulus to a pronephric kidney tubule by the loss of the peritoneal funnel.

Fig. 1.9. Glomerular border in the kidney of *Sphenodon*.

Fig. 1.10. Probable condition of a kidney unit in the chick. d; Nephric duct; eg. external glomerulus; gb, glomerular border; mb, Malpighian body; nc, nephrocoel; ns, nephrostome; nt, nephrotome; pf, peritoneal funnel; pf(h), peritoneal funnel homologue; sp, splanchnocoele; t, tubule.

large glomus external to the pronephros occurs in embryonic crocodilians and turtles, *Chrysemys picta marginata* and *Dermochelys* and *Lepidochelys*. In *Chrysemys* it lies against the pronephric tubules of segments 7 to 9. At the level of each pronephric chamber (remains of the somitic pedicles or stalks; a flattened internal wall abuts the external surface of the glomerulus to form a kind of Malpighian corpuscle. The glomus shows indications of segmental origin, which have not yet been totally incorporated into the pronephric chambers. In an intermediate pronephric-mesonephric zone, at the 10th segment, the glomus is partially enclosed by a chamber. An unenclosed external portion remains within the coelom, thus demonstrating glomerular gradiation from an external to internal type. This feature occurs in *Sphenodon.*

There is a large, highly vascular, continuous glomus (of fused glomeruli. In embryos of *Sphenodon punctatus* situated mesial to the pronephros. Pronephric chambers become Bowman's capsules, which lead into the splanchnocoele via ciliated peritoneal funnels and into the tubules via nephrostomes. The glomus and pronephric chamber may each be termed to form the Malpighian corpuscle. The pronephros (and its Malpighian corpuscle) eventually degenerates, together with the front of the mesonephros. *Sphenodon* has an internal glomerulus projecting into the Bowman's capsule, which further behind extends along the ventral border of the nephros, within the splanchnocoele, as a large vascular mass or glomerular border *Fisk* and *Tribe* believed that it secreted coelomic fluid. Similar glomerular tissue commonly occurs in other reptiles and it is termed the external glomerulus by more recent workers. *Fisk* and *Tribe* suggested that this tissue may be either the remains of internal glomerular tissue too extensive to be incorporated within Bowman's capsules, or a primary glomerular border which gave origin to the internal glomeruli, or again it arises separate from the glomeruli and these unite with it later in ontogeny.

The Mesonephros

The mesonephros originates from nephrotomes 11-23, which begin immediately posterior to the pronephros and extend along the embryo to a short distance anterior to the cloaca. A strand of unsegmented nephrogenic tissue separates from the somitic and lateral plate mesoderm but stays in continuity with the hindmost pronephric nephrotome, though it develops sometime after the formation of the pronephric duct. The series of primary mesonephric tubules is provided with Malpighian corpuscles but lacks open coelomic peritoneal funnels. Secondary and later tertiary and quarternary tubules originate from the nephrogenous tissue dorsal to the primary tubules; these open either into the primary tubules or directly into the Wolffian duct. More than one mesonephric tubule occur in each segment, as in turtles snakes, lizards and crocodilians. In lizards additional tubules develop only in the more posterior segments. The origin of additional tubules from extra-nephrogenous tissue agrees with the results of *von Mihalkovics* (1885), *Gregory* (1900) and *Schreiner* (1902), but does not support the results of *Braun* (1877), *Furbringer* (1878) and *Hoffmann* (1889). These authors claimed that they arose bydivision from primary tubules. This feature is probably not of prime importance.

It has been clearly stated that in the reptilian embryo the mesonephros, together with the pronephros, originates from the non-segmented, nephrogenous, intermediate cell mass and only secondarily segments to form repetitively arranged vesicles that differentiate further into S-shaped tubules. In terms of phylogenetic derivation, however, it is likely that the non-segmented nephrogenous tissue evolved from originally segmented nephrotomes. It is doubtful whether reptilian mesonephric tubules ever open into the coelom during their embryogeny; in contrast the pronephric tubules do so. The mesonephric tubules lack openings in *Trionyx, Chrysemys* alligators crocodiles and *Sphenodon*. Mesonephric tubules branch, have internal Malpighian corpuscles and are functional right through embryogeny until after hatching. The mesonephros finally transforms into epididymal structures in the male or disappears in the female. The histology and functional aspects of the embryonic mesonephroi of *Calotes versicolor* were investigated by *Malherkar* and *Joshi* (1974). Apart from the compactness and more numerous collecting ducts of the metanephros, the functional (renal) mesonephros and metanephros are generally similar histologically. *Alkaline phosphatase* was registered only in the brush border of the proximal mesonephric tubules, a similar location to that found in the mesonephroi of the chick and mammals.

Wolffian Duct

The pronephric duct which opens into the doaca becomes the evacuation duca of the mesonephros or mesonephric duct. One major problem in classical comparative renal embryogeny centred upon the germ layer derivation. Nowadays the germ layer theory is no longer "fashionable" nor indeed valid, but it did play an important part in early embryological research.

The pronephric duct was believed to arise from ectoderm in embryos of *Lacerta viridis, L. muralis* and *L. agilis, Anolis sagret* and *Sphaerodactylus notatus, Phrynocephalus* and *Trionyx sinensis* and *Mauremys japonica. Wiedersheim* (1890a,b) investigating crocodilians and turtles could not decide whether the duct arose from ectoderm or mesoderm. A mesodermal origin was supported by *von Mihalkovics* (1885) and *Schreiner* (1902). *Hoffimann (1889,* in *Lacerta agilis)* and Strahl (1886) did not find ectoderm participating in the formation of the duct. *Weldon* (1883, in *L. muralis*) described the duct as split off from the intervertebral region and grown posteriorly as a solid projection. *Dendy* (1999, in *Sphenodon*) generally agreed with the latter description. In *Trionyx* the duct was siad to form from the lateraltips of the pronephric tubules; the caudal end grew posteriorly by division of its own cells in some relation to the ectoderm *Kerens* (1907) on the whole agreed with Gregory, except that the duct remained free of ectoderm, the developmental process being identical with that of *Hypogeophis. Reese* derived the whole urinary system of alligators from mesoderm in *Chrysemys, Burlend* (1913) described the duct as originating from lateral plate mesoderm ; a groove grew caudally and was reinforced by mesoderm nipped off from the nephrotomes along the route. *De Walsche* (1929), likewise in *Chrysemyt,* found distal extremities of the pronephric tubules to join with a collecting tract (which crystallizes out in some way from pronephric crest) and the tip grew posteriorly by terminal proliferation, a process generally similarly that described by *Kerens* (1907). In *Sphenodon, Tribe* and *Fisk* (1940) showed the peripheral regions of the mesodermal pronephric tubules to grow posteriorly, develop a lumen, receive openings of the developing mesonephric tubules and finally open into the cloaca. No contribution from the surrounding tissues was observed. Recently *Pieau* (1965, in *Anguis fragilis*) described mesodermal nephric vesicles as developing lateral diverticulae, which fuse with their posterior neighbours to form an elongate cellular cord.

Thus, we see that the reptilian mesonephric duct arises from mesoderm as do those of other vertebrates. One may hazard a guess that it arises from an anterior mesodermal blastema, having a close relationship with the pronephros; certainly the duct and pronephric Anlagen are closely apposed in amphibians. The classical description of the origin of the duct, from anterior pronephric tubules in *Lacerta* may be a special; albeit primitive, condition and atypical, though the duct of turtles was described as originating in the same way. The duct probably grows-posteriorly towards the cloaca and receives mesonephric tubules as they form. Presumably inductive phenomena occur in reptiles as in amphibians and birds. The caudal tip of the Wolffian duct at first ends at the ventrolateral surface of the cloaca. Gradually, it is displaced dorsally, only slightly in *Testudo graeca* when it still finally opens cranio-ventrally, but in *Lacerta viridis, Anguis fragilis* and *Natrix tessellata* both ducts finally open on to the dorsal surface.

The Metanephros

It is the adult functional kidney of amniotes. It arises and develops similarly in all classes. Much of the classical work on reptiles and birds helped to establish the modern view of the duality of kidney origin.

These authors believed that the entire metanephros arose from branches of the ureter. *Braun* (1877) first described the origin of the reptilian ureter (and metanephros) from the upper surface of the *Wolffian duct*, a result which confirmed previous work by *Kupfer*. *Furbringer* (1878) thought it probable that the metanephros of *Anguis* and *Lacerta* embryos arose from two Anlagen, uréter and canal systems; the latter arose independently in the stroma above the mesonephros.

In a 20 mm embryo of *Lacerta* Anlagen of metanephric tubules show several S-shaped tubules in each segment, though these are still separated from the metanephric duct. In a 28 mm embryo the tubules are continuous with the ureter, which now opens separately into the cloaca near the *Wolffian* duct. The mesonephros and metanephros develop from the same intermediate cell mass and have the same ontogenetic and morphological values. The independent origins of the duct and tubules were clearly recognized by *Schreiner* (1902), who was strongly supported by *Kerens* (1907) from work on embryos of lizards, snakes and *Chrysemys*.

In crocodilians and turtles a continuous and similar *"Mutterboden"* of coelomic epithelium gives origin to the mesonephros and metanephros. In 10-12 mm crocodilian embryos the pronephric-mesonephric duct has opened into the cloaca and the metanephros includes convoluted tubules and glomeruli as does the mesonephros. A metanephric Anlage, dorsal to the mesonephros, but still without glomeruli, occurs in *Alligator* embryos of 9-10 mm. In larger embryos (14-18 mm) glomeruli are seen and the ureters join the *Wolffian ducts* before the latter open into the cloaca. *Tribe* and *Fisk's* (1940) descriptions of *Sphenodon* confirm earlier studies of reptiles. After degeneration of the pronephros and the anterior region of the mesonephros, a few outgrowths from the uretic bud appear in the metanephrogenic tissue, situated behind the mesonephroi. The metanephros eventually overgrows the hind region of the mesonephros and numerous metanephric collecting ducts soon open into the ureter. The still separated metanephric tubules develop Malpighian corpuscles and individually finally drain into the collecting ducts, which open into the anterior and posterior divisions of the ureter. The ureter has a separate cloacal opening close to that of the mesonephric duct, or to the *Mullerian duct* (oviduct) in the female.

The connection with a true ureter, the loss of urinary function in all other nephrogenous tissues, and the complete separation of the genital and urinary function in amniotes argue against the view that the metanephros is merely a specialized form of opisthonephros. The metanephrogenic blastema of amniotes may never have evolved in anamniotes, or be so retarded in its rate of development that it never arises in ontogeny. It may well be, however, that the key lies in the origin of the ureteric bud, the essential inductive role of which determines the origin and development of the metanephric tubules.

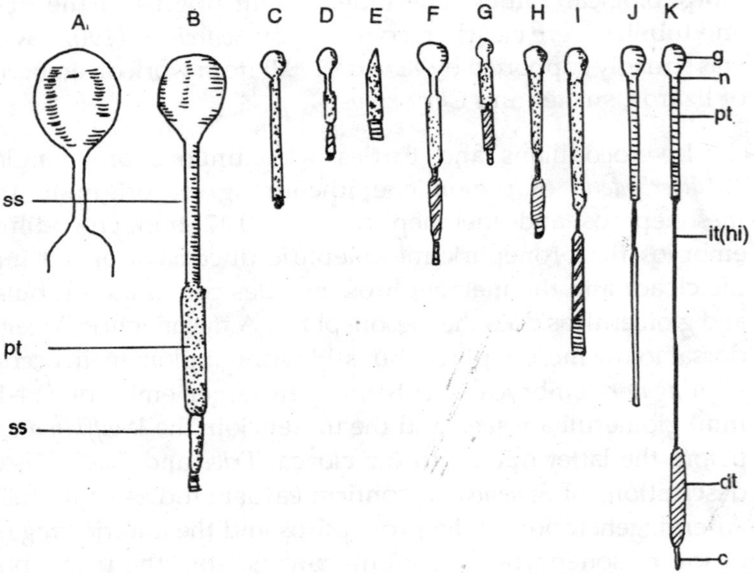

Fig. 1.11. Diagrammatic representation of uriniferous tubules (size relationships only very approximate) of different vertebrates. A cyclostome; B. elasmobranch; C., D., E. Teleost; F. *Rana catesbeiana* (frog); G. *Chrysemys picta marginata* Turtle) ; H., I. *Gallus domesticus* (bird); J., K. *Lepus cuniculus* (mammal). The giomerulus (g) (absent in some teleosts) is followed by a neck (n) and then a proximal (pt, intermediate (it), distal (dt and collecting tubular (ct) segments, A special segments (ss) situated between the glomerulus and the proximal segment and after the latter occurs in elasmobranchs and Henle's loop (hl) is typical of mammals.

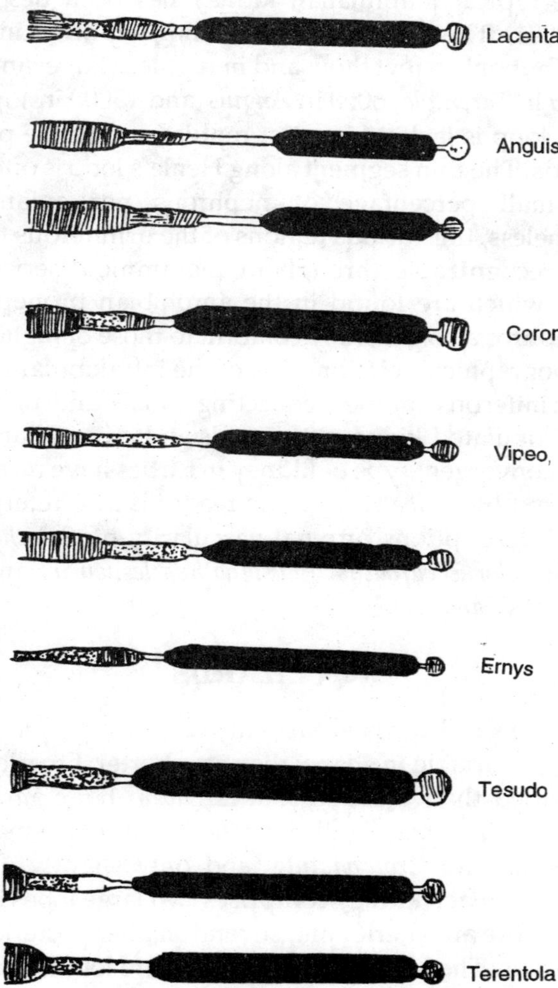

Fig. 1.12. Diagrammatic representation of uriniferous tubules of various reptilian kidneys. The glomerulus leads into the neck and thence the Hauptstuck and Ubergangstuck (black). Following in sequence are the ciliated region of the Schleifenstuck (white), then a non-cliated region followed by the Schaltstuck (dotted) and the Sammelrohr (lined). See text for other terminology, a and b represent ordinary and ureteral tubules in *Anguis* and *Tarentola* respectively, which are generally similar.

The typical mammalian kidney has been described by *Smith* (1951). There are about one million nephrons in a human kidney, but only a few thousand in reptiles. For example there are 1600 in *Tarentola*, 6000 in *Anguis* and 15000 in *Vipera aspis*. Henle's loop is lacking in most reptiles, as are the pelvis and pyramids. The thin segment along Henle's loop is only present in a small percentage of nephrons per avian kidney. Nevertheless, the various regions of the uriniferous tubule are clearly recognizable throughout the amniote series. Indeed tubules which are found in the amphibian pronephros and mesonephros also generally conform to those of higher groups. The topographical relationships of the intralobular artery and vein, uriniferous tubules, collecting canals and ureter in the seriate, lobulated kidneys of lizards and crocodilians and the radiate convergent type of kidney in turtles have recently been considered by *Guibe* (1970). The reader is also referred to the detailed descriptions of renal vascularity by *Splechtna* (1970) in *Amphibolorus barbatus, Physignathus lesueurii* and several species of *Chamaeleo*.

IN LIZARDS

The kidneys of lizards are usually of equal size, more or less lobed and variable in shape; they are shorter, broader and less sub-divided than those in snakes *Owen,* 1866; *Parker,* 1884). The kidneys are only slightly lobulated for example in *Holbrookia, Sceloporus,* and *Crotaphytut,* and only slightly notched in *Phrynosoma*. Each kidney comprises two large lobes in *Eumeces* and there are an anterior and several smaller posterior lobes in *Gerrhonotus*. The kidney of *Anguis* is more lobulated than that of other lizards. It is elongated, oval and flattened dorsoventrally as it is in *Anniella* pear-shaped in *Gekke,* somewhat elongate pearshaped in the agamids *Amphibolurus* and *Physignathus* and in Chamaeleo pardalis, *but uniformly elongate in C. brevicornis, C. jacksonii* and *C. fischeri*. The kidney is triangular and flattened with transverse furrows in *Hemidactylus*. The posterior region of the lobulated kidney is tapered in *Heloderma*. In practically

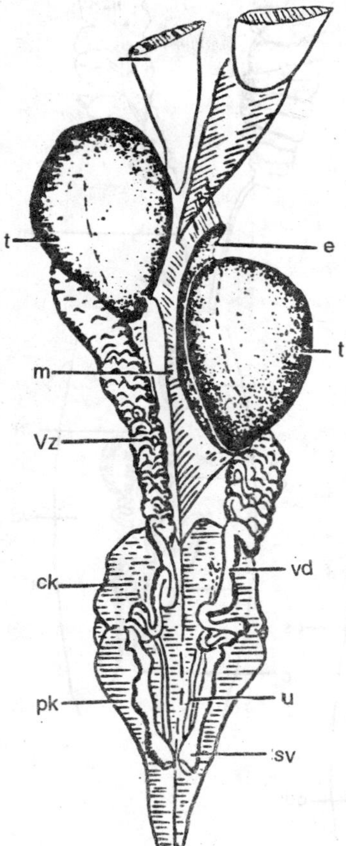

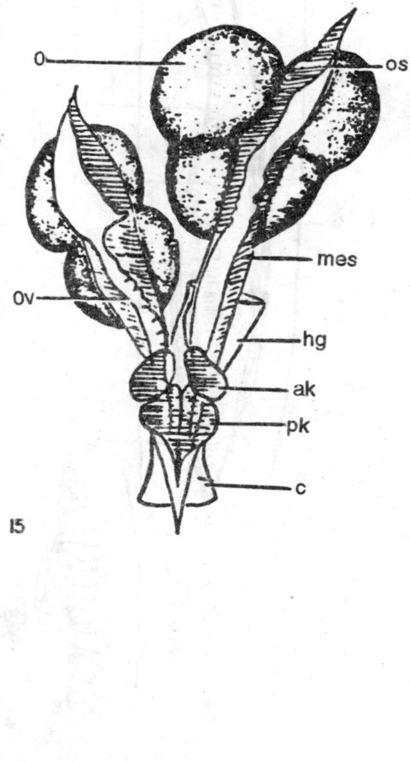

Fig. 1.13. *Sceloporus olivaceus*. Ventral aspect of male urinogenital system.

Fig. 1.14. *Sceloporus grammicus disparalis*. Dorsal aspect of female urinogenital system. ak, Anterior lobe of kidney; c, cloaca; e, epididymis; g, hindgut; k, kidney; l, lung; m, mesorchium; mes, mesometrium; o, ovary; os, ostium abdominale; ov, oviduct; pk, posterior lobe of the kidneys; sv, seminal vesicle; t, testis; u, ureter; vd, vas deferens.

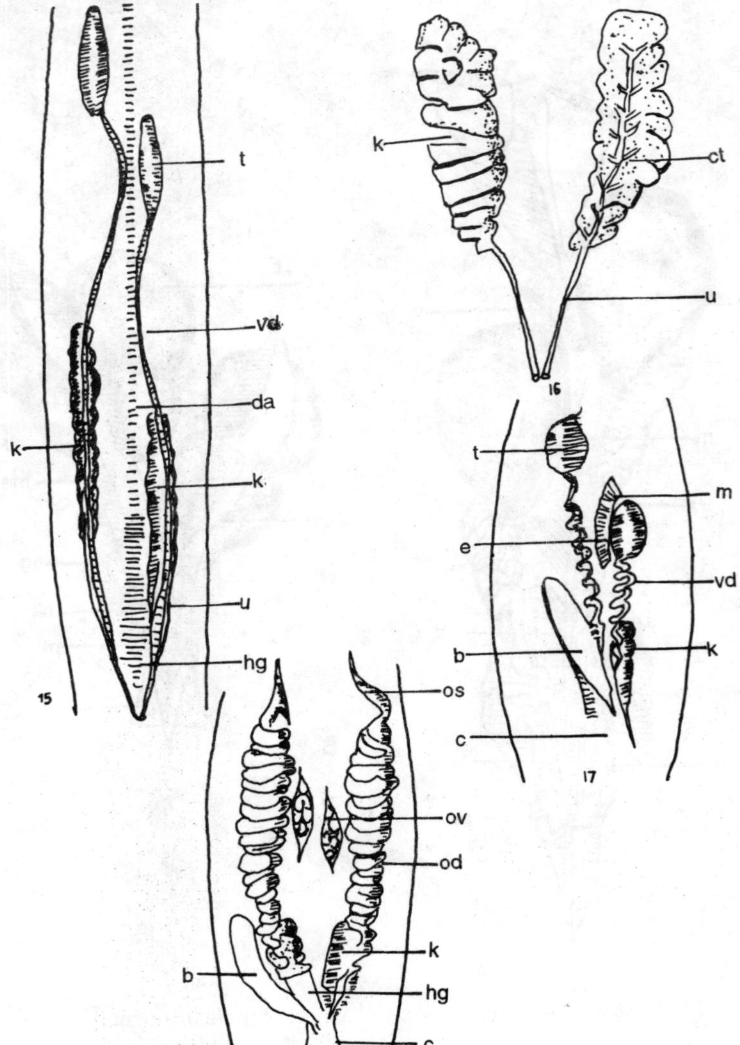

Fig. 1.15. Urinogenital system of male *Elaphe*.
Fig. 1.16. Kidneys of *Verinus indicus;* left kidney is turned over.
Fig. 1.17. Urinogenital system of male *Lacerta viridis*.
Fig. 1.18. Urinogenital system of female *Lacerta muralis*. c 1897). b, Urinary bladter; c, cloaca; ct, collecting tubules which open into the ureter; dorsal aorta; e, epididymis; hy, hindgat; k, kidney; m, mesorcrium; o, orsary; od, oviduct; os, ostium abdominale; t, testis; u, ureter; vd, vas deferens.

all cases the kidneys are situated in the posterior region of the abdominal cavity and extend beyond the cloaca where they may fuse *Reamey* reported median fusion only in *Lacerta agilis* and the kidneys are separate in *Varanus salvator* and in some chameleons. The kidneys of *Lialis burtours* are distinct, for though they extend back to the cloaca, the right is larger than the left and extends slightly further anteriorly, as does the kidney of snakes. The posteriorly located, lobulated kidneys of *Sphenocon* meet at their median surfaces, but never coalesce.

In lizards the wine carrying duct the ureter is usually short. It is recognized to course along the median or medio-ventral surface of the kidney along its entire length as in *Anguis*. The ureter of *Anniella* opens independently into the cloaca and gives rise to 2 short, blind, anteriorly directed diverticulum. In general, lizards and *Sphenodon* have an ipsilateral ureter and vas deferens opening conjointly into the cloaca, as in *Hemidactylus*. These structures may give rise to a short canal opening into a small cloacal papilla. In females the ureters and oviducts, of the two sides, open separately into the cloaca. In *Hydrosaurus*, however, the ureters unite and open dorsomedially into a papilla and in *Lacerta viridis* they open independently into the cloaca behind the vas deferens. There is indeed some variability in this feature even within the same genus; thus in *Sceloporus disparalis* and *S. poinsettii* the ureter and vas deferens open jointly into the cloaca, but in *S. olivaceus* the openings are separate. In the females oviducal and ureteral openings are separate. In the North American lizards described by *Brooks* (1906), the *ipsilateral ureter* and *vas deferens* share a common opening as do the ureter and oviduct. The right and left oviducal cloacal openings are fused in *Gerrhonotus* and *Cnemidophorus* but separate in *Sceloporus, Holbrookia, Phrynosoma* and *Crotaphytus*.

A urinary bladder is generally found in lizards. It has been reported in *Anguis, Iguana, Gekko* and the chameleon, though it is rudimentary in some Teiidae and absent in *Varanus salvator*. It usually opens ventrally into the cloaca. The bladder may be elongated and pear-shaped in *Heloderma* , or bilobed. A urinary bladder only occurs in *Eumeces* and *Gerrhonotus* among the various genera of lizards described by *Brooks* (1906). It is rudimentary in *Agama agama* or comprises merely a stalk in *Sceloporus grammicus disparulis* and *S. poinsettii.*

IN SNAKES

Snake kidneys are paired usually elongated and lobulated with the right situated in front of the left, though they may be nearly symmetrically disposed in *Calamaria Owen* (1966) called the lobules renules. There are five distinct lobes in *Lycodon* and about 20 to 30 of them in each kidney of the fox snake, *Elaphe vulpina* and in *Thamnophis.* Kidneys may be 13 cm long in *Natrix natrix* and transverse furrows separate the kidney into a series of further incompletely divided lobules. In contrast the kidneys are without lobules in *Tropidophis* and *Trachyboa.* The right kidney may be larger in *Thamnophis* ; it is larger in the male *Diadophis* than in the female.

The elongated ureters vary in length but are longer on the right side. They may lead posteriorly along the kidney's inner margin, ventromedially, either ventromedially or ventrally or again along its ventral lower border. Communication with the exterior is via a papilla on the dorsal surface of the cloaca; in most cases this papilla is distinct though close to the opening of the vas deferens or oviduct. The ureter and oviduct, however, share a common opening in *Natrix.* In most snakes the ureter dilates near its hinder end to form a small urinary reservoir, for snakes lack a urinary bladder.

IN CROCODILES AND TURTLES

In alligators and crocodiles the paired kidneys are lobote and have a highly convoluted surface. The left kidney is larger than the right. Large ureters originate from light coloured kidneys (covered by black tissue) to open separately into the upper wall of the anterior cloacal chamber. A dorsal division of the ureter serves the hinder fifth portion of the kidney; a ventral division drains the rest. Crocodilians lack a urinary bladder.

In turtles the paired kidneys are generally flattened, lobulated and usually symmetrical placed with their short ureters, are situated in the pelvic region of the body, as are those of the crocodiles and alligators.

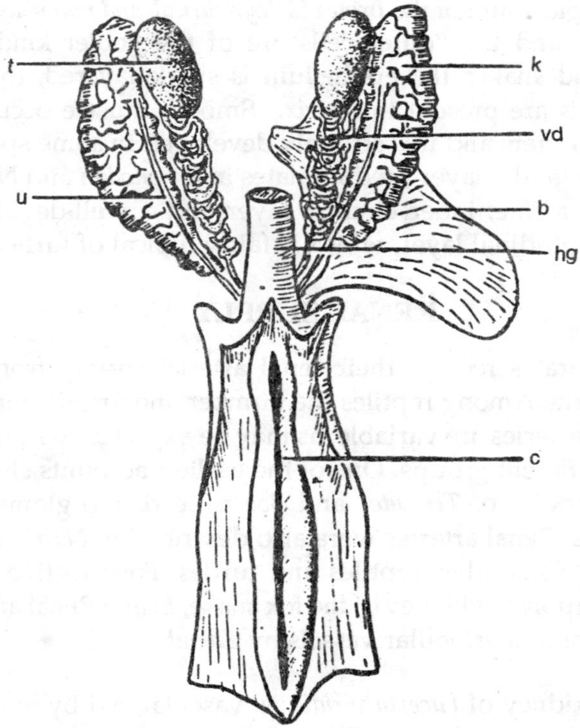

Fig. 1.19. Urinogenital system of male *Chelydra serpentina*: ventral aspect. b, Urinary bladder; c, cloaca; hg, hindgut; k, kidney; t, testis; u, ureter; vd, vas deferens.

However, the right kidney of *Mauremys japonica* is situated slightly in front of the left. The kidneys are oblong, broad and thick in *Testudo*, flattened anteriorly in *Chelonia mydas*, semioval in *Emys* or lozenge-shaped in *Chrysemys.* In the last genus the renal surface is made up of a series of shallow parallel grooves, sometimes anastomozing and arranged at right angles to the long axis. External convolutions of the renal surface resembling gyri were described in the female *Trionyx euphraticus.* The ureter is situated posteroventrally. Terrestrial and freshwater turtles have a large urinary bladder that leads into the lower wall of the cloaca.

The epithelium of the ureter is *"cylindrical and many layered"* in turtles, and the "outer cells are of the goblet kind." In lizards and snakes the epithelium is single-layered, though goblet cells are present in *Natrix.* Smooth muscle occurs in reptilian ureters and it is strongly developed in some species. An inner circular layer predominates in *Sphenodon* and *Natrix, Emys* has a fine, inner circular layer and a well-developed outer longitudinal layer, which is fairly typical of turtles.

RENAL SUPPLY

All vertebrates receive their renal arterial supply from the dorsal aorta. Among reptiles the number and arrangement of the renal arteries are variable, as may be expected, within and among different groups. One of the earliest accounts showed renal arterioles of *Testudo* and *Boa* to lead into glomerular capillaries. Renal arteries were also described in *Natrix natrix* and in various other reptiles and turtles. Four to five renal arteries supply the kidney of the fox snake, *Elaphe* Renal arteries were termed interlobular vessels by *Zarnik.*

The kidney of *Lacerta viridis* is vascularized by anterior, median and posterior arteries, as is that of *Agama agama* and the kidney is drained by a single vein. There are usually 4-5 renal arteries to each kidney in the agamids *Amphibolurus* and *Physignathus,* but chameleons have a variable number. Five renal arteries give origin to arteroles to the glomeruli and renal tubules in *Anguis.*

Among snakes the right kidney of *Xenopeltis* receives 2 renal arteries (the first enters the 2nd lobe and the second enters the 6th); the left kidney receives 1 artery only to the 2nd lobe. In *Thamnophis ordinoides* there are 7/p *Elaphe oxycephala* 6/5; in *Cerberus rhynchops* 2/1; in *Causus rhombeatus* 10/9; in *Elaphe quadrivirgata* 3/3 and in *Agkistrodon* 3 to 7/2 to 9.

There is a single renal artery to each kidney in *Eunectes notueus* and *Hmilhis scytale,* and the same condition seems generally applicable to the Boidae. In some support of Beddard's results. *Underwood* (1967) believed that the kidneys of primitive snakes tend to be supplied by a smaller number of renal arteries and those of more specialized snakes by a greater number. Thus, there is a single renal artery to each kidney in representatives of the Typhlopidae and Leptoryphlopidae, in *Anilius, Melanophidium* and most Boidae. Some Henophidia, however, differ, for example *Cylindrophis* Aniliidae) 4/5; *Eryx* 2/2;*Chersydrus* 2/3 and *Xenopeltis* 1/2. Few Caenophidia have only 1 renal artery, for example *Lycognathus,* but low numbers such as 1/2 and 2/2 are not uncommon as in *lyptodeira, Siphlophis* and *Tripanurgos.* High numbers of renal arteries occur in *Ophiophagus* 4/7, *Ptyas* 8/8, and *Causus* 8/12. A clear asymmetry is found in the unrelated genera *Naja* and *Homalopsis* both having 6/2. *Natrix natrix* is sexually dimorphic for the male has 4 or 5 renal arteries and the female 3 on each side.

Among turtles, the renal arteries of *Chrysemys* and *Chelydra* Penetrate the kidney mid-dorsally as interlobular vessels; smaller arterioles lead into glomerular capillaries. Efferent vessels leave the capillaries near the entrance of the afferents to anastomose with the venous network around the tubules. In *Clemmys* there are arteriae renalis cranialis, media and caudalis to each kidney.

All vertebrate except mammals and possible cyclostomes have a functional renal portal system. Foetal mammals were later included with groups provided with a renal portal system; while the system is retained in adult birds it is lost in adult mammals. A renal portal system typically is found to originate in the posterior region of the reptilian body passing from the pelvic veins to the kidneys. In *Lacerta viridis* the caudal vein bifurcates and the resulting pelvic veins to the kidney receive sciatic and femoral veins from the limbs. The arrangement is similar in *Agama agama*. In *Anguis fragilis* the renal portal system originates from caudal, intercostal, cutaneous and spinal vessels; the resulting venae renales afferens subdivide into capillaries distributed around the kidney tubules. Drainage is via venae renales efferens to the renal veins and thence to the posterior vena cava. In the tree dwelling lizard *Calotes mystaceus* some renal portal blood also appears to join hepatic portal blood leading to the liver.

In *Nartin natrix* the renal portal system arises in a manner similar to that of *Lacerta,* though the vessels are elongated in snakes. The system leads anteriorly together with the ureter along the ventral border of the kidney. In crocodilians afferent renal portal veins (renal advehentes) leave the epigastric shortly before fiburcation from the caudal vein, and likewise they receive pelvic veins on the way to the kidney. Venous sinuses occur between afferent and efferent renal veins in crocodilians and birds; thus some blood is shunted past the nephric capillary bed. However, in *Chrysemys picta marginata* blood traversing the renal portal system must enter a capillary network before reaching the posterior vena cava; it thus meets the requirements of a true portal system. In *Chelydra serpentina* this capillary network is supplemented by large indirect connections near the dorsal surface of the kidney, between the renal portal veins and the posterior vena cava. Here the renal portal system is partial or incomplete. Crocodilians and some turtles have only a partial renal portal system and in-any case only part of the hinder venous flow is canalized through it. This suggests that in reptiles this system has limited significance during life. Doubtless, however, it participates in permitting further kidney clearance of blood from the posterior regions of the body, though the process is subordinate to that of the renal arteral system.

Excretory Tubules

In reptiles the glomerulus which is a net work like structure of blood capillaries, is generally poorly developed. Its centre is of connective tissue with a reduced capillary content, in contrast to the birds whose capillaries are more numerous per gram body weight and smaller and have much syncitial tissue and mammals, which have numerous and large vascular glomeruli. The glomerular central mass of *Phrynosoma* is similar to that of birds except that the nucle are not as compactly arranged *Gampert* (1866) first described and figured the small oval or ellipsoidal glomeruli of snakes, the capillaries of which are located at the glomerular surface as they are in lizards. The emphasis is on fewer glomeruli of reduced filtration surface, to conserve water, and aglomerular diverticulae occur in snake kidneys. Lizards have the smallest glomeruli among reptiles, sometimes showing merely a few capillary loops and bifurcations. Turtles and crocodilians have "a fair glomerular surface and a solid or semi-solid urine." In testudines the glomerulus comprises a network of bifurcated and anastomosing capillaries, the loops of which are closely applied to each other to form a right ensemble. Occasionally capillaries traverse the core of connective tissue; this never occurs in snakes.

The following data give some indication of the numbers of glomeruli in each kidney and their size. The number of glomeruli per kidney is lower in smaller specimens of the adult iguana: 17000 (882 g specimen), 78000 (2200 g specimen), and 121000 (2500 g specimen. Presumably th larger animals had bigger kidneys and hence more glomeruli.

Studies on the normal development and from *unilateral nephrectomy* in the kidneys of mammals show that new functional nephrons only form in immature stages. For example in the newborn rat the number of glomeruli per kidney (about 10000) steadily increases at a declining rate, to about three times that number in the mature animal. Thereafter, the glomerular total is reduced to about one-half in senile forms. In teleosts apparently new nephrons can be recognized in animals at all sizes. The limited data on *Iguana* likewise would seem to suggest that new nephrons can differentiate in mature stages.

The range in number of glomeruli is considerable in reptiles, however, from 1000 glomeruli (total for both kidneys) for geckos, *Takydromus,* and *Eumeces* to 22000 in *Elaphe.* In *Clemmys* and *Natrix* the total for both kidneys lies between 13 000 and 16 000 respectively; this is generally less than in birds and mammals but more than in amphibians Glomerular size (diameter) is likewise variable, though the measurements given by earlier workers should not be treated too seriously, owing to faults of technique or shrinkage effects after fixation. In lizards, including among others *Phrynosoma* and *Iguana,* glomerular diameter ranged between 50 and 71 μm in snakes including among others *Boa, Natrix, Liopeltis* and *Vipera,* the size generally ranged between 60 and 120 μm. In turtles, including *Testudo, emys, Pseudemys, Caretta* and *Clemmys,* size was likewise variable, from 50 μm to 110 μm. An alligator glomerulus had a reported diameter of 50 μm.

One may only conclude from these inadequate data that on thewhole the lizard glomerulus is generally the smallest among reptiles.

Uriniferous Tubule

The reptilian nephron has been reported on by *Busch* (1855), *Bowman* (1842), *Leydig* (1857) and *Klein* (1881). The later described cilia in the constricted neck and also in the fine segment of the adult nephron. *Kolliker* (1945) likewise reported cilia in the neck of nephrons of embryonic lizards. The nephrons of Natrix, alligators and *Anguis* have been studied and described in details by a number of workers. in a series of detailed communications supplied substantial information on the structure ad histology of the uriniferous tubules of snakes.

Hestologically the Bowmans capsule made up of pavement epithelium, leads into a narrow, non-secretory neck segment composed of cubical cells with between one-third and one-half of them ciliated. The segment is usually short in snakes though it may be variable in length or larger in the male. The neck continues into a looped, glandular and secretory convoluted tubule the *"segment a brosse"* or segment a bordure strie—the tubulus contortus of mammals—with its microvillous brush border against the lumen. Short blind aglomerular diverticulae may originate here in some snakes. In section each cell of the convoluted tubule of snakes shows up to 50 urinary granules, though usually only 10 to 20 are seen. It is unlikely that they originate from the nucleolus. Abundant granules also occur in this segment in lizards and snakes including *Natrix*. Fat bodies and lipid droplets also occur. *Regaud* and *Policard* (1903a) further reported that some of the cells were ciliated though probably this is unlikely. The yellow *"graines urinaires"* or *"graines desegregation"* were previously described by *Leydig (1957) and* Heidenhain (1874) in snake kidneys and by *Solger* (1885) in alligators. On the basis of their varying contents of mitochondria and granules, of differing sizes, the proximal convoluted tubules of *Thamnophis* were considered to include three types of cells representing different stages in the secretory cycle of a single cell.

The next tubular region is the fine, non-secretory portion comprising (a) an initial ciliated region where about one-sixth of the cells are ciliated probably having a purely mechanical function, through *Regand* and *Policard* showed them to contain fine granules and lipid, which suggests a feeble secretory function as well, and (b) a terminal section of mucous cells (separated from the ciliated part by an intermediate region), probably akin to the succeeding sexual segment. In *Phrynosoma* the cilia of the "segment grele" were the longest found in those vertebrates studied. The succeeding segment is well developed in tortoises. The cells are without secretory granules though occasional fat droplets occur in the cytoplasm. Water is absorbed in this region and possibly diffusible substances also.

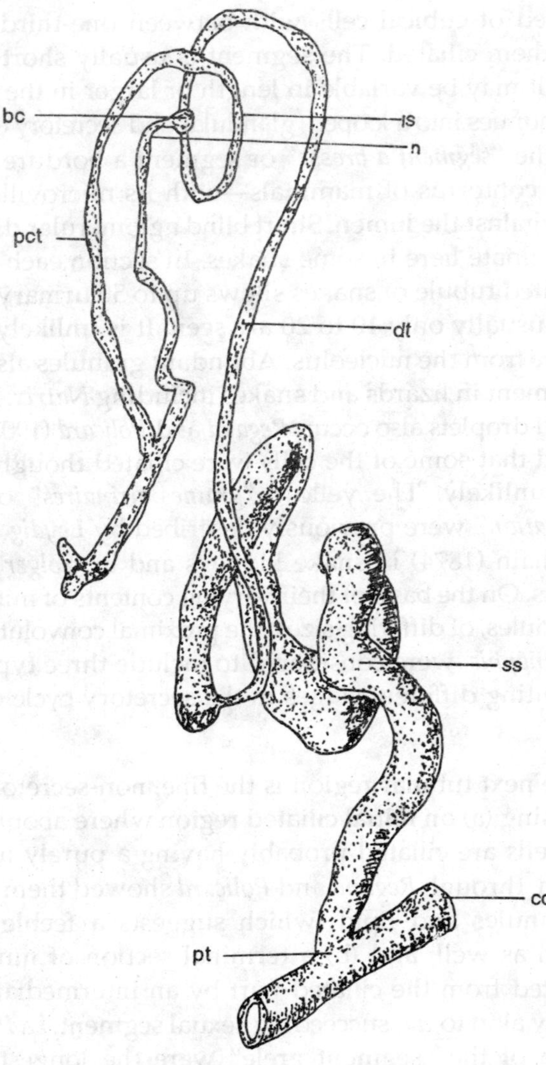

Fig. 1.20. **Uriniferous tubule of kidney of male *Themonthis sirtalis;* Diagrammatic illustration. bc, Bowmans capsule; cd, collecting duct; dt, distal tubule 1.9 mm); is, intermediate (fine) segment (0.3 mm) n. neck (0.07 mm); pct, proximal convoluted tubule (2.8 mm); pt, post-terminal segment (0.08 mm); ss; several segment (2.7 mm). Lengths in brackets. Total length of nephron 7/9 mm.**

Finally, a terminal segment leads into the collecting canals. This segment is hypertrophied in male snakes and lizards as the sex segment (*vide infra*). Apart from points of details of size and shape, the nephrons are generally similar in different groups of reptiles. Cilia in the reptilian nephron were supposed by *Marshall* (1934) to supplement the low filtration pressure, deduced from the small filtration area of the glomerulus and the low blood pressure. It is of interest that nephrons of birds and mammals (note, however, that *Klein* (1881) claimed their presence in mouse kidneys), where the blood pressure is high lack cilia.

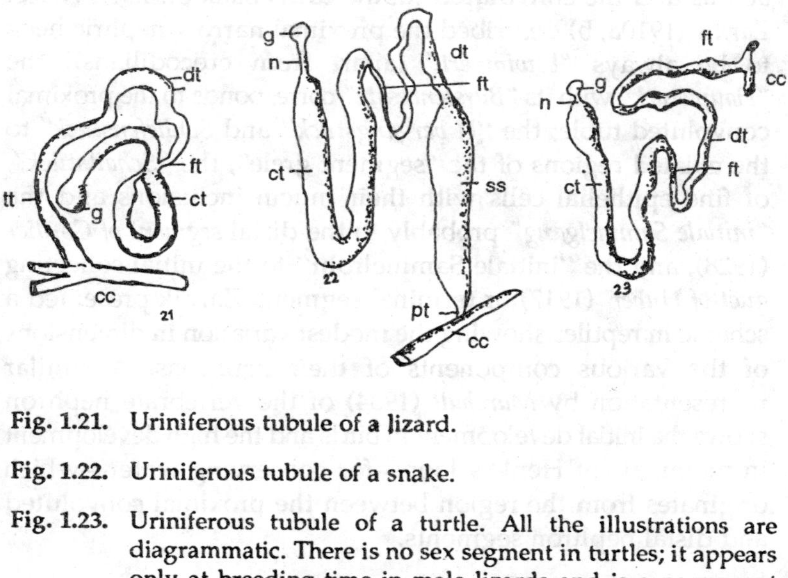

Fig. 1.21. Uriniferous tubule of a lizard.

Fig. 1.22. Uriniferous tubule of a snake.

Fig. 1.23. Uriniferous tubule of a turtle. All the illustrations are diagrammatic. There is no sex segment in turtles; it appears only at breeding time in male lizards and is a permanent feature of male snakes. cc, Collecting duct; ct, proximal convoluted tubule; dt, distal tubule; fts fine tubule (segment grele); g, glomerulus; n, neck of tubule; pt, post-terminal segment; ss. sexual segment; , terminal tubule.

The entire nephron is 3.3 mm long in *Chrysemys picta marginata*, 4.5 mm in *Alligator mississippiensis*, 4 mm in *Thamnophis sirtalis* and 2 mm long in a lizard (unnamed). In birds it is about 18 mm long, half the length of the largest in mammals.

From the glomerulus to the collecting tubule the distance measured was 8 mm in *Thamnophis sirtalis,* about one-quarter of the length of a human nephron, which was found by *Bishop* (1959) to range from 30 to 38 mm in length. *Huber* (1917) described the convoluted tubule to be U-shaped in the alligator, looped in *Thamnophis,* N-shaped in *Chrysemys,* lizards and birds and of elongate-U-shape in mammals (mainly Henle's loop). In general he confirmed *Zarnik's* (1910a,b) descriptions of reptilian nephrons and documented that neither alligators nor crocodilians have ciliated cells in the tubular necks. However, cilia were later discovered in the short neck segment of *Crocodylus* acutus and the convoluted tubule turns back on itself twice. *Zarnik* (1910a, b) described the proximal narrow nephric neck to be always *"bewimpert"* (apart from crocodilians): the *"Hauptstuck"* with its *"Burstenbesatz"* corresponds to the proximal convoluted tuble; the *"U bergangstuck"* and *Schliefenstuck"* to the ciliated regions of the "segment grele"; the *"Schaltstuck"*, of fine epithelial cells with their mucin inclusions and the *"initiale Sammelgang"* probably to the distal *segment of Cordier* (1928), and the "initiale Sammelrohr" to the initial collecting *duct of Huber* (1917) or terminal segment. Zarnik presented a scheme in reptiles showing the modest variation in dimensions of the various components of their nephrons. A similar representation by *Marshall* (1934) of the vertebrate nephron shows the initial development in birds and the high development in mammals of Henle's loop—for conserving water—which originates from the region between the proximal convoluted and distal nephron segments.

The kidney tubules and Malpighian bodies of reptiles are innervated by a complicated nerve net of fine fibres, located in th tubular parenchyma and for the most part closely associated with blood vessels.

Histochemically, Glycogen is present in the neck, proximal and sex segments, alkaline phosphatase in the Malpighian corpuscle, neck, proximal and intermediate segments and acid phosphatase in the distal, sex and post-terminal segments.

Succinic dehydrogenase activity occurs in the proximal, intermediate, distal and post-terminal segments, with the greatest activity in the two later segments. *Longley et al.* (1963) found mucoid substances frequently in the collecting tubules and excretory ducts of fishes, reptiles and birds but rarely in mammals and never in amphibians. In turtles mucins occur as coarse granulations, especially in the apices of the cells. The nephron of *Sphenodon* is rich in glycogen and poor in lipid. The significance of these findings in terms of tubular function is not understood.

The renal and collecting tubules of *Chamaeleo africanus, Python moiturus* and *Crocodylus porosuts* have moderate amounts of Δ^5-bβ hydroxysteroid dehydrogenase (HSDH), though there is weak activity in the sex segments of male chameleons and pythons. The kidney and collecting ducts showed intense activity of 17β HSDH, glucose-6-phosphate dehydrogenase and NADH diaphorase. *Gouder* and *Nadkarner* (1974) suggested that these enzymes convert ^{5}-3-hydroxysteroids to Δ^4-3β-ketosteroids and the sex steroids to 17 ketosteroids respectively, during steroid excretion.

Sexual Segment

In reptiles the sexual segment was first discovered by *Gampert* (1866) in *Natrix*, lying between the distal and collecting tubules; its cell contain numerous granular inclusions. *Regaud* and *Policard* (1903a) first termed this region the sex segment, of some sexual significance, and showed that it occurred only in the kidneys of male snakes and lizards but was absent in turtles of both sexes and in crocodilians. *Zarnik* confirmed its presence in snakes and in the lizard's *Anguis* and *Tarentola*, and *Reiss* (1923a) reaffirmed *Regaud* and *Policard's* (1903a) findings and clearly showed the sex segments of *Lacerta agilis* and *L. vivipara* to have a seasonal cycle, with glandular activity in summer followed by quiescence the following winter. Subsequently these principles were fully established in a variety of lizards and snakes though *Carlia rhomboidalis,* a tropical Australian skink, is usual among lizards in that its sex segment shows no seasonal variation in the diameter of its tubules. In this species reproduction is continuous throughout the year.

The sexual segment of male lizards and snakes originates from nephrons in generally similar extensive regions at the posterior end. It lies next to the last segment, or preterminally, in various lizards and snakes. It comprises the distal segment in *Crotalus adamanteus.* The sexual segment is described as being terminal in the lizards *Lacerta muralis, Phrynosoma cornutum, Cnemidophorus sexlineatus* and *Holbrookia* spp. various lizards and snakes and the snakes *Natrix natrix. N. maura, Thamnophis sirtalis, Diadophis, Liophis* and Xenolon. It comprises the intermediate segments and most of the collecting tubules in the amphisbaenians *Blanus cinereus* and *Trogonophis miegmanni.*

The sexual segment is situated after the distal but preceding the post-terminal segments in *Thamnophis.* It includes the preterminal terminal and collecting ducts in the lizards *Scincus* and *Chaltides* the terminal and collecting ducts in *Acanthodactylus* and in addition it extends from the pre-terminal segment to part of the ureter in *Takydromus tachydromoides,* the collecting tubules in *Sceloporous undulatus hyacinthinus* and *Anolis carolinensis* ; the middle and terminal collecting ducts in *Cnemidophorus l. lemniscatus* where the female has a small less differentiated region corresponding to the first portion; the secondary and tertiary collecting ducts and ureter in *Hemidactylus flaviviridis;* the collecting ducts and at least part of the ureter, which is found embedded in the kidney in *Carlia fusca* , the lizard *Anolis carolmensis* and the snakes *Leptotyphlops humilis, L. dulcis, Typhlops vermicularis* and *T. simoni.*

In snakes and *Varanus* the sex segment generally corresponds to the medial region of the distal convoluted tubule as studied by *Saint Girons.* In other squamates the collecting tubules form the main part. From the extensive reports on this morphologically variable renal component, it seems that in lizards and snakes it is usually situated terminally in the nephrons. In some others, it can also include the collecting ducts and the ureter.

The sexual segment of male lizards is involuted and cannot be distinguished from the adjacent tubular regions, for it is composed of similar mucous cells. In *Lacerta agilis* the diameter of the sex segment is 50-60 μm and the cells are about 20 μm high. In June the cells hypertrophy to three times the original height and they contain numerous secretory granules; secretion terminates in July and the sex segment is at rest in September. Similar levels of size and activity occur in the sex segment of the male *Sceloporus* in March to May. In *Takydromus* the columnar cells are 35-50 μm high in May to July and secretory granules occur, especially in their apical regions. Secretory activity is reduced in August and the sex segment regresses in the middle of September to become free of granules. In *Hemidactylus* its diameter when quiescent, in June to July, is 60-70 μm and cell height is 12-15 μm. In the active phase during October to May the segment diameter ranges from 100-250 μm and the cells, replete with granules, are 40-90 μm high.

Snakes do not show such drastic involution of the sexual segment and cellular hypertrophy is here termed continuous. However, though previous workers reported no apparent regression of this region during the sexually quiescent period, modest cyclical changes do seem to occur. *Volsoe* (1944), who remarked that hitherto, no seasonal variations of the ophidian sex segment had been described (in contrast to the situation in lizards), demonstrated seasonal changes in *Vipera berus;* a maximum tubular diameter and cell height in spring and a minimum in June to July, followed by a slow increase in size in autumn. The male sex segment of *Vipera* has an active spring period coinciding with the ripening of spermatozoa and a more or less inactive period during the remainder of the year, though variation is extremely small, compared with the condition in lizards. In *Natrix natrix* there are no obvious seasonal morphological variations in the kidney, but there is a difference in the abundance of intra-cytoplasmic secretory granules of the sex segment, which seems to show a low secretory activity in summer and the beginning of autumn; a cycle which is subtly related to the cyclic activity of the testis.

The sexual segment of *Vipera aspis* involutes in July. When the segment is active its diameter is 160-180 µm and that of the lumen 30 µm; the columnar cells, full of granules, are 60-70 µm high. At quiescence the segmental diameter is 100 µm, the lumen has disappeared and cell height is merely 40 µm. The diameter of the sex segment doubles during sexual activity. In *Thamnophis* it is claimed to enlarge three to five times which suggests that in some snakes at least more than modest seasonal variation in size may occur in this nephric component. According to *Gabe* and *Saint Girons* (1962) the sex segment is only slightly modified in *V. aspis, except* from June to the beginning of August, when some involution occurs. In Cambodian snakes hypertrophied kidney segments involute to a mucous state (as in lizards) when they are not reproductively active.

On the whole, however, snakes differ from lizards and do not show complete sexual quiescence and involution of the male kidney sex segment; the segment cells are always distinguishable from adjacent mucous cells or comparable ones from females and castrated males. The sex segment in the male ringneck snake *Diadophis* is remarkably enlarged.

At the peak of the breeding season, columnar cells of the hypertrophied reptilian sex segment have alveolar cytoplasm and a basophilic nucleus, and contain secretory granules. Externally the cells are enveloped by a thin layer of connective tissue. The sex segment of *Xenochrophis piscator is* surrounded by a thin layer of muscle that of *Typhlops vermicularis* is surrounded by a thick coat of smooth muscle, of outer longitudinal and inner circular fibres. A thinner sheath of circular and oblique fibres follows the sex segment within the kidney. Autonomic ganglia occur within the muscle layers of the within the kidney. Autonomic ganglia occur within the muscle layers of the ureter and collecting ducts.

In *Cnemidophorus l. lemniscatus* the cells of the first portion of the sex segment seem to include large secretory granules of varied opacities. The granular endoplasmic reticulum is abundant and there are numerous smooth vesicles and a well formed Golgi complex. Secretion is probably holocrine. In the succeeding regions of the sex segment the secretory granules are apical and have a denser core, probably derived from the Golgi. Secretion here is probably merocrine. In the female the region of the kidney corresponding to the sex segment of the male shows features generally similar in structures but less well developed. The collecting duct of the kidney was also investigated.

The detailed structure of the sex segment of *Natrix natrix* has been described. A sheet of contractile cells surrounds the sex segment. The intercellular spaces are occupied by collagen and between the contractile cells and the basal lamina of the secretory cells are two types of nerve cells, which are distinguished mainly in terms of the number of nerve cells, which are distinguished mainly in terms of the number of synaptic vesicles and membrane-bound granules which they possess.

It was reported that young immature male *Natrix maura* and a *Lacerta* sp. (with prespermatogenic testes and captured in May) had well-developed sex segments in their kidneys, that of *Lacerta* in full secretory activity. These reports may well be erroneous for in a young male *Lacerta muralis* this region was composed merely of hardly differentiated mucous cells showing no traces of secretory granules. No secretory granules occurred in the sex segment cells of an immature *Takydromus,* though the testis included a small quantity of spermatozoa. Again two immature male specimens of *Natrix* were still deficient in a hypertrophied preterminal segment, this region being as undifferentiated as in the female. *Anguis* one year old (1g in weight), shows no sexual dimorphism. By the time males weight 3.9g the cells of the collecting ducts and ureter are hypertrophied with PAS-positive granules beginning to appear in the collecting ducts.

The granules of the sex segment are either yellow or colourless, destroyed by acetic or alcohol strongly eosinophilic, they are discrete, spherical, highly refractile and contain protein. They measure 1.5-2 μm in diameter in *Crotalus*and 1-1.5 μm in *Thamnophis*. Their integrity is destroyed by Boutin's fluid and they include lipids, neutral glycoprotein and mucoprotein, tyrosine, lysine, cysteine and small quantities of tryptophane; possibly they have a proteolytic activity.

The enlarged sex segment of *Vipera berus* exhibits cyclica changes in the production and expenditure of cholesterol-positive lipids *Marshall* and *Wolff* (1957) and *Deb* and *Sarker* (1963) found unsaturated lipids and a high concentration of acid phosphatese during the breeding season of *Calotes versicolor,* the acid phosphatase disappearing at hibernation. In *Hemidactylus* a number of phosphatides are found in the secretion of the sex segment. The granules here are saliva-resistant and positive for PAS, phospholipid, choline and acid phosphatase. Furthermore, the sex segment seems capable of synthesizing its own phospholipids from suitable precursors under androgen stimulation. The histochemistry of the sex segment of eight genera of snakes was described by *Helmy* and *Hack* (1967). The male *Coluber* showed a high glycolipid content. Furthermore, the histochemical properties of the granules are not identical in the genera of snake examined.

The secretion of see segment is granular, eosinophilic and non-metachromatic, PAS-positive and saliva-resistant, rich in phospholipids, coloured by acid haematin of Baker, bereft of cholesterol, and is non-sundanophilic. Among other things it is poor in alkaline phosphatase, glucose-6 phosphatase, acetyl naphtholesterase-AS, acetylcholinesterase and succinodehy-drogenase, but rich in acid phosphatase, butylcholinesterase acetyl naphtholestemse and accomdoxyterase. In contrast the Malpighian corpuscles and the collecting tubules are rich in adenosine triphosphatase and the former include 5 nucleoti-dase and acetylcholinesterase.

During the reproductive phase granules are liberated from the apical regions of the cells, directly into the lumen proposed that the entire apical tip of the cells breaks down to liberate its granular contents into the lumen. After expulsion a new apical pole is rapidly reformed, granules accumulate again and the cycle is repeated.

The sexual segment regresses after castration in male lizards and snakes, either before or during reproduction. Hypophysectomy has a similar effect and injection of androgens stimulates it to enlarge. The sex segment therefore is under androgen control from the testis, the latter, in turn, being influenced by gonadotropic hormones from the pituitary. Nevertheless, the sex segment may also be influenced from other sources, for bovine growth hormone (but not prolactin) will stimulate its enlargement in the adult male *Anolis*.

Volsoe (1944) suggested that the secretion may help to separate semen from urine, by temporarily blocking the renal tubules and ureter during copulation, or that it assisted the total emptying of the ampulla and seminal grooves to secure retention of semen in the female cloaca and oviduct. Possibly it has a significance in mating by virtue of its scent. *Bishop* (1959) speculated that it may be a nutrient for sperm or that the acid phosphatase present could be an activating agent in the semen or that it may increase seminal viscosity. It may act as a primitive seminal fluid. *Sanval* and *Prasad* (1966) considered that the association of acid phosphatase and phospholipids in the sex segment, together with the esterases in the cloaca, and the reproductive tract of the female, may lead to enzymic degradation of phospholipids yielding the liberation of fatty acids and glycerylphosphory choline. These may serve as a source of energy for the sperm—a substrate for the survival of spermatozoa in the oviduct.

The sex segment supplies secretions which form a copulatory plug in *Thamnophis sirtalii, T. butleri* and *Natrix taxispilota,* which when present in the female's cloaca prevents rival males from coplating with the same female. Its function, however, is still not clearly understood. *Prasad* and *Reddy* (1972) suggested a homology of the sex segment with the mammalian seminal vesicle, based on its embryological origin, its relationship with the vas deferens, and its response to androgens. This view would seem to need more detailed investigation, before it can be supported with any degree of confidence.

The Structure of the Reptilian Kidney

The ultrastructure of the glomerulus *Phrynosoma cornutum* and *Trachydosaurus rugosus* has been described in details. The proximal and distal uriniferous tubules were likewise described in *Anolis carolinensis, Hemiditylus* spp., *Phrynosoma cornutum* and *Tropidurus* and *Crocodylus acutus. Kurosomi et al.* (1966) showed three types of mitochondria, which contained rod-like, round, oval or pipe-like inclusions, in the proximal and distal nephric tubules of hibernating and active *Elaphe quadrivirgata. Schmidt-Nielsen* and *Davies* (1968) studied *Chelonia mydas, Crocodylus johnsoni, C. porosus, Tiliqua scinocoides* and *Laticauda colubrina*. They related such changes in ultrastructure of the various segments of the tubules *expecially* in the intercellular spaces) to differences in the osmotic level of the tubular fluid. The information on the ultrastructure of the reptilian uriferous tubule was summarized by *Ericsson* and *Trump* (1969).

The pedicels of the podocytes, rest on a well-defined basement membrane with three layers, lamina rara externa. lamina densa and lamina rara interna and join the epithelium of the capsule and of the glomerular capillaries. The importance of the glomerular central cell mass was emphasized by *Pak Poy* (1959) and he likewise described the podocytes and pedicels which extend from the capsule to cover any exposed surface of the glomerular central mass at Bowman's space. The fine structure of the glomerular capillary wall with its attenuated endothelium is, apart from its position, no different from that of the central masscells. The latter may represent the vestigial mesenchymal Anlage out of which glomeruli develop. The cells of the proximal convolute tubule, which have a well-developed brush border, frequently show numerous pinocytotic vesicles at their luminal margin. The basal plasma membrane shows simple and complex infoldings, which likewise generally occur in vertebrate nephrons. *Roberts* and *Schmidt-Nielsen* (1966), however, claimed that the basal membranes of the proximal and distal tubules were not infolded in *Phrynosoma* and *Tropidurus,* but were in *Hemidactylus,* where they resemble renal tubule cells of mammals. The latter workers described cilia as well as microvilli on the proximal tubule luminal surfaces of the two xeric species. The cilia, however, probably belong to the intermediate segment. Blebs of the microvilli, thought by *Anderson* (1960) to be manifestations of a functional secretory cycle, as the contained material was assumed to be uric acid are probably fixation artefacts The surface plasma membrance of the ciliated intermediate tubule segment is not infolded and the cuboidal cells of the distal tubule show merelly a few short projecting luminal microvilli. The tubule is surrounded by conective tissue like the proximal and intermediate tubules; the plasma membrane has elaborate lateral interfoldings which interdigitate with those of adjacent cells.

The renal tubule cells of *Crocodylus* have been described, by a number of workers. Noting that they differed from those of other vertebrates in lacking basal infoldings. All other reptilian tubules had lateral intercellular spaces extending from the terminal bar to the basal lamina, with in interdigitating cytoplasmic projections extending from adjacent cells; a similar arrangement is found in the mammalian gall bladder. Fluid movements which proceeds across the proximal and distal tubules are correlated with the width of the lateral intercellular space.

The ultra structure of reptilian nephric tubules, shows fairly typical tetrapod nephrons. They demonstrate the brush border microvilli of the proximal tubules, large numbers of rounded, oval or irregularly shaped mitochondria round or slightly irregularly shaped nuclei either basally or centrally situated in the cell, pinocytotic vesicles between the mocrovilli and within the cell, a Golgi complex situated near the nucleus, various vesicular inclusions and lipid droplets, often surrounded by or in intimate association with mitochondria, and many individual ribosomes or polyribosomes in the cell matrix. Nevertheless, some profiles of renal components as reported by different authors could well be criticized for showing artefacts as a result of cell shrinkage. Kidney tubules (certainly amphibian ones would seem to be extremely sensitive to fixatives used in electron microscopy, especially; thus any exceptional ultrastructural profiles should be treated with caution.

THE GONADS

Development of the Gonads

The embryology of the reptilian gonads has been described by a number of workers Similarly. The sexual bipotentiality of the reptilian gonad during embryogeny has also been described. *Forbes* (1956) wrote: "If the gonad is to become a testis then the cortex regresses and the medulla forms the definitive medullary cords; if an ovary then the cortex thickens rapidly while the medulla degenerates into stroma." He believed that prolonged embryonic bisexuality in general was the rule in reptiles, though it now appears to be more typical of turtles and alligators. Precocious sexual differentiation of the gonads, however, occurs in *Thamnophis* at 16-19 days of incubation and in *Chamaeleo bitaeniatus* and *Lacerta siculd*. However in *Lacerta vivipara* there is prolonged sexual bipotentiality, but only in the female.

Evidence of normal heterosexuality is common in reptiles during embryogeny and in juveniles and adults. A transient Mullerian duct usually develops in male embryos of *Lacerta viridis, L. agilis* and many other reptiles *L. sicula* and in *L. vivipara* . However, Dantchakoft (1937) claimed that the duct regresses in male lizards to disappear by hatching time, as it does in *Thammnnophis* nor does it persist in older individuals of *Lacerta*. Nevertheless, a vestigial oviduct is recognized in adult males of various turtles in young males of *Emydoidea blandingii* and *Mauremys caspica leproca* for sometime after hatching, and in adult males of *Vipera* and *Natrix.*

A vestigial wolffian body and its duct were found in adult females of *anguis,* and *Lacerta, acanthodactylus, Agama, Uromastyx* and *chamaeleo Shufeldt* (1890) likewise found a parovarium (vestigial mesonephros) and a duct of *Gartner* in an adult female *Heloderma.*

At birth female *Agama agama* still have paired mesonephroi and their ducts, though the mesonephroi usually regress in older specimens. Adult female *typhlops* and *Leptotypholops* have large Wolffian ducts, some ending blindly. The right one accompanys the right oviduct and the left the left ureter. A female *Trionyx* possessed a rudimentary vas deferens and a female alligator 29 cm long had well-developed bilateral mesonephric ducts. Alligators 13 months old, have gonads with cortical (ovarian) and medullary (testicular) regions, and Wolffian and Mullerian ducts still exist simultaneously. In three immature female alligators, just under one metre long (2.5 years old), the *"medullary rest"* was composed of luminated testicular canals, containing germ cells; rete canals join them to the epididymis, which continues into the mesonephric duct (vas deferens).

In the turtle *Sternotherus odoratus* early gonadogenesis consists of a rapid proliferation of germinal epithelium on the ventromedial surfaces of the mesonephros to form germinal ridges, and the non-luminated sex cords first differentiate and grow inwards from the germinal epithelium. Blood vessels and stromal elements occur. At this indeterminate phase of bisexuality all embryos between 7.5 and 11 mm (carapace length) possess oviducts and a phallus at the same degree of development. A gonadal cortex and medulla (the potential ovary and testis respectively) coexist, though they are variable in development in different individuals. In specimens 11-14 mm long, a potential male or female gonad is distinguished by the relative amounts of medulla or cortex present in the gonad. During this period the primary germ cells develop slowly and they are equally distributed within the sex cords and germinal epithelium. In a female specimen 14 mm long, the separated medullary cords are beginning to degenerate. A potential testis has medullary cords still in contact with the cortex whose cells form seminiferous tubules in addition to those arising from the medullary cords. Regression of oviducts in the male or phallus in the female commences at about this time. A similar bisexual period occurs in *Alligator* embryos (77-88 mm head-trunk length at 60-70 days of incubation).

In the developing embryos of *Chrysemys picta marginata Allen* (1960a) called the mass of *cellular cords* destined to become *seminiferous tubules* the *sex cords;* the *rete cords* were those structures that formed canals uniting the seminiferous with the mesonephric tubules. *Rete cords* originate from evagination of Bowmans capsules and the basal portions of funnel cords (solid peritoneal funnels which are vestigial from the beginning). *Rete testis* (or rete ovarii) elements grow into the germinal epithelium, situated between the funnel cords and mesentery, and unite with the sex cords of germinal tissue. At hatching many young males (but not females) have luminated rete cords, though the latter are still recognizable in females. *Allen* (1906a) described distal portions of the funnel cords to be incorporated in the adrenal cortex. In *Sternotherus* rete cords originate in a similar manner. The complex in the male permits sperm to leave the testis via the rete, epididymis and vas deferens to the exterior. The rete, mesonephros and duct degenerate in the female. In the male the gonadal cortex disappears with concomitant development of the testicular tubules, which at hatching include spermatogonia. The intertubular region includes stroma, blood vessels and nerves, but no interstitial cells. New seminiferous tubules continue to develop from the peritoneum until sexual maturity. A female *Sternotherus* with a 16 mm long carapace possesses an ovarian cortex with numerous oogonia and oocytes; the medulla is reduced and now merely includes stroma, blood vessels and nerves. Only at hatching do the follicles enlarge and move towards the centre of the ovary. Females 9 to 11 years old are sexually mature, with large ripe ovarian oocytes and they deposit eggs. Mature sperm occur in the testis and epididymis of males 4 years old.

When the germinal ridges of a 15-16 day old embryo of *Thamnophis* have been observed it is found that each consist of a sightly thickened (2-3 layers) epithelium. Future *rete cords* join the ridges and mesonephric Bowman's capsules at regularly spaced intervals. By 19 days the gonads are distinctly ovarian or testicular. The testis now comprises a rudimentary cortex but primary seminiferous tubules extend throughout the gonad; few germ cells are present at this time. In potential ovaries a tunica albuginea separates the medulla and cortex. After 90 days the testis has no cortex; seminiferous tubules are well-developed and numerous spermatogonia are now undergoing development within their lumina. At birth the male lacks a *Mullerian duct* and the testes (the right is approximately 20% larger than the left) are completely differentiated, with abundant intertubular stroma and active spermatogenesis. At this time, the ovaries retain a loose central stroma. Each one is elongated, fusiform and transparent and the cortex has recognizable oogonia and oocytes. A tunica albuginea surrounds the irregular, sac-like vestigial *medulla*.

In *Anguis*, the embryos 5-6 mm long have Anlagen of gonads. At 6-7 mm longitudinal projections of coelomic epithelium cover a mass of undifferentiated cells derived from it. Differentiation into either ovaries or testes occurs in embryos weighing 75-100 mg, and sexes are clearly distinguishable in embryos 39-41 mm snout-cloacal length (75-80 mm total length). Ovaries are well differentiated and the cortex contains germ cells. A vestigial medulla of small tubules is still present, and rete tubules, often retained in the female, penetrate the hilus. The testes, of luminated seminiferous tubules often with germ cells within, still retain a rudimentary cortex; as yet there are no interstitial cells. *Raynaud* concluded that testicular canals proliferate by the reorganization of cells of the undifferentiated gonad. The sex cords may retain contact with the germinal epithelium preserving their ancestral relationship. A central longitudinal testicular canal receives openings of the epithelial cavities (future seminiferous tubules); this structure is found in fishes and in some Gymnophiona, but it soon rapidly regresses in *Anguis*. In both sexes of *Anguis* rete originate and develop in a way generally similar to that of *Chrysemys* and *Sternotherus* *Raynaud* and *Pieau* showed in an 80 mm long embryo that cellular elements from Bowman's capsule join coelomic epithelial proliferations. Segmentally arranged bridges of cells penetrate the undifferentiated basal region of the gonad, thereafter to fuse into a continuous epithelial rete lamina. The lower half separates into future rete cords, and the upper half into a reduced, narrow, longitudinal ribbon along the median wall of the mesonephros. In males at the end of foetal life, the enlarged *Wolffian duct* is convoluted; the rete cords and the longitudinal canal are luminated. Of the mesonephric tubules 47% become vasa efferentia, the others become vasa aberrentia and some elements of the paradidymis. In females the *epoophoron* (homologue of the male epididymis), the *Wolffian duct*, and *rete ovarii* ultimately regress. *Wiedersheim* (1890 a,b) concluded that the ostium abdominale of the Mullerian duct of turtles and crocodilians is probably derived from a readymade second pronephric nephrostome, a view supported by *Balfour* (1885) for the chick. The *Mullerian duct* grows caudally near the pronephric duct, to open into the cloaca by cellular proliferation from an originally solid anterior cell mass. According to *Wilson* (1896) the ostium abdominale of embryo crocodilians originates from the thickened epithelium around the pronephric nephrostomes. When the latter disappear the groove like *Mullerian anlage*, anterior to the mesonephros, becomes a duct on the internal surface of the mesonephros.

In the alligator (9-10 mm crown-rump length) the *Mullerian epithelium* is generally located between the *Wolffian canal* and the angle formed by the lateral edge of the mesonephros and the body wall. It also grows caudally by the extension of a solid rod of epthelial cells, which secondarily develops a lumen. In contrast *Moens* (1911) suggested that the duct of crocodiles develops from coelomic epithelium along its course. Forbes' results, however, were later supported by *Forsburg* and *Olivecrona* (1963) in alligators, for they found the tip of the Mullerian duct seemed to grow caudally in close association with the Anlage of the Wolffian duct. At the tip scattered granules were present during growth, which presage the degeneration of the duct in males.

The *Mullerian ducts* of *Sternotherus* arise simultaneously with the gonadal sex cords. Lateral to the *Wolffian duct*, on either side, primordia form a tubal ridge on the dorsolateral wall of the mesonephros, similar to the arrangement in alligators. The first groove-like depression forms the anterior coelomic aperture and thence separates from the ridge, to grow caudally towards the gonads. In the female it opens into the cloaca after hatching.

Among the lizards the walls of the oviduct of *Lacerta agilis* become thickened due to enlarged albuminous glands. *Braun* (1877) noted that the *Mullerian duct* originates as a groove in the peritoneal epithelium and continues posteriorly as a solid rod of cells between the peritoneal epithelium and the Wolffian duct. The *Mullerian ducts* of lizards and snakes arise from flattened depressions in the sub-peritoneal tissue of the Wolffian body. The subsequent groove on the lateral edge of the mesonephros opens proximately into the coelom—"*der Mullersche Trichter*" or "*Abdominal Offnung.*" Growth is by cellular proliferation along the whole length, and there are no connections between the Mullerian and Wolffian ducts.

The *Mullerian Anlage* of *Anolis carolinensis* develops into two parallel longitudinal folds, which join to form a tube; later the tube extends caudally to the cloaca. The Mullerian ducts of *Xantusia* ad of *Chamaeleo* likewise originate when the gonads are undifferentiated and after the disappearance of the pronephros. Anterior to the mesonephros there is a short thickening or plaque of cells in the coelomic parieto-pleural epithelium. The cells of the Mullerian "pavilion' proliferate and extend caudally at the external surface of the mesonephros alongside the Wolffian duct. There is an anterior ostium abdominale in the female, in which caudal development is more rapid than in the male. Near the cloaca the Mullerian ducts unite to form a uterus; in the male they regress before reaching the gonad. As soon as the gonad starts to differentiate in a 6.5 mm embryo of *Xantusia*, the Mullerian ducts regress in the male but enlarge in the female. Caudally the band becomes a narrow ribbon and its upper part forms the *Mullerian duct*. Cranially the plaque becomes the ostium; it grows caudally (below the raised narrow epithelial ribbon) alongside the mesonephros. The median and ventral condensations were described in *Lacerta agilis* by *Bertelli* (1897). According to *Raynaud* the primordium of the *Mullerian duct* has no relationship with the raised epithelium in the dorsal angle of the mesonephros.

In early embryos of *Anguis* (90-100 mg weight) the ostia are narrower in the male. In females (100-120 mg weight) the Mullerian duct cells hypertrophy and differentiate, and they have almost reached the cloaca. In somewhat older males (120-140 mg weight) duct development and cellular height are a variable. The duct terminates before the urodaeum and probably never reaches it. *Pycnotic nuclei* occur in the duct of such embryos when 160 mg in weight, and the duct finally regresses.

Thus, we see that, the *Mullerian* plaque of *Anguis* formed from tissues medial to the pronephros, and the ostium tubae originates here or at least in the region between the pronephros and mesonephros. Furthermore, the bases of the vestiginal pronephric vesicles are incorporated into the epithelium of the Mullerian plaque and the small invaginations from which the latter develops may correspond to pronephric nephrostomes.

This would imply that the Mullerian ostium is derived from a pronephric (or mesonephric) nephrostome(s) in ontogeny and phylogeny. *Rayaud et al.* (1970) described the relationship between length and degree of histological differentiation of *Mullerian duct*, and the corresponding stage of testicular differentiation in the male embryos of five species of reptiles, at the time of the arrest of the duct's caudal progress. In these species the *Mullerian ostia* form at the same stage of somatic development. However, the testes commence to differentiate before *(Natrix)*, simultaneously with *(Lacerta viridis, L. sicula)* or after *(Anguis, Testudo)* the formation of the ostium tubae. *Mullerian ducts* are short in *Natrix, Lacerta viridis* and *L. sicula*. They reach the caudal part of the mesonephros and cloaca in *Anguis* and are complete in *Testudo*. Thus, the earlier the testis differentiates, the sooner the arrest of the Mullerian duct. These authors describe the histological pattern of degeneration at the tip of the Mullerian duct, and the cytotoxicity is referred to testicular secretion, for testosterone propionate elicits similar degeneration in this region of female *Anguis* embryos. Whether hormonal action is direct or, as is more likely, it triggers secondary activities such as *autolysis* or *phagocytosis* was not considered.

In cyclostomes the *Mulleriar ducts* (oviducts) are absent; probably they have not evolved in agnathans, but these forms have a well-developed pronephric and mesonephric system. In craniates *Mullerian ducts* presumably have evolved subsequent to the renal system. If indeed the ostium abdominale is of nephrostomal origin and the *Mullerian plaque* incorporates pronephric tissue cells within it, then this would imply that caudally proliferating cells of the *Mullerian* duct have a renal derivation in phylogeny. Germ cells of both sexes in Amphibia and Amniota may therefore utilize renal ducts for agress. External fertilization occurs in the former or within the *Mullerian ducts* in the latter. According to this argument the vas deferens (*Wolffian duct* or *mesonephric duct*) and oviduct (*Mullerian duct*) are therefore homologous structures of "*coelomoduct*" derivation.

THE REPRODUCTIVE SYSTEM OF ADULTS

Male Gonads

Like other vertebrates the lizards and snakes have paired testes the right testis is generally situated anterior to the left, though in *Eumecs obsoletus* and *Crotaphytus collaris* the left is anterior. The testes are almost symmetrically disposed in *Agama tuberculata* The right testis is anterior to the left in all amphisbaenians. Tests are particularly elongate in snakes and snake-like lizards. Generally speaking the testes of reptiles are similar to those of birds and mammals and they are vascularized by spermatic arteries and veins. Among lizards *Hemidactylus* is typical in having abdominal testes surrounded by a tunica albuginea of connective tissue, and attached dorsally to the body wall by a fold of peritoneum, the mesorchium. The convoluted seminiferous tubules are interspersed with interstitial cells, blood vessels and *lymphatics*. Loosely coiled seminiferous tubules of the creamy-white testes of *Anniella* are easily seen with a hand lens. In *Scincus* the compact tunica of fibrous connective tissue has an inner vascular surface and is bounded externally by a serious coat, which is continuous with the mesorchium. The tunica of *Lacerta* has a smooth muscle constituent with parallel connective tissue fibres and blood vessels.

The oval white testes of *Xantusia* become greyish in summer. Their tunica comprises 6-7 layers of connective tissue with a few scattered fibroblasts, and its septa separate the tubules. Intertubular spaces contain blood vessels, fibroblasts and interstitial cells. Each seminiferous tubule includes an outer basement membrane, Sertoli cells, and a syncitium of germinal elements. The typically ovoid testes of *Lacerta agilis* and *Eumeces fasciatus*, which are about 6 x 3 and 3 x 5 mm respectively, are generally similar to those of *Xantusia*. Testes of chameleons are black in colour.

Leptotyphlops humilis *and L. dulcis* have paired spindle shaped testes located at the front of the caudal third of the body. They resemble strings of closely abutting small lobes (the largest at the centre), 2-7 lobes per testis, with from 4-6 the most common. The left testis has fewer lobes, though occasionally the same number or even more may occur. In a fully active testis the lobes are 0.5 mm long and 0.5-2 mm wide. There are 4.15 distinct ellipsoid lobes to the testis of *Leptotyphlops phillipsi,* the testicular units each having one or more seminiferous tubules giving origin to 4-5 short Testes of *Typhlops simoni* are generally similar in appearance (5-7 mm long; 1 mm in diameter; and 4 or 5 lobes per testis). In all these species each lobe is encased by a thin tunica albuginea of mesothelium, supported by connective tissue (3-5 μm thick) and 1 or 2 layers of smooth muscle run obliquely and longitudinally.

In the case of *vipera berus* the testes are elongate, cylindrical, yellowish-white organs, situated in the caudal third of the body. The right is larger but they vary in size according to the reproductive cycle. Each is covered ventrally by the peritoneum and encased in a tough, thin, transparent tunica, 3-5 μm, thick of elastic fibres. No tunica septa occur between the tubules. In no advanced snake is the testis multipartiate or lobulated. In *V. berus* about 20% of the testicular volume is composed of intertubular (interstitial) tissue. Large interstitial cells occur singly or in groups of up to 100 cells, as do fibrocytes and other connective tissue elements. The volumes of the fresh testis are 1250 mm^3 (right) and 980 mm^3 (left) and the average diameter of a seminiferous tubule is 200 μm. Each tubule is invested by a membrana propria (2.4 μm thick) of circular collagen and a few elastic fibres. The epithelial cells lining the tubules vary seasonally in appearance and number *Vipera berus* has a continuous annual spermatogenetic cycle and different regions of the tubule reach the same stage of development. Simultaneously. Sertoli cells are always present, especially near the basement membrane; their number is fairly constant as new ones are created as old ones degenerate. The testis of *Natrix* is only half as large as that of *Vipera,* despite the larger body size of *Natrix*. A thin tunica, usually without smooth muscle, and the absence of septa are characteristic of sauropsid testes, in contrast to the condition which occurs in mammals.

In the turtle *Emydoidea blandingii* the yellow and testes are bound to the kidneys by the mesorchium, the latter extending mediolaterally over the epididymis and kidney and being continuous with the body wall and the dorsal mesentery *Bojanus* (1819) had earlier described the testis of the turtle *Emys* to be situated alongside the kidney (the ovary is situated more anteriorly) The testes of *Trionyx spiniferus* are elongated; those of *Emys* elliptical in shape. The testes of immature *Crocodylus palustris* are small, white, rod-like bodies comprising seminiferous tubules and interstitial cells, held together by a thick tunica. In section tubule walls (one cell thick) surround a lumen filled with a fibrous network. Sertoli cells and spermatogonia line the lumen. At this stage there is no indication of spermatogenesis and spermatocytes, spermatids, and spermatozoa are not yet recognizable. Spermatrogenesis in the tubular testis of *Sphenodon* includes the various germ cell stages arranged as in mammals. Sertoli cells but not interstitial cells were recognized.

The sexual cycles of reptiles have been widely studied. It should be stressed, however, that between say North American, tropical and Australian forms, for example, the cycles are not strictly comparable in terms of seasonal changes in pattern, unless the times and lengths of the seasons and hence temperature supervening, are taken into account. Many lizards emerge from hibernation with their testes at maximum size (probably they use sperm produced mainly during the previous autumn), but testicular size is greatly reduced in summer. This group includes *Sceloporus undulatus,* whose maximum testis size in November is seventeen times the minimum in July, *S. graciosus, S. occidentalis, Phrynosoma douglassi,* and *Uta stansburiana* of south-eastern Arizona. The last regression of the testis is complete by mid-August, and from September the seminiferous tubules develop rapidly, to reach maximum size in December-January and then decline after February.

In other lizards the tests are smallest in late autumn to spring and they are maximal in size from spring to early summer. This group includes among others *Anguis* and *Lacerta, Eumeces laticeps, Agama impalearis, Anolis carolimensis, Anolis* lizards of Puerto Rico and the Virgin Islands, *Uta stansburiana stejnegeri* of western Texas, *Urosaurus ernatus,* the testes of which are maximum in size in June *Cnemidophorus hyperythrus, Uma notata, U. inornata, U. scoperia, Sceloporus orcutti,* and *Dipsosaurus dorsalis, Acanthodactylus erythrurus lineomaculatus, Sceloporus malachiticus, Sitana ponticeriana* and *Calotes nemoricola* of Kashmir, India. *Calotes versicolor* of India. *Holbsnokia texana* and *Cnemidophorus gularis, Uromastyx hardwickii* of West Pakistan.

1. Urinogenital System

It is of interest that different species of the same genus show differences of testicular size relative to hibernation. The rates of *Phrynosoma solare,* which are minimal in size at hibernation in November to March (each 40 mm^3 in volume; 5.7 mm long x 5.7 mm in diameter), are largest during the May breeding season (15 x 11 mm) and then decrease in size in July and August. In *Xantusia vigilis* there is an autumn—winter spermatogenesis followed by a spring spermiogenesis. At the height of the breeding season each testis measures 5 x 4 x 3 mm (weight 33 mg), an increase over the winter volume of nearly nine times. Tobular diameter increases from 50 to 350 μm and the lumen, which is closed at hibernation, reaches a diameter of 30-60 μm in the mature testis. Regamey (1935) showed the testis of *Lacerta agilis* to be small at the end of June (diameters of tubule and lumen are 190 and 40 μm respectively). Against the wall are *Sertoli cells* and spermatogonia but mainly spermatocytes and only few spermatids and spermatozoa. Spermatogenetic activity commences in July when the diameter of the seminiferous tubules increases to 220 μm. Spermatids and spermatozoa are abundant during October and November.

Spermiogenesis is in full activity in the spring and the seminiferous tubules, though little changed in diameter, are replete with spermatids transforming into spermatozoa.

Seasonal cyclic changes have also been reported in the North American *Eumeces fasciatus* is somewhat similar is its testes are enlarged in January and there are spermatozoa in the tubules in March to June. Spermatogonial proliferation also begins in April and continues slowly until tubule repopulation is achieved in November. All spermatozoa are eliminated from the testes by June in *E. latiscutatus* of Japan. *Scincus* (end of June to beginning of July) and *Chalciaes* (April) have restricted breeding seasons. The testes are quiescent for about nine months of the year in *Scincus*, but for a slightly shorter time in *Chalcides*. Just before breeding, the testes of *Scincus* increase enormously in size; in *Chalcides* maximum size occurs at the end of March. Testes of both genera dramatically reduce from August until autumn, when enlargement recommences; maximum size is reached the following spring.

During the spring mating season, the testes enlarge in the south Algerian *Uromastyx, Varanus* and *Scincus. Acanthodactylus* also breeds in autumn and, in contrast to the other three genera, has testicular tubules wich do not involve. Sexual activity is practically continuous except for quiescence in July and August. In *Varanus komodoensis* testicular activity is highest in June and July. In *Chamaeleo pardalis* spermatogenesis is completely arrested in July and August (winter in Reunion Island) and the testis is involuted. It starts again in September and the efferent ducts are full of sperm from November to June. *Sceloporus jarrovi* of southeastern Arizona mates in the fall when the testes are their maximum size. Ovulation and fertilization occur in November and December, but embryonic development stops at the blastoderm stage till March; it then recommences and the young are born 10 weeks later. The mode of development is related to the food supply.

In *Emoia cyanura* and *E. werneri* reproduction is probably continuous throughout the year, with a minimum in May and June (shortest days) ana a maximum in November and December (longest days). Even during May and June spermatozoa are abundant in the epididymis of the male of *E. werneri.* Testes of *E. cyanura,* however, are smaller at this time. Likewise *Anolis acutus,* endemic in St Croix, Virgin Islands, reproduces throughout the year though reproductive potential is highest during June to October, when the level of rainfall is maximal and lowest in January to April, when rainfall is minimal. Nevertheless, even during the latter period more than 50% of the females are ovigerous. Continuous breeding also occurs in the *Cyrtodactylus malayanus, C. pubisulsus, Draco melanopogon* and *D. quinquefasciatus.*

There is no seasonal variation in the *spermatogenic cycle* of the geckos *Hemidactylus, Cosymbotus* and *Gehyra* of Java or of *Hemidactylus flaviviridis* of India. In *Carlia fusca,* a tropical skink, testicular size is maximal just before peak sperm production and storage in October and at a minimum after sperm evacuation in February and March. Furthermore, in contrast to the significant seasonal size variation in lizards of the temperate zone, only slight (if any) changes in size occur in the testis of *Carlia rhomboidalis,* an Australian skink which inhabits areas of heavy rainfall; reproductive activity is the continuous throughout the year. Likewise seminiferous tubules of *Amerva* spp. of Costa Rica, never involute during the year. In *Scincella lateralis* of North America, breeding occurs in March to August, and there is irregular hibernation during the winter. Gonadal size and hence reproductive activity of *Uma scoparia* would seem to be directly related to the supply of available food, insects, which inhabit annual plants. The prime factor therefore is thelevel of rainfall.

Among snakes the testes of *Thamnophis radix* are largest in July to October and smallest in December to May. *Thamnophis elegans terrestris* and *T. brachystoma and Leptotyphlops humilis, L. dulcis, Typhlops vermicularis* and *T. simoni* show parallel changes. In *L. humilis* in March the diameter of the seminiferous tubules is 180 µm; by mid-April is has risen tc 260-270 µm and there are large lumina, which reach a maximum of 250-275 µm. In July tubular diameter is again reduced to 140-190 µm and the basement membranes are shrunken. Again in *Diadophis punctatus* from Florida testes are smallest in the winter to spring period; spermatogenesis then begins and maximum size is reached in the summer. The pattern is repeated in *Natrix natrix* of Europe, for testes are minimum on emerging from hibernation in March and maximum (4-5 times as large) in August and September. Maximum size occurs slightly earlier (April to early May) in *Naja naja* of southern China, when tubules are 300-350 µm in diameter and free spermatozoa abound within their lumina. Testes regress from May to the end of August; tubular diameter is then 90 µm and they are spermatogenically inactive. Spermatogenesis begins again from the end of August to the end of September, and after winter accelerates to reach the peak again from March onwards.

Little seasonal variation occurs in the testis of *Vipera berus*. Volume increases about twofold in March just before mating, mainly due to increase in tubular diameter (from 130-120 µm) and length as well as intertubular space. The diameter of the seminiferous tubules of *V. aspis* of the Pyrenees doubles between May to June and September, and thereafer reduces again. The volume of the testis of *V. berus* is minimal immediately after mating in May. In *V. berus* of southern Britain testicular size decreases in July, but a second development occurs in autumn before hibernation.

Among turtles the seminiferous tubules of *Mauremys caspica* increase in diameter threefold in August (up to 450 µm), and the epididymal canals when full of sperm in December are similarly enlarged, compared with those in September. Testes of *Chrysemys picta* weighed the most in March after hibernation and after loss of sperm to the epididymis decreased in size in April. Likewise the tests of the Panamanian *Pseudemys* are enlarged from July to December, coincident with spermatogenesis. Regression occurs thereafter, though the epididymides stay large when full of sperm. Sperm were found in the vasa deferentia of adult males of *Macroclemys temmincki*, of Louisiana, throughout the year, which suggests that there is an extended breeding season, as found in *Chelydra serpentina*. However, the testes of *C. s. serpentina* from Tennessee have an active phase in summer and autumn and an inactive phase in winter and spring. The testes are smallest in spring the then larger from July to September; they reduce in size in October when the epididymis becomes full of sperm. The condition is similar to that in *Clemmys* and *Sternotherus (vide supra)*.

Reproduction is more frequent in tropical or sub-tropical forms, as may perhaps be expected, and such species often have a second breeding season in the autumn or even breed throughout the year. Presumably breeding is greatly influenced by temperature, though probably light and rainfall, which secondarily influence food supply, are appreciably involved. Such stimuli doubtless act through the medium of the endocrine system in a manner not yet understood.

"Prenuptial" spermiogenesis occurs in a number of species in spring, before copulation, and *"postnuptial"* spermiogenesis in summer, after it. In the former spermiogenesis usually starts in the spring, after hibernation, and ends before mating; in the latter spermatogenesis begins in spring and spermiogenesis finishes at the end of summer. Spermatozoa are thence stored in the vas deferens for use the following spring.

The Accessory Ducts

In several species of snakes like *Vipera berus* and *Natrix natrix* there is a narrow band-shaped epididymis on the dorsomedial surface of the testis, connected to it by a mesentery through which pass blood vessels and ductuli efferentia that join the seminiferous tubules to the epididymis. The junction between the testicular tubules and the ductuli efferentia-occurs along the entire length of the testis, though *Morgera* (1905) reported that in *Coluber* and *Elaphe* they arise only at both ends. *Volsoe* (1944) believed that the central ones were probably overlooked, owing to their shortness. *Natrix* has up to 33 connections that anastomose and form a network in the mesentery (incipient rete testis). The ductuli efferentia lead into the ductuli epididymides. There is a distinct junction between the stretified cells of the seminiferous tubules and the single layer of flattened epithelium of the ductuli efferentia. Likewise there is sharp separation between the latter and the high columnar cells of the ductuli epididymides. Some of these cells are ciliated, and during the breeding season, may be filled with secretory granules. *Volsoe* (1944) could not confirm *Alverdes'* (1928) claim that smooth muscle surrounds the ductull epididymides. The latter ductuli continue into a ductus epididymidis along its entire length. The non-ciliated cells of the latter form a convoluted tube with a wide lumen; the cells are 50 *μm* high and 40 *μm* wide; they are enveloped by smooth muscle. In the breeding season the lumen is filled with sperm. So-called basal cells, of unknown function, occur amid the epithelium of the ductus epididymidis. The latter leads into the proximally convoluted was deferens, which straighens distally to continue caudally along the lateral surface of the kidney and accompany the ureter to their common opening into the cloaca. Distally the ureter has an ampulla which fills with sperms during the breeding season.

In *Thamnophis elegans* the nasa efferentia (ductuli efferentia) likewise penetrate the mesorchium along the entire testicular length. The short first portion of the dectuli epididymidis is secretory for part of the year; the distal region includes ciliated cells. The larger ductus epididymidis is also secretory and, as in *Vipera berus,* is not ciliated. The non-secretory ductus deferens (vas deferens) is surrounded by smooth muscle and its main function is sperm storage; it is convoluted in its hinder region in *Thamnophis brachstoma*. The arrangement in *Vipera aspis* is generally similar to that in *V. berus* and *Thamnophis.* There is a shorter non-ciliated region called ductulus epididymidis 1, of columnar cells 20 μm high with secretory granules; secretion

TABLE 1.1

Tetis-epididymis complex of reptiles. Terminology of various components by different authors

	Testis			Accessory Ducts of Testis		
SAURIA						
Henry (1900)				Petite tubes	Greoss tubes	
Morgera (1905)	Tubuli seminiferi	Vasa efferntia		Tubi medii Tubi piccoli	Canale deli Epididimo	Defernete
Alverdes (1928) (and snakes	Tubuli contorti	Hodenaus fuhrgange	Hauptkanale Verbindungsstuke,	Vasa efferentia epididymidis Haupstucke	Ductus	Vas deferens
Badir (1958) *Scincus, Chacides*	Tubuli seminiferi	ductuli efferentes	longitudinal canal	Ductuli epididymides I Ductuli epididymides II	Ductus epididymidis	Ductus deferens
OPHIDIA						
Van dne Brock (1933)	Tubuli seminiferi	ductuli efferentes		Ductulus epididymidis	Ductus epididymidis	Ductus deferens
Volsoe (1944)	Tubuli seminiferi	ductuli efferentes	marginal canal	Ductulus epididymidis	Ductus epididymidis	Ductus deferens
W. Fox (1952)	Seminiferous tubules	ductuli efferentes		Ductuli epididymides	Ductus epididymidis	Ductus deferens
Saint Girons (1957)	Tubuli seminiferi	canal efferents (ductuli efferentia)		Ductuli epididymides I Ductuli epididymides II	Ductus epididymis	Ductus deferens
RHYNCHOCEHALIA						
Gabe and Saint Girons (1964)	Seminiferous tubules			Ductulli epididymides I Ductuli epididymides II	Ductus epididymis	Ductus deferens

occurs at the end of June and July. A large region, called ductulus epididymidis II (diameter 65 μm), has ciliated cells 15 μm high and is not secretory. The ductus epididymidis (diameter 260 μm) has a large lumen and epithelial cells 55-60 μm high, which are little modified during the year though they are smaller during June and July. The individual variation in size of the convoluted, non-ciliated and non-secretory vas deferens (200-250 μm in diameter, height of cells 15-50 μm), probablyd reflects the varying quantity of sperm which fills it.

Among lizards, *Scincus* has four anterior and two posterior ductuli efferentia; the latter fuse, branch and then join those in front which likewise have fused in the mesorchium. This anterior network leads into a longitudinal canal by nine separate openings and other ducts end blindly. The epithelium of the longitudinal canal is similar to that of the ductuli efferentia. The tubules of the ductuli epididymides which originate all along the canal are lined with high columnar cells some of which are ciliated. The resulting networkk then leads into the ductuli epididymidis II, lined with cells of smaller diameter though cilia are still present. The tubules of the ductuli epididymides lead into the unbranced, convoluted ductus epididymidis, which is lined by cylindrical cells and surrounded by circular muscles, and then finally into the vas deferens. The vas joins the ureter near the cloaca. *Chalcides* was shown to have 9 ductuli efferentia which join the seminiferous tubules and form a networkk. At least 12 ductuli efferentia join ductuli epididymides I.

Lacerta has only 1 ductulus efferentis, while *Anguis* has 5 to 9 that open into a longitudinal canal. The ensemble constitutes a rete testis, although the epididymis proper is derived from about a dozen vasa efferentia, opening at one end into the epididymal canal and at the other into the cranial region of the longitudinal canal. A single duct joins the testis to the epididymis in *Lacerta muralis, L. agilis, Sceloporus* and *Eumeces*. The last is thus out of line with the other skinks (*vide supra*). *Reynolds* (1943) reported a longitudinal canal in *Eumeces*. Cilia occur in the epididymis of lizards as in other reptiles. *Van der Stricht*

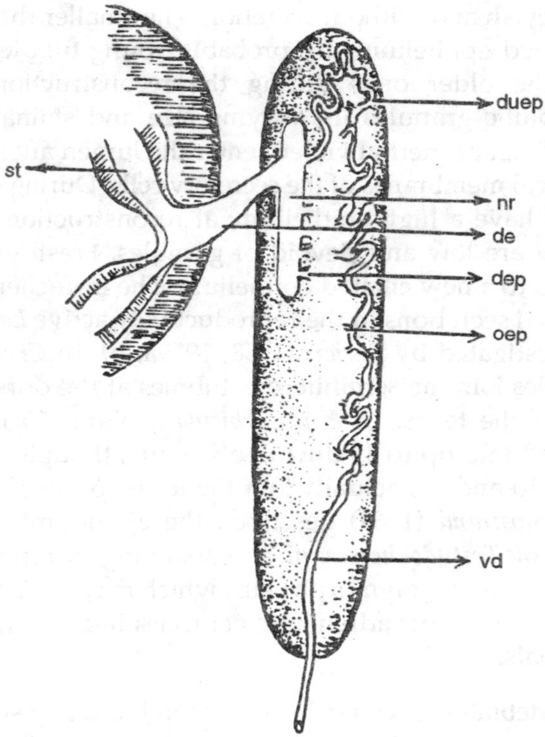

Fig. 1.24. *Lacerta*. Testis-epididymis junction de, Ductus efferens; dep. ductuli epididymidis; du.op, ductus epididymidis; st, seminiferous tubule; vd, vasa deferens.

(1893) described cylindrical, ciliated and non-ciliated secretory cells in thin and thick epididymal canals of *Lacerta vivipara*, as well as the discharge of saffranophilic granules from secretory cells into the lumen, where they mix with spermatozoa. According to *Henry* (1900), however, van der Stricht's descriptions relate merely to the secretory phase, for Henry described resting, presecretory, secretory and reconstructive phases succeeding one another, in the epididymis of various adult *Lacerta, Hemidactylus* and *Anguis*, as well as of *Vipera*. The epididymis is large when in full secretion (April to June) and contains a milky fluid. At rest (August to September) it is small, greyish and without secretion. The smaller thin canals, with ciliated epithelium, are probably young tubules, which replace the older ones during the reconstruction phase. Saffranophilic granules (of varying size and stainable with gentian violet or methyl violet), enter the lumen after rupture of the apical membranes of the secretory cells. During secretion the latter have a high epithelium; at reconstruction (June to July) they are low and devoid of granules. Fresh growth of cells leads to a new ciliated epithelium. The histochemistry of epididymal secretions in the reproductively active *Lacerta* has been investigated by *Furieri* (1958, 1959a, b). In *Chelydra* the rate tubules join the seminiferous tubules at the dorsomedial surface of the testis. In *Emys orbicularis* about 15 testicular canals lead into up to 45 ductuli efferentia, though only two, numbers 10 and 11, actually join the testis to the glomerular tubules. *Siostrand* (1969) described the epididymis and vas deferens of *Testudo hermanni* as containing norepinephrine confined to small polymorph cells, which may have the same function as the short adrenergic neurones in the genital tract of mammals.

In vertebrates reduction in number and resulting shortening of serially segmental organ components, generally denotes specialization: this feature is clearly seen in renal tubules. By this criterion snakes and turtles retain a primitive arrangement in having a large number of tubular ductuli efferentia. Lizards have a reduced number or only one. Turtles have numerous anastomoses forming a rete testis, similar to that found in mammals; unlike the situation on the latter, however, the rate is situated outside the testis. The rete cords (ductuli efferentia) are derived from peritoneal funnel canals. Transient funnels occur during embryogeny in marsupials and the mouse, but in other mammals these vestiges do not appear, and the rete cords are prolongations of the seminiferous tubules that open into the mesonephric tubules.

Interstitial (Leydig) Cells

Interstitial cells occur between the testicular tubules of most mature reptiles; however not seen in *Sceloporus undulatus*. Interstitial cells are also rare in *S. occidentalis, Xantusia vigilis*, in which they occur in small masses but are not recognizable during gonadal development and *Acanthodactylus erythrurus lineomaculatus*. *Evans* and *Clapp* (1940) did not mention them in *Anolis carolinensis* but *W. Fox* (1958) reported small groups each of three or four cells in *Anolis*, where they are hypertrophied in the intertubular space. They were not recognized in *Mabuya multifasciata*, nor at any stage of the reproductive cycle of *Uma notata, U. inornata*, or *U. scoparia*. *Forbes* (1961) failed to recognize interstitial cells in the testes of immature alligators but groups of 3 or 4 cells were noted in immature two-year-old male specimens of *Crocodylus palustris*. There was no sign of interstitial components in *Sphenodon*.

In general, the size, number, and appearance of the intersitial cells very during the reptilian reproductive cycle (*vide infra*), which may explain why they are sometimes missed. Their shape is polyhedral, elongated, or oval; occasionally they are syncitial. They have a large rounded nucleus and vacuolated or alveolar cytoplasm, which may include pigment granules, crystalloids and fat globules. Interstitial cells are binucleate in *Varanus*. Cellular elongation could well due to pressure exerted by the seminiferous tubules, which should be borne in mind when measurements are reported. Interstitial cells occur singly or in small groups, near blood vessels, among the seminiferous tubules. This applies to turtles (*Risley*, 1938b in *Sternotherus*, groups of 6-8 cells); crocodilians; lizards, who described the interstitial cells to be scattered in small masses in *Anguis* and geckos but in more voluminous masses in turtles and snakes and snakes.

In the quiescent testes of *Bungarus coeruleus*, during August and September, interstitial cells occur in small groups between the seminiferous tubules. During the period of testicular spermatogenic activity, in October and November, they are dispersed and seem to be few in number.

Numerous individuals have described secretory activity, or quiescence of interstitial cells, in terms exemplified by various criteria such as: (a) syncitial arrangement or changes in volume, (b) size and vacuolation of the cytoplasm or changes in granulation, (c) nuclear appearance, and (d) amount of tissue. Some of these criteria offer neither strictly nor reliable evidence of secretory activity. However, many oi these authors reported multiple cellular changes, including changes in numbers during the reproductive cycle.

The lizards *Lacerta agilis, L, vivipara* and others, exhibit a chronological relationship between cyclic activity of the interstitial cells and the secondary sex characters. In *Phrynosoma solare* the interstitial cells are syncitial at the end of the breeding season but have definite cells boundaries during the period of hibernation. Cellular volume is greatest at the height of breeding but minimal when spermatozoa are produced. The interstitial cells of *Hemidacrylus* are most numerous in August during the time when the testes are at rest and the seminiferous tubules shrunken, and fewest when seminiferous tubules are most active. In *Uromastyx* the interstitial cells hypertrophy in October and their cytoplasm is vacuolated. Secretory activity occurs after hibernation, in the spring and continues until sexual quiescence in the summer. The interstitial cells of *Varanus* are small in August, where they are crowded between the testicular tubules. Spermatogenetically inactive seminiferous tubules of *Varanus bengalensis* and *Mabuya macularia* have large compact masses of interstitial cells between them. At the height of spermatogenesis the tubules are expanded and the interstitial cells are dispersed and appear fewer. The interstitial cells of *Anguis* increase in size during spring; in April their cytoplasm is poor in fat but rich in fuchsinophilic granules. By June the granules have disappeared and the cells are charged with fat. The cells are small in July and August and a fatty surcharge in autumn diminishes in winter when the cells are inactive.

Herlant (1933) showed the cycle of secondary sexual characters to proceed in parallel with changes in size, number and constituents of the interstitial cells; this was similarly noted in *Anolis carolinensis*, in which they reach maximum size in July, perhaps owing to reduction of pressure by the seminiferous tubules. They are smaller and scarcer, with diminished activity in autumn, when they cannot be distinguished from fibroblasts. *Uta stansburiana* shows maximum cellular developed in mid-February; the nucleus is spherical and cytoplasm vacuolated. Regression occurs during the period April to June, and the nuclei are small and pycnotic by early August. However, the interstitial cells of *Lacerta agilis* show maximum activity in April to June; they are then charged with granules. By July the granules have disappeared and the cells regress. Similarly, interstitial cells in *Eumeces fasciatus* attain maximum size in March and April just before the seminiferous tubules which reach it at the end of this period. *Bourgat* (1968) reported reduction in the amount of interstitial cells tissue of *Chamaeleo pardalis* during the warm (mating) season. Probably the testes were then at maximal size and the interstitial cells were therefore regressing. Other evidence is more difficult to interpret. Peak enlargement of the interstitial cells of *Sceloporus occidentalis* occurs from November to February, with minimum size in the spring and summer; testicular volume is greatest in February-March (after hibernation) and least in July and August. In *Carlia fusca* activity of the interstitial cells and the size of the seminiferous tubules are reduced in February and March while the reproductive activity of the testis is at its peak from December to January. *Bons* and *Saint Girons* (1963) reported the interestitial cells of the amphisbaenians *Blanus* and *Trogonophis* to involue in June after secretory activity; the testes involute in July. In all these cases, however, the secretory activity of the interstitial cells seems to precede the peak activity of the seminiferous tubules and the former regress before the latter.

It seems that the interstitial cells generally of different species of lizards vary, especially seasonally, in number, size and cytological content. Maximum size occurs during the breeding season and the lipid content, which is high in autumn, is reduced during winter hibernation and then replaced by secretory granules.

In *Cnemidophorus tigris* and *C. burti* a unique layer of sub-tunic albugineal cells coexists and is in continuity with typical intertubular interstitial cells. *Callisaurus,* an iguanid, has no sub-tunical layer. The tunica and intertubular interstitial cells show maximal development in early summer when spermiogenesis is at a maximum. In some regions the tubular interstitial cells disappear soon after spermiogenesis is completed, the tunical cells remained during all the months the animals were observed. Likewise the *Cnemidophorus lemniscatus* has abundant interstitial cells that form a continuous collar between the tunica albuginea and the peripheral seminiferous tubules. This collar increases in volume and the cells hypertrophy from April to October; it is reduced and the nuclei are pycnotic from December to February. Highest activity of the interstitial cells coincides with the highest values of day length, temperature and rainfall. Sperm production seem to occur throughout the year without any appreciable variation. A tunica layer was described in *C. tigris.* It was postulated that its cells were identical to those of the intertubular interstitial cells. The peripheral layer is absent in males of *Callisaurus, Klauberina, Sceloporus* and *Uta,* but does occur in some 18 teiid lizards (17 spp. of *Cnemidophorus* and *Ameiva undulata;*). The number of tunical cells does not vary seasonally but there is seasonal variation in the number of intracellular granules. The cells are secretory in June and in the later period of reproduction. Deposition of granules occurs in late April and May at the first copulation and depletion in July at the last. Young adults have fewer tunical cells than older animals and there is quantitative intranspecific and interspecific variation. On the basis of the histochemical reactions of various enzymes studied the circumtesticular sub-tunical sheath in *Cnemidophorus tigris* is considered to be functionally equivalent to the interstial Leydig cells. Both groups of cells show positive reactions for hydroxy dehydrogenase, 17-βdehydroxysteroid-dehydrogenase, NADPH diaphorase, and glucose-6-phosphate dehydrogenase. Such correspondence supports the hypothesis that the circumtesticular cells have an endocrine function.

Snakes show abundant interstitial cells during hibernation. In *Thamnophis radix* these cells are large and vacuolated, with distinct boundaries, from October and through the hibernation period to the spring, but are small and dispersed, though still secretory, in July and August. Similar changes in size were reported in *T. brachystoma*. In *T. elegans* interstitial cells hypertrophy in the spring and follow a similar pattern to those of *T. radix;* some slight reduction in size, but not in numbers, ensues in winter. In *Naja naja* the nuclear size of the interstitial cells is maximal in spring and reduced in summer; a new generation then develops to enlarge at hibernation. The interstitial cells are largest-during the breeding season, though in some snakes they may enlarge at other seasons as well. The annual cycle of the interstitial cells of *Natrix natrix* would appear to be similar to that in *Thamnophis*.

Interstitial cells of *Typhlops vermicularis* and *T. simoni* from Israel are sparse in April and May, when there is full spermatogenetic activity; the sex cycle begins in spring.

The large interstitial cells of *Vipera berus* and *V. aspis* vary in size even in the same individual. The nucleus is round and vesicular: the cytoplasm is abundant with a distinct cell membrane. Their total volume never exceeds 1% of the testis, an amount which bears no relation to the size or age of the mature animal. Maximum development, in terms of numbers and appearance, occurs simultaneously with the final stages of spermatogenesis, and the interstitial cells regress to a minimal size in August. This is followed by a slow increase in size and numbers until hibernation, while final growth and proliferation ensue in the following spring. No changes occur in volume or secretory activity of the interstitial cells of the tropical Indian snake *Xenochrophis piscator* throughout the year. The interstitial cells of the turtles *Chrysemys, Emys* and *Testudo* have numerous granules whose "richness" does not change between May and December, so that *Herlant* (1933) concluded there was no seasonal variation, a view also reached by *Risley* (1938b) for *Sternotherus odoratus*. However, *Bulliard* (1921) had earlier noted that the abundant interstitial cell of the tortoise were more plentiful in winter than summer. In *Emys orbicularis* the interstitial cells are small during quiescence and their lipid granules are only light stained. During sexual activity both the number of interstitial cells and the number of their granules increase. The relative proportion of small interstitial cells increases again at the end of the breeding season; interstitial cells probably secrete lipoidal substances. *Combescot* (1954a, b) and *Combescot* and *Guyon* (1955) showed that the interstitial cells of *Mauremys caspica leprosa* are well-developed and have "attained full expression" from March to May, when they are distinguished as large round blocks of cells. The regress in June, become rare in August and enlarge again in November and December; they lipid containing areas have the appearance of a large collar surrounding the seminiferous tubules, which *Lofts* (1968) believes to be homologous to the tubular boundary cells of some fishes. *Mauremys* first mates in April and May and spermatogenesis terminates at the end of August. Consequently turtles do show a seasonal cycle of activity, and *Herland* (1933) later modified his conclusions and reported that the interstitial cells of *Testudo graeca* reach peak development in spring and start to involute from May to June, when they become filled with lipochrome pigment.

According to *Frankenberger* (1928) the interstitial cells perform a trophic function; diminution in their number is due to their transformation into connective tiissue or endothelial cells, and new interstitial cells develop at their expense. In contrast *Blount* (1929) and *Dutta* (1944) specifically denied reversion to connective tissue cells. However, the view of *Bouin* and *Ancel* (1903) that the interstitial cells serve as a source of male sex hormone has gathered strong support. This is proved from experimental castration, leading to the involution of the secondary male sex organs and from injections of mammalian or synthetic androgens, which stimulate the development of these organs.

Ultra Structure of Interstitial Cells

The ultrastructure of the interstitial cells of adult *Lacerta vivipara* and *Anguis fragilis* characteristics of cells which elaborate steroid. The cytoplasm includes a smooth endoplasmic reticulum of small or larger vesicles; numerous free ribosomes and polyribosomes abound, and mitochondria are plentiful in a density related to the sex cycle. During the spring, the interstitial cells of *Lacerta vivipara* show a well-developed system of vesicles and vacuoles of the smooth endoplasmic reticulum and of the Golgi complex, which are possibly related to androgenic production. Vacuoles are less prominent in summer, after breeding, when the smooth endoplasmic reticulum comprises a network of tubules often associated with lipid.

Leydig cells of *Lacerta sicula* subjected to a low temperature (4°C) stopped their secretory activity and regressed; warmth (28°C) stimulated their cellular development and secretion.

The testes of *Bothrops jararaca* and *Crotalus terrificus* contain androgens; oestrogens and progesterone are present in the ovaries of these forms. The testes of *Uromastyx hardwickii* show maximal levels of cholesterol when they are regressed and minimal levels when in full spermatogenic activity. *Melampy* and *Covazos* (1954) showed that the interstitial cells of *Phrynosoma cornutum* contain ketosteroids, phospholipids and variable quantities of cholesterol, and the lipid content of the interstitial cells of the testis of *Vipera-berus* is also variable. After the onset of the breeding period cholesterol-positive lipids are gradually depleted; they presumably contribute to the production of androgens. After seasonal exhaustion, new interstitial cells develop by mitosis and some specimens of *Vipera berus* show autumnal sexuality reflected by further development of the interstitial cells. *V. aspis* also has two periods of sexual activity and shows similar effects. During hibernation or summer sexual quiescence, blood androgenic levels are low and interstitial cells are largely filled with acidic or anisotropic lipid droplets. Such droplets practically disappear during spring and autumn sexual activity.

Nevertheless, *Hahn* (1964) still quoted *Forbes* (1961) that testosterone had yet to be established as derived from the interstitial cells of reptiles, though these cells are probably its source.

Incubation *in vitro* of testicular germinal epithelium, sperm, and the interstitium from *Natrix fasciata pictriventris* showed histochemical localization of 3-β-hydroxysteriod dehydrogenase in the interstitial cells and slight activity in the germinal epithelium. The testis was able to synthesize testosterone and convert pregnenolone into progesterone. The interstitial cells are probably the site of this steroidogenesis.

In *Naja naja* testicular androgen was found to include testosterone, androstenedione, deoxycorticosterone and aldosterone. Their levels rise from late August to September and remain elevated, except for a drop during winter hibernation, until the end of the breeding season in early May. The reduction of the lipid level of interstitial cells probably reflects an increasing androgenic secretory activity, the accumulation of cholesterol-positive lipids a reduction in secretory activity. The interstitial cells of *Chrysemys* spp., *Varanus niloticus* and *Natrix natrix* and of *Lacerta muralis,* and *L. vivipara* have a strong activity of steroids 3-β-ol-dehydrogenase; this is feebly shown in newly hatched young of the last. The gonads of *L. vivipara* show activity of the same substance (more intense in females), which disappears after hatcing. In *Lacerta* the dehydroxydehydrogenase (3β-dehydroxysteroid-dehydrogenase) appears just after sexual differentiation in the ovarian medulla, as well as in the testicular sex cords of the embryo and the interstitial cells of young animals. Such enzymatic activity occurs also in the interstitium of the testis of *Sceloporus occidentalis.* The enzyme disppears from the ovary of *Lacerta vivipara* and *Vipera aspis* before birth.

Lofts and *Boswell* (1961), and *Lofts* (1968) made a strong case for a clear relationship between accumulation and depletion of the lipid substance of the interstitial cells and the active breeding periods of reptiles. In *Naja naja* the spring accumulation of cholesterol-positive lipids in the interstitial cells is succeeded by a depletion simultaneous with the maximum production of androgens and the development of secondary sex characters, as they noted in birds. The transient phase is followed by one in which the interstitial cells again become heavily charged with cholesterol-positive lipids. The cells eventually disperse, are phagocytosed by macrophages and a new generation starts to accumulate lipids during hibernation. When spermatozoa leave the seminiferous tubules of *Mauremys caspica* to be stored in accessory ducts, the interstitial cells show a similar cycle but out of phase with the spermatogenetic cycle. Thus maximum secretory activity of the interstitial cells and their denudation of lipid occur in May, just before spermatozoa are expelled from the epididymis and when the seminiferous tubles ** spermatogenetically inactive.

Thus, we see that the interstitial cells of reptiles mature and are secretory simultaneously with, or more probably just preceding, maturation or final spermiogenesis in the seminiferous tubules. Near the end of spermiogenesis the interstitial cells usually regress together with the testicular tubules. Male interstitial cells probably secrete androgens, synthesized from cholesterol-positive lipids, and their secretory cycle generally synchronizes not with spermatogenesis, but with the development of the secondary sexual characteristics. In females oestrogens from the ovary likewise influence maturation of the secondary sex characters, which in both sexes might well be expected to develop before the occurrence of final spermiogenesis, actual copulation, and thus fertilization. interstitial cells and gonads mature under the influence of pituitary hormones, and mammalian gonadotropins (except for luteinizing hormone, LH) were shown, in the lizard *Scincella lateralis,* to increase the number and size of interstitial cells, with consequent cytoplasmic granulation, when the interstitial cells were few and the epididymis and the renal sex segment were quiescent.

The phasing of maturation of the interstitial cells, secondary sex characters and spermiogenesis could well result from the liberation of pituitary interstitial cells stimulating hormone (ICSH) before those other gonadotropic hormones which stimulate testicular tubules. Alternatively, the hormones may be released simultaneously, and the ICSH be more potent and faster in action ; or again the interstitial cells are quicker than the testicular tubules to respond to normal stimulation from different gonadotropins.

Sertoli Cells

Sertoli or *sustentacular cells* have been reported in the reptilian testis, against the basement membrane inside the seminiferous tubules of turtles, crocodilians, lizards and snakes. The Sertoli cells of *Uta stansburiana* are 'binucleate or possibly multinucleate; when seminiferous tubules regress much of the lumen is filled by Sertoli cells; *Uma* sp. is similar. In regressed aluminate seminiferous tubules of *Dipsosaurus dorsalis* Sertoli cells and spermatogonia still persist at the periphery. In *Uta stejnegeri*, Sertoli cells at maximum size have triangular or polyangular nuclei. These cells are syncitial in the lizards *Lacerta muralis*, *Tarentola mauritanica*, *Auguis fragilis*, *Carlia fusca*, and *C. rhomboidalis*, as well as in snakes and turdes, and they often contain fatty droplets and siderophilic granules (especially in lizards during sexual activity, *Herlant*, 1933; also in *Phrynosoma cornutum*.

Kehl and *Combescot* (1955) conclude that the sertoli cells of lizards are practically unchanged throughout the year, nor do these cells seem to change in size during the sexual cycle of *Eumeces laticeps*. However, *Lacerta agilis* shows maximal development of the Sertoli cells in the spring, *Uromastyx* shows quiescence in June, and from July to the end of winter *Anguis* has merely a few compressed cells lying against the tubular walls, though these cells divide again in March. *Cnemidophorus* has abundant Sertoli cells from April to July. After completion of spermatogenesis in late July, their numbers decrease and some appear to be evicted through the epididymis. Likewise Sertoli cells disappear in the amphisbaenians *Blanus* and *Trogonophis* during August. The syncitial, triangular or polygonal Sertoli cells of *Xantusia vigilis* extend as fibrous strands from the basement membrane of the seminiferous tubule towards the lumen. The syncitium, which is fairly extensive in May and June, practically fills the lumen and phagocytoses the remaining sperm, finally to decrease by October. In *Anolis* during sperm evacuationo and the reorganization of germ cells, the Sertoli syncitium remains as a thin peripheral ring round the tubule. In contrast, in the snake *Thamnophis elegans,* as in *Xantusia,* it fills the lumen at this stage. In *Anolis* and *Thamnophis* Sertoli cells move away from the basement membrane and orientate so that their long axes are perpendicular to it. Cells become compressed and vesicular and, when the spermatogenic cycle recommences in September or October, the Sertoli cells often appear flattened against the tubular basement membrane. Their numbers are fairly constant and delineated cells are probably new ones.

Volsoe (1944) suggested that the number of Sertoli cells is relatively constant in *Vipera berus;* as old ones are destroyed, new ones are derived from primary spermatogonia and all intermediate stages between the two are recognizable. However, *Miller* (1963) drived spermatogonial and oogonia ultimately from primary germ cells, and Sertoli cells from germinal epithhelium. Emphasis was also laid on their nuclear irregularity and size, as they are often larger than adjoining spermatogonial nuclei. In *Vipera aspis Saint Girons* (1957) did not exclude an annual cycle in Sertoli cells, with a maximal amount in March, though there is great individual variation in different specimens. In the lizard *Anguis* and in amphisbaenians they are smallest and lowest in number during spermatogenesis, and *Saint Girons* (1963b) claimed evidence of intermediate forms between spermatogonia and Sertoli cells. Nevertheless, in *Cnemidophorus Goldberg* and *Lowe* (1966) found no intimate association between these two cell cetegories. The derivation of Sertoli cells is still not clear.

Dufaure (1971) described the ultrastructure of Sertoli cells of *Lacerta vivipara*. They include small mitrochondria, fee ribosomes and a moderately well-developed smooth endoplasmic reticulum, a Golgi complex and micro-tubules. Lipid, glycogen and possibly lysosomal bodies are present. There is seasonal variation in glycogen content. .Hypophysectomy leads to a retention of metabolites concentrated around the lipid. The Sertoli cells probably have a phagocytic role and eliminate remaining sperm after the spermatogenetic cycle, or perhaps a nutritional one as well, assisting in some way the development and differentiation of sperm. Probably they salvage lipids and steroids and assist endocrinal economy. *Lofts* (1968) has generalized that in non-mammalian vertebrates, especially birds and reptiles, seminiferous tubules regress and Sertoli cells accumulate cholesterol-positive lipids at the end of the breeding season. The cells practically occlude the tubular lumen, owing to its reduced diameter. Residual bodies, are mainly composed of lipoidal substances, RNA, mitochondria, and Golgi material. They are phagocytosed by the Sertoli cells with ingest most of the substances except for the lipids, which coalesce and lie within them (seen in the rat Sertoli cells by electron microscopy). The process is similar in *Naja naja*. In non-mammalian tetrapods cholesterol-rich lipoidal material rapidly accumulates in the Sertoli cells after evacuation of spermatozoa from the seminiferous tubuies; it decreases as spermatogenetic activity resumes in the germinal epithelium. In contrast to other reptiles, the tubules of *Naja* are never completely free from sudanophilic material, for lipid droplets begin to accumulate in the basal regions of the Sertoli cells before the final traces of post-nuptial lipids have disappeared. *Gottfried et al.* (1967) show that *in vitro* the testis of *Naja naja* synthesizes, from endogenous precursors, dehydro-epi-androsterone, androstenedione, 17α dydroxypregnenelone, and oestrone. In an addendum to their paper *Lofts et al.* report that testosterone is the major androgen, though its productoin fluctuates with the season. Furthermore, seminal tubules dissected free from the testis and incubated *in vitro* with ^{3}H progesterone, can synthesize 17α-hydroxyprogesterone, androstenedione and testosterone, which suggests that the synthesis is due to intratubular elements, possibly Sertoli cells. *Lofts* (1968) suggested that in addition to their phagocytic activity, Sertoli cells are stimulated by the pituitary FSH to act as a link between adeno-hypophysial stimulation and tubular spermatogenesis. Oestrogens, progesterone and cholesterol are associated with tubular sudanophilia in many vertebrates, and these substances are known to be precursors in the biosynthetic chain of events leading to the production of testicular androgens or oestrogens. FSH may, therefore, influence Sertoli cells by eliciting a Sertoli cell hormone involved in spermatogenesis and influencing the ordered, spatial and chronological sequence of germinal events, commencing with spermatogonia at the tubular periphery and proceeding via spermatocytes spermatids, and spermatozoa inwards towards the lumen. *Miller* (1963) suggested that Sertoli cells and granulosa cells surrounding the oogonia are non-gonocytic, and probably derived from coelomic epithelium. In contrast gonocytes drive exclusively from primordial germ cells. This suggestion is of considerable interest as follicular granuulosa cells are well known for their endocrine response to adeno-hypophysical gonadotropins.

The Gonads of Females

There is a pair of ovaries enveloped by peritoneum, and suspended by a mesovarium comprising storma, nervous and vascular elements. Like the testes they exhibit seasonal changes and typically are of greatest size at the height of the breeding season, when the ovules are enlarged just before ovulation, and smallest afterwards. The ovaries of turtles are symmetrically positioned. In lizards and snakes, however, they are usually situated asymmetrically, with the right one anterior to the left. Indeed, the right ovary is usually larger than the left in the snakes *Thamnophis radix, Xenochrophis vittara, N. sipedon, N. rhombifera* and *Ptyas mucosus*. Asymmetry likewise extends to the oviducts. In *Lacerta vivipara* the left ovary extends back to the pelvic cavity.

Some lizards have been reported to have practically symmetrical ovaries. Here belong *Crotaphytus collaris, Eumeces obsoletus, Gerrhonotus liocephalus infernalis* and *Cnemidophorus sackii gularis, Lacerta muralis,* and *Sceloporus poinsettii* and *S. olivaceus*. In *S. orcutti* the left ovary is anteriorly positioned in about half of 48 females, the right in 15% and the rest had symmetrical ones. In *Dipsosaurus dorsalis* 40% of 52 females showed the right ovary anterior, 2% the left, and the remainder had symmetrical ovaries.

The oblong ovaries of the lizard *Xantusia,* in the dorso-posterior region of the body, are composed of a single layer of flattened cubical cells surrounding the stroma ovarii, which in all reptiles include collagen fibres, scattered fibroblasts and blood vessels and an ovarian lymph cavity, lined by a squamous epithelium. Ovaries of *Carlia rhomboidalis* and *Acanthodactylus erythrurus lineomaculatus* appear similar. There are two genital beds of small isolated masses in each ovary; for examle in *Carlia* one is anterior, the other behind, on the dorsal ovarian surface. Each mass is 10-60 µm thick and of 1 to 5 layers of cells. Each is composed mainly of primary oogonia and ooxytes in various states of division. During most of the year immature ova (2-20 Whitish/opaque bodies, 1-1.3 mm in diameter) are surrounded by a vitelline membrane, formed by a striated and homogeneous layered zona pellucida. Various corpora atretica occur throughout the year, for all but one or two of the ova become atretic. Yolk deposition occurs shortly before ovulation with maximum size occurring during April-May, as in *Uta stansburiana*. A ripe ovum is 7-8 mm in diameter and usually one or two are found in each ovary during the year.

In *Lacerta sicula* five arbitrarily separated groups of previtellogenic oocytes are present throughout the year. The criteria of grouping depended upon their diameter, single or multilayered granulosa layer and the presence of pyriform (nurse) cells. Yolk deposition occurs 3-6 weeks before ovulation.

In *Phrynosoma solare* ovarian volume in 100 mm^3 at the beginning of June and follicles are 1.5 mm in diameter. By the middle of June the volume is 3600 mm^3 and follicles are 7-10 mm in diameter. Ovulation begins in mid-July andhas ended by the middle of August.

In *Hemidactylus flaviviridis* the small whitish ovaries have only 2 or 3 follicles during the period from November to February. Ovarian volume and follicular number increase in spring, and in April to May one of the 6-8 follicles greatly enlarges and ovulates, to enter the oviduct. Ovaries rapidly regress in June to October and become minimum in size when the testes are maximum. *Chamaeleo lateralis,* of Madagascar is of interest in that the left ovary produces more oocytes than the right.

Among snakes the elongated, sac-like ovaries of Natrix sipedon and *Thamnophis radix* have membranous walls. The ovaries include an irregular lymph-filled central cavity surrounded by a loose, semi-transparent stroma as in *Natrix rhombifera,* in which oval creamy-while, relatively avascular oocytes contrast with yellow, vascular atretic follicles and corpora lutea. In *Vipera lebetina* growing follicles are situated in the ovarian cavity on a germinative stem, from which oocytes are derived.

Mature females of *Crotalus viridis* which are *post-partum,* have only small ovaries, for they gave birth to young the previous year in autumn. Ripe females that have not yet bred have large ovaries and there is a two-year cycle in this species.

Aspects of ovarian morphology of snakes are mentioned by Fraenkel and *Martins* (1937) and by *Fraenkel et al.* (1940) in

Xenodon merremii and other Brazilian snakes (*Crotalus terrificus: Bothrops jararaca* and *B. alternatus*). Asymmetrical sac-like organs of unequal length and surrounded by connective issues, include stigmata of corpora lutea.

Gonadal arteries from the dorsal aorta vascularize the ovaries and oviducts (and male gonads too) in various New World Crotalinae. Ovarian veins drain the overies and open into the post-caval veins.

The loose membranous ovaries of *Emydoidea blandmgii* spread over the kidneys and posterior regions of the lungs. Connective tissue (medially) and mesovaria support them and the mature eggs. Lateral and anterior mesovarial extensions are continuous with the mesotubaria which support the oviducts. In *Terrapene carolina*, the ratio of ovary to body size, is highest in May when 2-8 enlarged follicles are usually present. The ovary becomes smaller from June to July after ovulation, when 2-5 eggs are found in the oviduct; it reaches its peak size again the following spring.

In *Mauremys caspica* follicular enlargement occurs throughout March and April and ovulation in April and May. Enlargement of the oviducts parallels that of the ovary. Ovulation coincides with the discharge of spermatozoa from the epididymis. During February and March ovaries expand and the seminiferous tubules of the testes have regressed. Maximum ovarian size coincides with minimum testicular size; during ovarian quiescence in May to September, the testes show recrudescence of spermatogenesis and the seminiferous tubules enlarge. The epididymis stroes motile sperm, which presumably are active for up to 6 months before mating and fertilization of ova; for most of this time ovaries are in a state of regression.

In vertebrates the chief source of oestrongens is the ovary. The locus is mainly cells of the theca interna. In lizards during oestrus, the theca cells hypertrophy and become full of sudanophilic inclusions; they are strongly positive for 3-β-hydroxysteroid dehydrogenase in polarised light are birefringent, features which disappear at anoestrous. The activity of the thecal cells parallels the development of the secondary sex characters.

The ovarian cycle in reptiles is probably influenced in some way by the environment (light, temperature), acting via the hypothalamus. Furthermore, by feedback, oestrogen and progesterone influence steroid-sensitive areas in the hypothalamus, the latter thence activating the release of pituitary gonadotropins, which influence ovarian growth, steroid synthesis, ovular development and ovulation. Growth hormone would also seem to be involved. Variuos mammalian gonadotropins stimulate reptilian gonads and are probably of similar nature to those occurring naturally in reptiles. Prolactin and progesterone are anti-gonadal substances in reptiles and inhibit gonadal development. Sex harmones in the ovaries of *Lacerta sicula* were investigated by *di Prisco et al.* (1968).

Removal of the largest follicles from the larger or both ovaries of *Anolis carolinensis,* results in the smaller follicles accelerating their growth rate. This work raises the whole question of control of follicular size in reptilian ovaries. The authors suggest that a number of factors are concerned in the process, such as vascularity, the larger follicles being exposed to more circulating gonadotropin; or cells of larger follicles may well be more sensitive to gonadotropin; larger follicles may secrete substances inhibiting or desensitizing smaller follicles to the circulating gonadotropins. Again follicular growth may be related to the surface area of the oolemma involved in micropinocytosis of yolk.

Smith et al. (1972) described the following ovulatory cycles for anoline lizards: *allochronic cycles,* during which ovulatoin proceeds alternately from paired ovaries throughout the breeding seasoni; autochronic cycles during which ovulation occurs simultaneously from both ovaries; and monochronic cycles, during which only a single ovary is functional. The first two categories are further sub-divided into monoallochronic, polyallochronic, monoautochronic, and polyautochronic cycles respectively. The latter is the rule in iguanid genera, except for *Anolis, Chamaeleolis, Chamaelinorops,* and *Phenacosaurus,* which are monoallochrronic.

The Oviduct

There is a pair of oviduct. The anterior oviducal opening into the coelom of lizards and snakes, the *"ostium abdominale,"* or *"trompe de Falloppe"*, was well known. The coiled oviduct, suspended along its length by the mesatubarial fold of the mesentery, the *"Haltband"*, or *"cordon ligament"* joins the ipsilateral ureter in many lizards, to lead by a common opening into the cloaca. The oviducts of snakes join to form a median uterus, which has a wall thicker than the rest of the oviduct. Published descriptions of the histology of the oviduct noted cilia in the ostial region in *Lacerta* and *Anguis* and the mucous cells lining it in *Sphenodon* (Osawa, 1898). In *Lacerta viridis* and *Coluber viridiflavus* there are tubular glands with slender ducts leading to the oviducal lumen. Connective tissue, circular muscles and a serosa surround the mucous membrane. In the anterior longer region glands secrete albumin; the more posterior region includes shell secreting glands. Glands increase in size during the breeding season.

Lizards in general have paired oviducts, each comprising a think-walled, pleated Fallopian tube, which is lined by ciliated and mucous cells. Single alveolar glands of cuboidal epithelium open into the lumen by short ciliated ducts. A thicker-walled uterus (with thicker muscles), has poorly developed uterine glands in *Hoplodactylus pacificus,* though these glands are better developed in the viviparous lizards *Anguis fragilis* and *Lacerta vivipara*. Uterine glands are large and branched in Australian viviparous genera. Oviparous species generally have better developed uterine glands than viviparous ones. A short, narrow, almost glandular vagina is lined mostly by ciliated cells. The pleated walls are lower and less frequent than in the uterus and goblet cells are fewer; they were not found by *Giersberg*

(1922). Mucus is secreted for lubrication. Granular secretions from the uterine glands serve for the production of shell membranes, especially in oviparous forms.

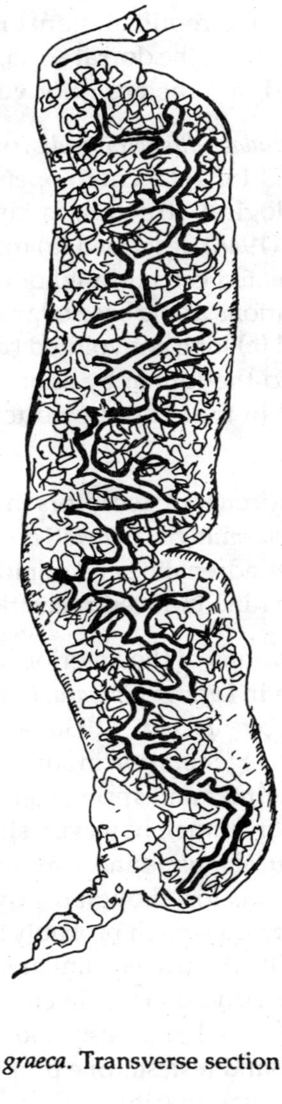

Fig. 1.25. *Testudo graeca*. Transverse section through the oviduct.

The Fallopian tube is surrounded by collagen fibres (there are no elastic or reticular fibres) and capillaries; externally lie circular muscles and then longitudinal muscles and a peritoneal layer. A small sphincter muscle separates the Fallopian and uterine tubes. In general the reptilian uterus is supplied with blood by a uterine artery from the dorsal aorta, and drained by uterine veins which lead into the posterior vena cava.

The oviducts of *Hemidactylus flaviviridis*, originate as wide anterior ostia and extend for 1-1.5 cm to open separately into the cloaca. Their histological structure is similar to that of *Hoplodactylus* but *Dutta* (1946) described (a) unicellular glands which open between ciliated cells throughout the oviduct, (b) glandular cell accumulations at the base of grooves, especially in the funnel region and (c) long convoluted branched tubular glands, which open into the lumen by short epithelial ducts and are best developed in the central oviducal region where albumin is secreted.

There are α and β adrenergic receptors in isolated uteri of the viviparous lizard *Liolaemus gravenhorsti* and ovoviviparous *L. t. tenuis*. In both species adrenaline and noradrenaline elicited contraction and isoprenaline induced relaxation.

Histological descriptions of the anterior funnel, Fallopian tube, uterus, and vagina in *Lacerta vivipara*. *L. agilis* and *Anguis fragilis* were given by *Jacobi* (1936). *"Schleimzellen"*, *"Wimperzellen"*, oviduct degeneration in summer and autumn, and its regeneration in winter and spring were also described. The region of the anterior funnel, however, shows no annular changes, nor do the oviducal muscle and serosa in *Cnemidophorus inornatus* or *C. neomexicanus* for, though the oviducal muscles expand, they are thinner and stretch probably as a result of the enlargement of the oviducal mucosa. Similar changes to those occurring naturally were induced in specimens of *Cnemidophorus* treated with oestrone, oestriol and oestradiol and, to a lesser extent, with testosterone and testesterone proprionate. Ovaries showed in increase in vascularity with these treatments.

Castration of non-gravid *Sceloporus cyanogenys* leads to a reduction in oviducal size, and likewise in *Lacerta sicula* prevents annual growth.

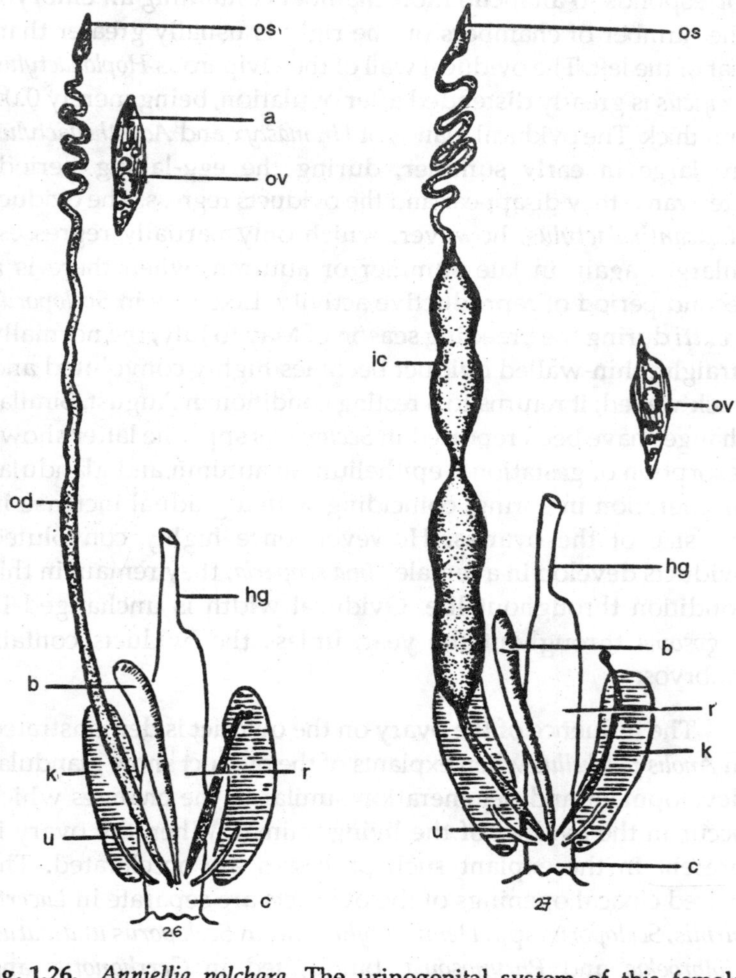

Fig. 1.26. *Anniellia polchara.* The urinogenital system of this female limbless lizard when non-gravid. The left oviduct is vestigial.

Fig. 1.27. *Anniellia polchara.* The urinogenital system of the gravid female, a, Adrenal; b, urinary, bladder; c, cluca, hg, hindgut; ic, incubation chamber; k, kidney; od, oviduct; os, ostium abdominale; ov, ovary; r, rectum; u, ureter.

Fertilized *Lacerta vivipara* and *Anniella* show a series of uterine swellings (transitory structures probably found for the first time by *Giacomini*, 1891, in *Chalcides*), each of which corresponds to an incubation chamber containing an embryo. The number of chambers on the right is usually greater than that of the left. The oviducal wall of the viviparous *Hoplodactylus pacificus* is greatly distended after ovulation, being merely 0.04 mm thick. The oviducal glands of *Uromastyx* and *Acanthodactylus* are large in early summer, during the egg-laying period; afterwards they disappear and the oviducts regress. The oviduct of *Acanthodactylus*, however, which only partially regresses, enlarges again in late summer or autumn, when there is a second period of reproductive activity. Likewise in *Sceloporus orcutti* during the breeding season of May to July, the normally straight, thin-walled oviduct becomes highly convoluted and thick-walled; it returns to a resting condition in August. Similar changes have been reported in *Sceloporus* spp. The latter shows resorption of gestational epithelium in autumn and glandular proliferation in spring, coinciding with a gradual increase in the size of the ovaries. However, once highly convoluted oviducts develop in a female *Uma scoparia*, they remain in this condition throughout life. Oviducal width is unchanged in *Lygosoma* throughout the year, unless the oviducts contain embryos.

The influence of the ovary on the oviduct is demonstrated in *Anolis pulchellus* when explants of the oviduct show glandular development and degeneration similar to the changes which occur in the oviduct of the living animal. When the ovary is present in the explant such processes are accelerated. The paired cloacal openings of the oviducts are separate in *Lacerta viridis, Sceloporus* spp. *Hemidactylus* and in *Sceloporus undulatus, Holbrookia* and *Phrynosoma*, but united in *Gerrhonotus* and *Cnemidophorus*. The two latter genera and *Eumeces* have a muscular sphincter at the posterior end of the oviducts. *Anniella* has a non-functional, aborted, and degenerate left oviduct, a narrow luminated tube that is either blind in front or has merely a tiny ostium. The right oviduct is large and functional. The histology and histochemistry of the oviduct of *Sphenodon punctatus* have been described by *Gabe* and *Saint Girons* (1964). On the whole the structure is typically reptilian with numerous uterine glands, as in turtles. There is an extremely short vaginal tube.

In *Thamnophis elegans* the terminal ciliated infundibulum leads into a convoluted tube, which has alveolar glands that serve as seminal receptacles. Alveolar glands are bound into groups of three to six by connective tissue, and lead into the lumen by ciliated ducts. In lizards structures comprising ciliated pits or tubules closely resemble the seminal vesicles of snakes, though *Fox* and *Dessauer* (1962) did not find sperm in them. No trace of seminal vesicles or receptacles was recognized in the female *Sphenodon*.

As the infundibulum of *Thamnophis* flares out towards the ostium it becomes thin-walled and ciliated; mucous-like cells are still present. The uterus is lined by a tall epithelium, which is columnar, intermingled with ciliated and non-ciliated cells. Flask-like tubular and alveolar glands hypertrophy in the spring, at which time the entire reproductive tract becomes extremely vascularized. The non-glandular vaginal pouch comprises a straight, narrow tube several millimetres long and a thick-walled bulbous cloacal opening about 1-2 mm more posterior. The two regions, which differ in the thickness of their muscular and connective tissues, are lined by ciliated and non-ciliated columnar cells. As in lizards the entire duct is surrounded by connective tissue, blood vessels and circular and longitudiinal muscles and suspended by the mesometrium.

The left oviduct is absent in the Australian *Typhlops polygrammicus T. ligatus* and *Helminnthophis frontalis, T. braminus, T. angolensis, Leptotyphlops humilis, L. dulcis,* and *L. Phillipsi,* and vestigial in *Anomalepis flavapices, A. dentatus,* and *Helminthophis flavoterminatus* in which it does not reach the posterior border of the left kidney and in seven out of ten species and subspecies of *Tantilla*. However, the oviducts are equally developed and functional in all species of *Liotyphlops* and in *Typhlophis squamosus*.

Enlargement of the oviduct of snakes during the breeding season occurs as in lizards. In *Vipera aspis* the infundibulum varies in dimensions and the "tuba" or "trompe" in structure according to season. The tuba is extremely pleated and in females at oestrus the epithelium is 17-21.µm high. Cilia, mucous cells, and deep tubular glands, the ducts at which are lined with cilia, occur. At maximum development of the oviduuct the lumen is greately reduced, cilia have disappeared and the glandular diameter has increased. After ovulation the glandular epithelium rapidly degenerates and is invaded by eosinophilic leucocytes. During anoestrus the oviduct is greatly reduced in size (epithelium 5 µm high) and the glands are barely distinguishable. In the case of the triennial cycle of *V. aspis* the regressed condition of the oviducts is retained until the winter of the second year after previous breeding, when the oviduct begins to enlarge again. The uterus and vagina one enlarge to a maximum, however, every third year in a pattern generally similar to that of *Thamnophis*. Oviducal size depends on secretions (such as oestrogens) from the ovary, which in turn vary in function with the size and activity of the follicles, the whole influenced by hormones liberated from the pituitary. In *Crotalus* for example, as the ovary reaches maximum size so does the oviduct, which becomes more convoluted in the process.

The luminal margin of the aglandular oviducts of a young *Testudo graeca* (plastral length 10 cm) is lined by ciliated cells surrounding areas of mucous epithelium. At 13 cm incipient glands have invaginated the chorion to form culs de sac. By 17 cm complex glandular cells include granules 2 µm in diamter. The mucosa includes intermediates between ciliated and calciform cells, which implies development of the latter from the former, a view not supported by *Dutta, Kehl's* (1944) description for Testudo was generally similar to that of Argaud

(1920). At 20 cm the voluminous festooned oviducal lumen is lined by mucilaginous and ciliated cells and well-developed glands. In Mauremyys caspica leprosa (plastral length 8 cm) cilia and glands are absent in the 2 cm long oviduct; at 11-13 cm plastral length, the festooned oviducal mucosa includes ciliated and incipient epithelio-glandular cells. In a specimen of 13 cm plastral length the oviduct (9 cm long) is similar to that of *Testudo graeca;* granules are present in certain acini. The oviducts are voluminous in the sexually mature. Mauremys. A short, anterior, aglandular ciliated region is followed by a portion (two-thirds of the total oviduct), which includes cells with secretory granules. A short ciliated region 1-2 cm long continues into a final section that opens into the cloaca. Here glandular acini, separated from the epithelium by connective tissue, secrete the egg shell. The coiled oviduct (length about 12 cm) of adult *Emyaoidea blandingii,* closely bound to the ventral kidney surface by the mesotubarium, enters the anterior cloacal wall ventrolaterally, but dorsolaterally in the adult Trionyx euphraticus. In Terrapene carolina, anterior ostial cells (9 µm in height) are sparsely ciliated but vacuolated in January, and they are heavily ciliated and 127 µm high in May. Together with non-ciliated cells they line grooves; the cells in grooves of the posterior ostial region are non-ciliated. The remaining pseudo-stratified columnar cells, either ciliated or not, are 28 µm high in March and 37 µm high in July. The albumin-secreting region of the oviduct is lined by cells similar to those of the post-ostium region. Giersbert (1922) reported seasonal changes in the oviduct of turtles; the oviduct almost doubles in thickness during sexual activity and its structure is very similar to that of birds.

In adult alligators the oviducts open into the coelom some way anterior to the ovaries are convoluted anteriorly but become straight ventral to the kidneys. Shortly after hatching (11-12.5 cm headtrunk length; 82-96 days of incubation), the *Mullerian ducts* are fully formed in both sexes and supported from ostium to cloaca by a broad meso-salpinx. Peritoneum covers dense connective tissue, but no muscle fibres occur. The oviduct rapidly denerates in the post-hatching male. No essential differences have occurred in specimens 18 months post-hactching. Forbes believed that the mucosa later becomes ciliated and specialized for secretion of egg membranes, and that a recognizable musculature develops, elicited by hormones from the sexually mature ovary. This view was based upon experimental administration of oestrone to immature alligators, which developed hypertrophied oviducts, oviducal glands, well formed muscle and ciliation at the cloacal extremities. It is unclear whether incipient oviducal myoblasts occur and can be distinguished from fibroblasts of connective tissue in young post-hatching forms; presumably muscular differential occurs extremely late in these forms.

Callard and *Leathem* (1967, 1970) showed that in the oviduct of *Coluber, Elaphe* and *Natrix*, the activity of oviducal glycogen, alkaline phosphatase and β-glucuronidase is correlated with ovarian steriodogenesis, probably of ovarian oestrogen and progesterone. In *Coluber* oviducal alkaline phosphatase activity increases during the reproductive cycle; the β-glucuronidase level increass during this time in *Elaphe* (as in *Natrix*) but the activity of alkaline phosphatase decreases. Glycogen levels of *Coluber, Elaphe* and *Natrix* are high at the beginning of reproduction, but they then decrease. The mucus secreted by the uterine secretory cells in *Thamnophis sirtalis* is an acid mucopolysaccharide, that of the chameleon also reacts strongly with PAS and alcian blue.

The mechanism by which sperms ascend the oviduct to fertilize ripe ova was first explained by *Parker* (1928, 1931) for pigeons and *Chrysemys picta*. In the turtles the entire oviduct is lined by ciliary cells, interspered by glandular cells. Eggs descend the oviduct by muscular propulsion (peristalsis) supplemented by ciliary activity. *Cuellar* (1970) described a ventral oviducal migration, during oogenesis is lizards, with the infundibular ostium intimately enveloping the pre-ovulatory ovarian occytes. The ovaries are thus virtually isolated from the general coelom, so it is unlikely that the occytes loosely migrate from the ovaries into the ostium.

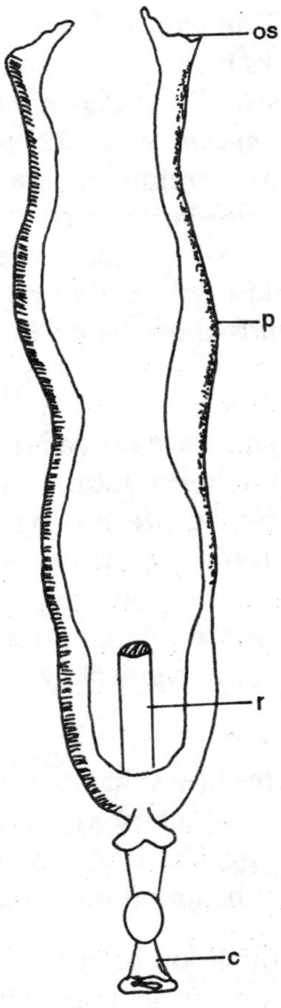

Fig. 1.28. *Chresemys picta*. Oviducts spread out. The dots along the margin (see on the right) represent the area of the pro-ovarian ciliary tract. c. Cloaca; os, ostium abdominale; pe, pro-ovarian ciliary tract; r, rectum,

Parker showed that sperm ascend the *isthmus* a (short intermediate region between the uterus and the albumin-secreting region), probably by antiperistalsis against the downward ciliary current. Thence they move to the *infundibulum* by the action off a pro-ovarian longitudinal ciliary band, 2 mm wide, the stroke of which is towards the ovary and the reverse of the abovarian cilia. The externally recognizable, vascular edge of the oviduct marks the pro-ovarian band. *Crowell* (1932) confirmed these results in *Chrysemys picta, Pseudemys scripta elegans, Phrynosoma cornutum* and *Sceloporus undulatus*. In *Phrynosoma* the albumin-secreting region is only 2 cm long (less than one-quarter of the total oviducal length; the region is one-half the oviducal length in the turtles), the pro-ovarian ciliary band is about I mm wide. *Mauremys japonica* has two ciliated pro-ovarian bands these lie on the mesotubarial and the opposite sides.

In contrast mammalian sprmatozoa are transported mainly by muscular contraction of the oviducal wall. These contractoins are probably mediated by the effect of oxytocin on the uterus. The sperm rapidly pass from the cervix to the site of fertilization. The role of cilia is questionable, but they are possibly of more importance in lower amniotes. Among mammals there could well be some active flagellar movement by the sperm themselves.

Sperm Retention in the Oviduct

A survey of literature shows that numerous reptiles retain viable sperm within the oviduct for a considerable time before actual fertilization of the egg. Ovulation, mating and fertilization may be separated by a substantial interval.

Delay in actual fertilization of eggs, as distinct from the mating process, has been termed *"amphigonia retardata"* by *"Kopstein* (1938). Retention of sperm in the oviduct of female snakes was emphasized by *Klauber* (1956) in *Crotalus*. Live sperm were found in the uterus of *Agkistrodon contortrix mokasen* for at least 11 days after mating; in *Storeria dekayi* sperm were retained for three months. Mating occurs in *Bungaris fasciatus* and *Naja naja* several months before ovulation. *Thamanophis sirtalis* mates in autumn and fertilization is delayed until the following spring. Motile sperm were removed from the uterus of non-pregnant females for up to 3 months after isolation from males; these sperm retained the capacity to fertilize other females for at least a month. *Causus rhombeatus* had live sperm in its uterus after 5 months, *Rhabdophis subminiata* for 4 months, *Boiga multomaculata* for 8 months, and *Natrix vittata* for up to 1.5 years.

Young were born from isolated females of *Trimeresurus popeorum* at least a year after mating; embroys of *Drymarchon corais* were obtained after 52 months and from *Thamnophis couchi hammondi* after 53 months. Females of *Leptodeira annulata polysticta continued to lay fertile eggs* (with an area vasculosa) for up to 5 years after the last copulation.

In turtles a high percentage of fertile eggs may be obtained from female *Malaclemys* for 2 years without annual copulation, though fertility levels tend to fall later. In fact, a single mating of this species shows isolated females laying fertile eggs for at least 4 years. *Ewing* (1943) claimed the same for isolated females of *Terrapene carolina*. In the *lizards Chamaeleo jackosnii* and *C. pumilis, Atsalt* (1953) noted sperm retention in the oviduct. In *Uta stansburiana* fertile eggs were found 81 days after mating. Delayed fertilization does not occur in *Lacerta muralis*.

Rahn (1940b) suggested that the survival of sperm in the uterus may well be a safety factor enabling eggs to be fertilized in the absence of males at the time of ovulation or when ecological conditions in some way prevent preovulatory mating. Mating in the following spring with simultaneous fertilization of eggs from two batches of sperm, may thus be quite possible.

In the biennial reproductive cycle of *Crotalus v. viridis* eggs develop rapidly in summer of the second year and remain in the ovaries during winter hibernation. In the following spring eggs are fertilized by sperm derived from matings in the previous summer and autumn and stored in the interim in special vaginal pouches. Sperm confined to the anterior pouch do not penetrate beyond the uterus, the glandular secretion of which may well prevent access until the time of ovulation, when final migratory ascent of the oviduct occurs. In *Thamnophis elegans terrestris* and *T. brachystoma*, however, sperm are stored at the base of the infundibulum in seminal vesicles, which lead via ciliated ducts to the oviducal lumen. The sperm remain in the vagina for several weeks after mating in the spring, and thenceforth they ascend the oviduct to remain in the seminal vesicles for storage.

In *Thamnophis elegans terrestris* sperm from autumn matings are retained during winter in the posterior two-thirds of the oviducts and they do not reach the seminal vesicles until February or March. In spring matings sperm reach them in April. These receptacles, found in groups at the corners of folds in the oviducts, also occur in the blind snakes *Typhlops* and *Leptotyphlops*. Sperm of *Vipera aspis* remain in the vagina (after autumn mating) during winter, and then proceed to the seminal vesicles where they fertilize eggs the following spring. Among lizards, iguanids have sminal vesicles containing sperm in the anterior segment of the vagina, while gekkonids (*Phyllodactylus homolepidurus* and *Coleonyx variegatus*), have them in the region of the oviduct between the uterus and infundibulum. Usually they are situated near the mesentery and smooth muscle. The tubular arrangement of the seminal vesicle is similar to that described by *Saint Girons* (1962b) for the chameleon and by *W. Fox* (1963) for *Anolis carolinensis* and consists of branched tubes, formed by a single layer of epithelial cells and lined internally by cilia. Amphisbaenians, as far as is known, lack seminal vesicles.

Spermatozoa were identified in seminal receptacles of females of *Terrapene carolina* isolated from males for 14 months. These seminal receptacles, which are described for the first time in turties, are morphologically unspecialized tubules of albuminous glands found in the caudal part of the albuminous regions of the oviduct. However, it may be as well to express caution, particularly for extreme cases of delayed fertilization and the occurrence of fertile eggs several years after the last copulation. In view of the fact that many groups of lizards and even one snake have been shown to be parthenogenetic.

The Corpus Luteum

A yellow coloured *corpus luteum*, or *yellow body* appears as the ovarian components in the raptured follicles following ovulation. It may be retained in modified form for a variable period in some reptiles. The corpus luteum is essentially a mammalian organ. Similar structures, however, have been found in other tetrapods and were seen long ago in such reptiles as *Trionyx*, *Lacerta agilis*, and *Anguis fragilis* and *Chalcides chalcides*. Indeed the corpora lutea of reptiles and mammals are clearly homologous. The corpus luteum of *Chalicides* can be distinguished from the flase yellow bodies (corpi lutei flasi) or atretic bodies; the latter bodies are ovules which, in various reptiles, are never shed into the oviduct but degenerate before reaching maturity. In *Lacerta vivipara* corpora lutea and atretica are only difficult to distinguish at the period just before ovulation. According to *Bragdon* (1952) in *Thamnophis* and *Betz* (1963) in *Natrix*, atretic follicles are hyperaemic and more flaccid than developing follicles; the more fluid yolk is paler and the follicular wall thinner and more easily ruptured. Atretic follicles are replaced by stromal tissue and leave a persistent scar on the ovarian surface. Nevertheless, in the oviparous *Lacerta viridis* *Cunningham* and *Smart* (1934) described and figured in atretic follicle instead of a corpus luteum, though their illustration showed two corpora lutea. Atretic follicles are common in most reptiles, but rare near the end of gestation in ovaries of *Lacerta vivipara*, *Tarentola*, *Hoplodactylus* and *Chrysemys*. *Guraya* described scars after final resorption in the ovaries of snakes. In *Hoplodactylus*, *Thamnophis*, and *Bungarus*, the zona of the atretic follicle disappears and the follicular epithelium becomes a festooned layer of gians cells bulging into the yolk. Thecal fibroblasts thence penetrate between the giant cells and enter the yolk of the ovum where numerous polymorphonuclear leucocytes and lymphocytes are situated near small capillaries. Yolk is first absorbed by the giant cells and then by leucocytes;

finally, the former are reduced in size and disappear. The vestigial follicle is marked by thecal fibroblasts. *Lovez* (1906) had earlier provided a somewhat similar description of follicular atresia in various crocodilians, turtles, lizards and snakes. Generally (especially in turtles) resorption of the vitellus occurred from the periphery inwards; in crocodilians it occurs from the centre outwards.

In *Xantusia vigilis* corpora atretica are present throughout the year, but all but one or two of the 20-40 follicles, which develop annually, become atretic. Large macrophages from the granulosa region are involved in their degeneration as in *Lacerta vivipara*, in which the atretic follicles are usually larger than unovaluated follicles, and in *Scoloporus occidentalis*. There are indications that immediately after the breeding season in *Anolis carolinensis* the corpora atretica suppress ovarian sensitivity to environmental and hormonal stimulation. Their removal increases ovarian responsiveness to gonadotropin five-fold.

Lacerta vivipara has two to five yellow-orange, reniform-shaped corporlutea per ovary, a number equal to (or occasionally greater than) that of embryos in each oviduct. Correspondence in the number of corpora lutea and oviducal embryos or eggs was also noted in various lizards and snakes. The difference in the numbers of corpora lutea per ovary and of the ipsilateral oviduoal eggs or embryos has been used as evidence of ovular migration to the contralateral oviduct. Most workers agree that, as the corpus luteum develops, the interior of the discharged follicle fills with hypertrophied cells of the membrana granulosa.

In *Hoplodactylus pacificus* luteal cells are invested by fibroblasts from the theca externa; septa penetrate amid luteal cells but there are no capillaries. The theca interna does not contribute to the development of the corpus luteum, nor are there capillaries or even connective tissue elements among the luteal cells of *Hemidactylus flaviviridis*, though the corpus luteum is highly vascular in Phrynosoma. The viviparous skinks *Spheromorphus quoyii, L. quadridigitatum (L. quadrilineatum) Egernia whitii*, and possibly *E. cunninghami*, are rather like *Hoplodactylus*, though there is only superficial fibroblastic ingrowth. The oviparous Amphibolurus nuricatus has fibroblasts from the theca interna, and blood vessels penetrating amid luteal cells. However, only connective tissue fibres extend between the latter. Corpora lutea of the viviparous *Leiolopisma weekesae* and *L. entrecastenauxii* are generally similar to those of *Amphibolurus*. In *Lacerta vivipara*, blood vessels occur within the corpus luteum and fibroblasts of the theca interna grow between luteal cells, as they do in *Calotes versicolor*. The degree of luteal vascularity of *L. vivipara*, however, is no measure of its physiological activity.

In *Hoplodactylus* as in *Mabuya* the follicular opening is plugged by follicle cells after ovulatoin, and the theca thus does not overgrow unlike the condition usually found in reptiles.

In snakes, a lizard (*Phrynosoma*), and a turtle (*Chelydra*) the hypertrophied granulos cells first invade the follicular cavity, and the infiltrating supporting fibres arise from the theca interna; finally blood vessels penetrate. The corpus luteum is highly vascular in the viviparous snake *Thamnophis* and oviparous snakes *Lampropeltis* and *Diadophis*. In *Enhydrina* the corpus luteum forms from similarly hypertrophied follicular cells, but ingrowths from the theca are not accompanied by blood vessels, nor do fibroblasts penetrate between individual cells. In contrast blood vessels accompany thecal ingrowths and fibroblasts penetrate between luteal cells in *Hydrophis*. The related snake *Cereberus rhynchops* has merely superficial ingrowths of thecal fibres and no invasion of blood vessels. The *Vipera aspis* within 3 weeks luteal cells from proliferated follicular cells fill the entire follicular cavity, and thecal fibroblasts and capillaries have commenced invasion of the luteal tissue. After 2 months the corpus luteum comprises a mass of elongate luteal cells surrounded by fibroblasts and capillaries, and the theca externa is reduced in size. Connective tissue fibres invade the interior. Degeneration begins near the end of the period of gestation, in early September, and proceeds slowly after parturition. A degenerate corpus luteum is recognizable amid fibroblasts and connective tissue in mid-winter; its vestige is seen after a year, probably to remain for the animal's lifetime. Vestigial corpora lutea (corpora albicantia) also persist throughout the life of the Japanese lacertid *Takydromus tachydromoides*. However, the corpora lutea of *Cnemidophorus hyperthrus* disappear soon after oviposition. Snakes are similarly variable. In *Natrix rhombifera* the corpus luteum disappears completely in the spring following the year of reproduction or it may last for 2-3 years in *Thamnophis sirtalis*.

In general the development of the corpus luteum of turtles is similar to that of squamates. In *Terrapene carolina* the corpus luteum develops in June when eggs are in the oviduct. Soon after ovulation cells of the granulosa, isolated in cords 50-200 µm wide (so-called granulosa-lutein cells), and of the theca, situated externally in groups of one to three cells, fill the collapsed follicle. Both types of cell contain isotropic lipid droplets. Capillaries, do not penetrate the central luteal cell mass. Artophy occurs by mid-August and follicular atresia by July or August. *Pseudemys* is probably typical of turtles in which corpora lutea regress and disappear within a few weeks after ovulation, though final disappearance in *Chrysemys picta* coincides with ovulation of the new clutch of eggs.

The true relationship between the corpus luteum and (a) viviparity or (b) egg rention within the oviduct in reptiles is still not clearly understood, though *Millar* (1959) claimed a definite correlation between the corpus luteum, longevity and egg laying or retention, a relationship generally recognized by *Chieffi* and *Botte* (1970) and *Varma* and *Guraya* (1973). In oviparous reptiles the corpus luteum regresses soon after oviposition; it is retained in so-called ovoviviparous and viviparous forms. In specimens the gestation of which is about 3 months, the corpus luteum starts to degenerate during the last month.

In view of the time which occurs between ovulation, egg

deposition, and the development of luteal tissue, as well as the high degree of vascularization of the corpus luteum in numerous oviparous reptiles, the corpus luteum may well be associated with oviducal egg retention as well as with viviparity.

In lower vertebrates hypophysectomy leads to atresia of the yolky egg follicles, and gonadotropins elicit and maintain vitellogenesis. *Agama* shows an increased number of yolky follicles after FSH injection. In the adult female *Dipsosaurus dorsalis* in the spring, pituitary gonadotropins initiate ovarian oestrogenic synthesis, indicated by the accompanying increase in oviducal weight. Oestrogen elicits hepatic synthesis is specific calcium-binding phospholipoprotein (globulin), a yolky precursor which circulates in the blood to the ovary where it is absorbed by the follicles under gonadotropic stimulation. Similar changes in plasma protein are induced by a single injection of oestradiol 17β. However, after hypophysectomy in female lizards, oestrogen injection has no stimulatory effect on the level of phospholipoprotein, or plasma calcium, suggesting that the pituitary is somehow involved in their biosynthesis or induction. The granulosa cells may well transfer yolk to the oocyte by pinocytosis, presumably originally withdrawn from the fat body reservoir via the liver.

It is still doubtful whether a corpus luteum is essential for the maintenance of gestation in reptiles, and if so how it exerts its effect. According to *Cunningham* and *Smart* (1934) degeneration of the corpus luteum is inhibited by the presence of fertilized ova in the oviduct:; they consider that hormone-like substances are passed into the blood-stream and lead to the persistence of the discharged follicles. The latter either directly or indirectly prevent further ovulation or ovular development. Likewise the corpus luteum may influence follicular atresia.

In various *viviparous* snakes *ovariectomy* early in the gestation period elicits resorption of embryos; during the middle of the gestation period it causes the embryos to be born dead; at late gestation the operation has no effect and parturition is normal. Hypophysectomy early or in the middle of gestation yields similar results. Injection of crystalline progesterone into ovariectomized snakes does not prolong gestation, but injection of post-pituitary extracts late in gestation elicits premature parturition. *Clausen* believes that the reptilian corpus luteum is more important in the early stages of gestation, and later possibly other tissues supply endorine secretions (analogous to the mammalian corpus luteum and placenta). He indicates that in reptiles pregnancy is controlled by a series of hormones with the corpus luteum initially involved. On the whole these results were supported by *Fraenkel et al.* (1940) who concludes that the corpusluteum "protects" pregnancy until a well advanced period of gestation. The importance of progesterone and the corpus luteum in the maintenance of gestation of the ovoviviparous *Chamaeleo p. pumilis* was emphasized by *Veith.*

It has been observed that living embryos of caatrated pregnant females of *Thamnophis* and *Natrix* survive for at least 25 days, though in some previous experiments two out of six embryos aborted. However, *Bragdon* found that hypophysectomy of *Thamnophis* and *Natrix* has no effect at any stage of gestation, nor does it influence the capacity of granulosa cells to digest yolk in atretic follicles. Furthermore, a corpus luteum can develop in the absence of a pituitary. Hypophysectomy of *Thamnophis* after ovulation merely leads to a decrase in the cell size of the corpus luteum (smaller cytoplasmic nuclear ratio and denser cells). Possibly the corpus luteum of *Agama* is dependent on the pituitary immediately after ovulation and later is independent of the pituitary. *Bragdon* found that snakes ovariectomized in the middle quarters of gestation have normal young which reach term, though there is some interference with parturition. He concluded that the pituitary and ovaries (including the corpus luteum) are not essential for normal pregnancy in reptiles, or at least not in these viviparous snakes. These results are generally supported by *Panigel* for *Lacerta vivipara* and by *Badir* for *Chalcides ocellatus*. Gestation of *Sceloporus cyanogenys* is likewise unaffected by gonadectomy; parturition, however, is delayed and foetal death is probably due to the exhaustion of the volk supply. It is possible, therefore, that the corpus luteum may have different degrees of importance in different species of reptiles, and the independence of embryos in the oviduct, and their survival, may improve the longer a corpus luteum is retained. This subject needs further investigation.

Progesterone is found in the corpus luteum of the snakes *Bothrops jararaca* and *Crotalus terrificus* and Δ^5-3β hydroxysteroid-dehydrogenase in that of the pregnant *Natrix*. There is an intense activity of 3β-ol-dehydrogenase and a positive reaction for 3 and 17 ketosteroids in the corpus luteum of *Lacerta sicula*. These findings are supported by *Klicka* and *Mahmoud* (1973) as the corpus luteum of *Chelydra serpentina* appears to function as a steroidogenic organ converting cholesterol of progesterone under the influence of the pituitary hormones. *Progesterone* activity in the plasma of female *Natrix sipedon* rises following ovulation and reaches its highest level at term; minimal amounts occur in the plasma of males. The functional signifiance of these results has yet to be elucidated.

The ovarian cycle of lizards, including its relationship with the corpus luteum, has been reviewed by *Callard et al.* (1972b). In the ovoviviparous *Sceloporus cyangenys* substantial progesterone is produced by the corpus luteum under the control of pituitary hormone which, for the duration of gestatoin and *post partum*, inhibits follicular development and ovulation. It acts directly on the hypothalamus and also peripherally by antagonizing vitellogenesis. *Yaron* (1972) suggests that in viviparous reptiles ovulation deactivates those enzymes responsible for the conversion of progesterone to oestrogen, leading to an increase of the former in the ovaries and plasma. Within the post-ovulatory gonad, the steroid-converting capacity is limited to the lutein and thecal cells of the corpus luteum. Such hormonal changes are favourable for the rention of the foetus within the oviduct of ovoviviparous and viviparous forms.

Thus, we see that the structure of the reptilian corpus luteum is generally similar to that of mammals. It develops in a variable manner from the ovarian follicle, after ovulation when follicular and thecal elements, fibroblasts and capillaries invade the follicular cavity to form luteal tissue; lipid is present. In contrast to the condition of atresia, granulosa cells do not become macrophagic but first hypertrophy and probably poliferate to form luteal tissue. The degree of involvement of the variable components is not related anatomically or taxonomically to the level of viviparity of different genera. Vestiges of the corpora lutea may remain throughout life in the ovaries of some lizards and snakes, though scars are usually temporary and disppear. The origin of the corpus luteum is under the control of the pituitary, though the degree of its dependence is not well understood. It probably functions as a steroidogenic organ and produces progesterone, but its functional relationship to gestation is unclar. Probably, it is not essential to maintain pregnancy in viviparous forms, at least in the later stages of gestation.

The corpus luteum is produced by the same processes which occur in atresia, and if the corpus luteum exercises some physiological effect on gestation then it may have evolved before the placenta in viviparous reptiles and only subsequently become endocrinal. Reptilian viviparity doubtless is a feature evolved independently in different groups, with the ultimate development of an omphaloplacenta and thence allantoplacenta. Bearing in mind the relatively variable and simple nature of the reptilian placenta compared with those found in mummals, the reptiilian corpus luteum and may still represent a liable, rather intermediate stage of endocrine function, the importance of which could well differ quite profoundly in different groups in which it has not yet attained the high degree of specialization and functional significance seen in eutherian mammals.

Fat Bodies

Fat bodies are basically the food storing structures formed in animals. Abdominal fat bodies occur in all examined reptiles from the temperature zone, as well as in some semi-temperature anoline lizards. Fat passes to the river where it is processed to form yolk for the developming occytes inthe ovary. In females of *Uta stansburiana* fat bodies decline in winter and early spring when yolk is deposited in overian follicles; males show a similar fat body cycle though their fat bodies are considerably smaller than those of the females. In *Cnemidophorus sexlineatus*,fat bodies decrease in size after emergence from hibernation are small in May, and have disappeared by mid-June. They enlarge in July and in September have reached practically maximum size. Similar results were obtainded in *C. neomexicanus* and *C. inornatus* non-reproductive females of *Sceloporus magister*, however, showed no consistent differences in size, though a graph did show a reduction from June to July after egg laying. Possibly sampling methods were not sufficiently controlled.

Hahn and *Tinkle* (1965) show that expiration of the fat bodies in proestrous females of *Uta stanburiana* results in the inhibition of follicular development; extirpation in early oestrous females results of increased follicular atresia and retardation in the rate of yolk deposition. Ovariectomy prevents rapid lipid mobilization from the fat bodies, which occurs is sham-operated controls. Therefore, ovarian hormones may directly or indirectly regulate lipid metabolism from the fat bodies. In *Ameiva* spp. the size of the fat body is correlated with follicular development; the relationship is also found in *Agama, Dipsosaurus. Sceloprous* and *Cnemidophorus*. Furthermore, males of *Ameiva festiva* and *A. quadrilineata* increase the weight of their fat bodies after orchidectomy; administration of testosterone prevents this. The fat reserves of males may thus be related to energy demands during the breeding season. Again in seven species of Caribbean *Anolis*, both sexes undergo a conspicuous increase in size of the fat bodies when gonadal activity is reduced. In winter when egg production ceases fat bodies of females increase in size; they shrink when egg production commences again. However, in *Vipera berus* fat bodies are generally well-developed even after hibernation and are reduced only after prolonged starvation, or in females immediately after pregnancy when they may disappear completely. In *Crotalus horridus* they are also reduced in gravid females and extremely so after parturition. Perhaps fat bodies serve as a reserve storage of food, as well as a safety reserve in times of dietary hardship. There is no seasonal variation in the size of the fat bodies in *Homalopsis buccata*, of Kuala Lumpur, Malaysia, which has no really definite breeding season. Fat body size has no temporal pattern in the male *Anolis acutus* of St Croix, Virgin Islands, though in the female it is smaller during the dry season of December to April when reproductive potential is lowest. Fat bodies are absent in the viviparous *Hemiergis peronii*, though present in another (oviparous) Australian skink *Morethia boulengeri* and are also missing in the Javanese geckos, *Hemidactylus frenatus*, *Cosymbotus platyrus* and *Gehyra mutilatus*.

Pseudohermaphroditism

Embryonic or juvenile bixexuality often called *hermaphroditism* or *transient heterosexuality* is well known in reptiles. Cases are described in a previous section. Sexual development and differentiation are presumably mediated through the endocrine system and genetically determined. Any genetic change in germ cells, endocrine imbalance, or pathological condition in the gonads of an embryo may, therefore, be expected profoundly to influence the development of the urinogenital system. Thus, male and female specimens frequently retain, to a lesser or greater degree, reproductive components atypical of their sex.

Frequently more extreme and abnormal examples of pseudohermaphroditic system are recognized. A male *Lacerta agilis*, seemingly normal in appearance, possessed a normal pair of oviducts, which opened anteriorly by funnels into the coelom and posteriorly into the cloaca. Again a pair of blind Mullerian ducts was found in an *Amphibolurus muricatus* and *Lacerta savicola*, but they were more vestigial in the latter. A fully developed unilateral oviduct occurred in one specimen of *Lacerta viridis* and paired oviducts in another. Likewise one adult *Mauremys caspica leprosa* had a single aborted. Mullerian duct, dorsal to the epididymis, and another had a pair of oviducts which were normal except in the posterior region.

An adult female *Chalcides ocellatus* had ovaries (containing eggs) as well as testicular canals opening into short, blind ductuli efferentia; one of *Gerrhonotus* had two vestigial Wolffian ducts and three out of 24 specimens of *Chrysemys* had rudiments of the epididymis and the vas deferens on either side, while a fourth specimen had a vas deferens.

The best studied turtles is an adult male *Testudo graeca* having a well-developed epididymis, vas deferens and oviduct (each with a ciliated ostium) on each side. The specimen possesses a single penis. A right testis contains spermatozoa and the left ovotestis including an ovarian component with well-developed yellow eggs and a testicular region with seminiferous tubules. In another male there is a pair of normal testes and a rudimentary thin-walled Mullerian duct on each side, both of which open into the coelom anteriorly while the slightly larger left one opens into the cloaca. An *Emys orbicularis* was said to have a left ovotestis (with ovocytes) undergoing spermatogenesis, and *Hansen* (1943) reported an adult male *Pseudemys scripta troostii,* the paired gonads of which comprised four-fifths ovarian and one-fifth testicular tissue by volume. Large yolky eggs were present and the whole was enveloped by a mesovarium. Potentially functional epididymis, vas deferens, and oviduct were recognized on each side. Doubtless the specimen could have functioned as a female and laid eggs; presumably it could also have produced sperm and was thus a true hermaphrodite. Again *Risley* (1941a) reported that an almost normal looking adult male *Chrysemys picta marginata* had an ovotestis, male ducts, and an oviduct, with pathological alteratoins to the accessory sexual ducts on the left side. A female juvenile *Malaclemys terrapin centrata* had a medullary tumour in the left ovary, while inhibited normal regression of the medullary cords into hollow ovarian sacs.

Among lizards and snakes, a *Lacerta viridis* possessed a pair of ovotestes. Each gonad comprised normal seminiferous tubules separated by interstitial stroma, and there were ovarian appendages that contained large, fully developed ova. On each side an epididymis led into a vas deferens, which united with its ipsilateral ureter to open into the posterior region of the cloaca. Paired copulatory organs were present. Oviducts with glandular walls and open coelomic funnels were about one-third the normal length. The most perfect bisexuality in reptiles was claimed by *Bons* and *Bons* (1969) for a Moroccan lizard *Ophisaurus koellikeri,* which possessed complete sets of male and female sexual structures on each side of the body, including external hemipenes. Hypertrophied renal sex segments were also recognized. From histological examination both sets of organs appeared functional.

An agonadal adult garden lizard, *Calotes versicolor,* externally resembled a female and had no hemipenes, but Mullerian ducts and a dorsal crest were present. An intersexual form of *Bothrops insularis* of the Quemada Grande of the South Atlantic, 40 miles south of Santos, showed 36% of them to be females with unilateral or bilateral hemipenes; 21% to be females and probably sterile, without hemipenes. The discoverers coined the term "arrenoidism" (females with a unique male character) for this condition. Intersexual females of *Pseudoficimia frontalis,* a Mexican colubrid snake, had well-developed hemipenes, the first case of adult female snakes claimed universally to have such primary male characters. Young females have epoophorons and vas deferentia and young males have oviducts.

THE GERM CELLS

Germ Cells are the Cells Undergoing Ganetogenesis

Waldeyer (1870) originally propounded the view that the definitive germ cells of vertebrates originated in the gonad. In contrast *Nussbaum* (1880) claimed that they originate outside the gonadal germinal region and migrate to it early in ontogeney. *Allen,* who studied reptiles, supported *Nussbaum,* and first

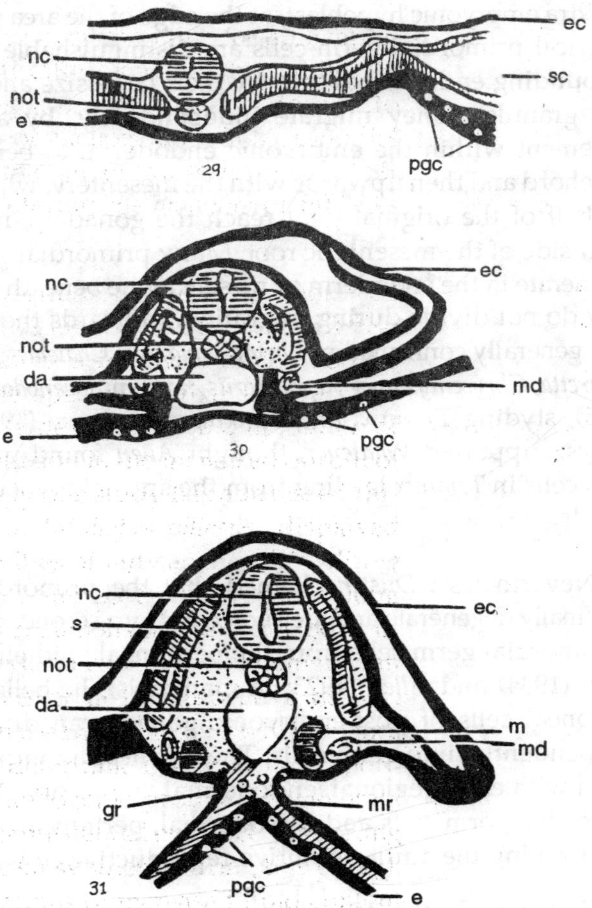

Fig. 1.29-31. *Chrysemys picta marginata.* Migration of primordial germ cells, in embryos from the extra regionoal entoblast. da, Dorsal aorta; e, endoderm; ec, ectoderm; gr, gonadal region; m, mesonephros; md, mesonephric duct; mr, root of mesentery; nc, nerve cord; not, notochord; pgc, primordial germ cells; sc, splanchnocoele; s, somite.

described the primordial germ cells of *Chrysemys picta marginata* as arising in a posteriorly directed horseshoe shaped zone of the extra embryonic hypoblast, at the edge of the area pellucida. Spherical primordial germ cells are distinguishable from the surrounding endodermal cells by their large size and contain yolk granules. They migrate independently, by amoeboid movement within the embryonic endoderm, to beneath the notochord and then upwards with the mesentery, where about one-half of the original total reach the gonadal Anlagen on either side of the mesenteric root. Many primordial germ cells degenerate in the endoderm or mesenchyme beneath the aorta. They do not divide during migration. In lizards these results were generally confirmed in *Phrynosoma* and *Chalcides*; in turtles in *Caretta*, *Chrysemys* and *Sternotherus*, and in *Sphenodon*. *Simkins* (1925), styding *Trionyx*, and *Simkins* and *Asana* (1930), with *Calotes*, supported *Wadleyer*, thought *Allen* found primordial germ cells in *Trionyx* leading from the splanchnopleure to the gonads.

Nevertheless, *Dustin* claimed that the primordial germ cells finally degenerate and are succeeded by a second generation of primordial germ cells from the peritoneal epithelium. Like *Risley* (1934) and *Allen* (1923; in mammals), he believed that peritoneal cells of post-embryonic gonads can differentiate independently, into germ cells. There is here an implied dual origin with extra-regional, endodermal, migratory, amoeboid primordial germ cells and mesodermal, peritoneal, epithelial cells forming the adult definitive reproductive cells.

Berenberg-Gossler (1914) believed that the primordial germ cells of *Lacerta agilis* are late-forming mesodermal cells, some of which reach the gonads to become primordial germ cells; others contribute to the Wolffian duct. This view has never been corroborated and should probably be discarded. According to *Risley* (1933a), the primordial germ cells of *Sternotherus* lose their yolk during migration and develop a large attraction sphere (also called *"Nebenkern"* or *"Mitochondrialkorper"*, *Dufaure* and *Hubert*, 1965), which lies next to the nucleus and represents mitochondria and yolk granules. The primordial germ cells divide rapidly on arrival in the gonad. In *Sphenodon,* as in other reptiles, the primordial germ cells arise in the endoderm of the yolk sac. They lie in the area opaca all round the embryo, though chiefly in the anterior area. This arrangement is found in birds though probably secondarily acquired by them. From this condition the primordial germ cells are found to originate in the posterior extra-embryonic area in *Lacerta* spp. and in the anterior area in *Vipera aspis.* Migration of the primordial germ cells in *Sphenodon* is first amoeboidal and thence by passive movement in the blood vessels to reach the gonads. *Tribe* and *Brambell* regard the dual migratory mechanism as intermediate between the interstitial, amoeboidal process in most reptiles and the passive vascular transportation of birds. However, migration of primordial germ cells is a variable phenomenon in reptiles. It is vascular in *Chamaeleo bitaeniatus* and *Mabuya megalura, Anguis fragilis, Vipera aspis* and *Thamnophis.* The latter concluded that all reptiles the primordial germ cells of which are transported vascularly have them first located in the extra-embryonic region and predominantly anteriorly. Migration in interstitial in *Chalcides ocellatus, Lacerta vivipara* and other *Lacerta* spp.

Asymmetry of the urinogenital system of snakes, seen in the posterior displacement and reduction in the length of organs on the left side, is also expressed in the distribution of the primordial germ cells. Greater numbers occur on the right side of reptiles and on the left in birds.

Experimental results now available for vertebrates support the view of Nussbaum (1880), that primordial germ cells not only give rise to all the definitive genocytes of both sexes but also play a role in the organization of the gonads. The reptilian gonocytes derive only from the primordial germ cells has received support from *Miller* (1963), and this view has been later reiterated for man by *Falin* (1968). In most reptiles oogonia may persist until after sexual maturity, then they divide to form oocytes.

Reptilian primordial germ cells have been described by light and electron microscopy. The presence of osmiophilic granules in the cytoplasm (more numerous than in birds) and a newly discovered specific paranucleolar mass alongside the nucleolus, permit easy recognition. The paranucleolar mass is not recognizable in the primordial germ cells of *Anguis* but it is clearly visible in those of *Lacerta vivipara, L. agilis,* and *Vipera aspis.* It is negative to PAS, colours green with methyl green pyronin and it is probably of histone nature.

The ultrastructure of the cytoplasm of the primordial germ cells of *Lacerta vivipara* includes ribosomes and polyribosomes, the latter being more numerous in older cells, though always less profuse than in somatic cells. Likewise there is a progressive development of the smooth endoplasmic reticulum and the Golgi network. Mitochondria are small, poor in cristae, and grouped together, particularly in the region of the attraction sphere. Yolk bodies ultimately disappear, and the lipid droplets correspond to the osmiophilic bodies previously described. The granular nucleolus is ring-like with a less dense central zone, which includes fibrillar structures. A structure similar to the paranucleolar mass occurs in the adult spermatogonia. *Hubert*, considered that this permits one to establish continuity between the primordial germ cells of the extra-embryonic entoblast and the definite germinal cells of the adult. This view, however, does not allow for the independent acquisition of a paranucleolar mass in any cells which could possibly be derived from peritoneal epithelium. The amoeboid form of the primordial germ cells of embryos of *L. vivipara* possesses cytoplasmic microtubules. The micropodia are of characteristic ultrastructure; they are devoid of mitochondria, lipids and yolk but contain small vacuoles and fibrils.

Study by autoradiography reveals that there is little synthesis of RNA in the gonocytes of *Laccerta vivipara;* the migrating gonocytes do not synthesize DNA. However, as sexual differentition occurs, thymidine is actively incorporated into the gonocytes of the genital region just before intense mitosis proceeds. There is only sparse information on the histochemistry of primordial germ cells. A strong PAS reaction occurs in the vitelline inclusions in *Emys, Pseudemys* and *Chamaeleo*. No acid phosphatase activity occurs in the cells *L. vivipara*, and there are no reports of any alkaline phosphatase activity in such cells in reptiles.

Endocrine Glands and the Urinogenital System

There have been numerous rather recent publications on the pituitary and thyroid glands of reptiles and their endocrinal relationships to the urinogenital system. Reptilian pituitary cells have been classified by modern histological staining procedures, and correlation between their appearance and changes in various components of the reproductive system have been investigated. The evidence suggests an intimate relationship between the number and histological appearance of specific cellular components of the pituitary, especially of the adenohypophysis, and the sexual cycle of reptiles.

From extensive research among reptiles *Saint Girons* (1970a, b) described adenohypophysical rostral gamma (LH) cells; large mediolateral or ventrally situated beta (FSH) cells; small, conical, widely dispersed but mainly posterior delta (thyrotrophe) cells; posterior acidophiles; and medial and mediorostral X or epsilon (somatotrophic) cells. These five categories seem generally similar to those described earlier by *Gabe* and *Rancurel* (1958). During the reptilian breeding cycle, the LH and FSH cells may become degranulated and vacuolated (assumed to reflect secretory activity), or become granulated and lose their vacuolation. Thus in both sexes of *Acanthodactylus erythrurus lineomaculatus* at the beginning of the reproductive cycle the beta and gamma cells of the hypophysis degranulate and reduce in volume. There is an assumption of casuality between such cellular activity, mediated either via secreted gonadotrophic hormones to elicit gonadal alterations or via the gonads which, through androgens or oestrogens, elicit bodiliy or behavioural changes.

Nevertheless, there is still profound disagreement between different workers about the actual cells concerned in initiating particular endocrinal function. *Licht* and *Rosenberg* (1969) for example, have questioned the validity of some of *Saint Girons'* findings. From partial hypophysectomy and bioassay in *Anolis carolinensis*, *Licht* and *Rosenberg* (1969) believe that the gamma cells secrete corticotropin, the beta basophiles thyrotropin, delta basophiles gonadotropin, carminophilic acidophiles prolactin and the alpha orangeophilic cells somatotropin.

Cytologically the pars distalis of *Anolis carolinensis*, described as gonadotropic, was investigated by *Pearson et al.* (1973), in sexually active and castrated specimens. Similar investigations by *Forbes* and *Dent* (1974) showed gonadotrophs to elongate and degranulate in castrated *A. carolinensis*. Increased numbers of microtubules and smaller "intermediate filaments" (100 A° in diameter) orientate parallel to the long axis of the cell, which may facilitate cellular elongation. The intimate relationship between the intermediate filaments and the secretory granules may be associated with granular movement during secretion. The hormonal influence of the pars anterior on the maintenance, development, and function of the gonads and their ducts has been tested experimentally by hypophysectomy. Conversely the influence has been studied by implanation into hypophysectomized specimens of pituitary tissue or by intraperitoneal injections of powdered hypophysial extracts or of urine from mare, hog, cattle and sheep, or of the commercial preparations of such gonadotropins.

After hypophysectomy, the gonads (and interstitial cells, are reduced in size and spermatogenesis is inhibited. Egg laying, however, is unaffected and the Sertoli cells appear to be unchanged. In contrast *Chiu* and *Lynn* (1971) found that hypophysectomy of the snake. *Arizona elegans* impaired egg laying. Implantation of pituitary tissue into *Thamnophis* leads to an increase in the volume of testicular tubules as well as that of the interstitial cells. Injections over varying periods of the following substances, into specimens at the quiescent period of the reproductive cycle elicited growth, differentiation and hypertrophy of testicular tubules, interstitial cells, epididymis, vas deferens, ovaries and oviducts; whole pituitary extract into the alligator and *Malaclemys;* hypophysial extracts into *Uromastyx*; pregnancy urine into lizards including *Anguis*; hog anterior pituitary extracts pregnant mare's urine into *Phrynosoma;* gonadotropin (equine) into hypophysectomized *Thamnophis;* anterior pituitary extract (gonadorme) into *Lacerta*, antuitrine (anterior pituitary-like substance) into *Anolis* and *Eumeces*; ovine gonadotropins FSH and LH into *Anolis* and ovine FSH and LH in saline-into hypophysectomized *Hemidactylus, Anolis, Uta;* and *Sceloporus*. Such injection increases the volume of individual eggs rather than their number in` *Phrynosoma* and stimulates them to mature and ovulate in *Lacerta vivipara*. In *Sceloporus dyanogenys* the number of follicles reaching maturity is increased after injection of PMS (pregnant mare's serum; *Callard et al.*, 1972a). Secondary sexual structures including the sex segment of the male kidney are also stimulated. presumably secondarily via gonadal hormones under pituitary control. FSH is far more potent in eliciting ovarian and oviducal enlargement and some ovulation in female hypophysectomized lizards and *Reddy* and *Prasad* (1970a) conclude that in *Hemidactylus* FSH alone may maintain testicular function and also stimulate production of androgens. This result was subsequently confirmed in snakes and turtles by *Licht* (1972a, b). In *Lacerta vivipara* eggs are never expelled prematurely (oviposition) after administration of anterior lobe extract. Similar treatment in juvenile alligators elicits premature disappearance of the mesonephric glomeruli, though a vestigial Wolffian duct is still retained in treated females. In *Xenodon* ovulation of the recipient is prematurely stimuluated after homografting the hypophysis. Recent evidence by *Lisk* (1967) shows the pituitary-gonad axis of *Dipsosaurus dorsalis* to be influenced by the hypothalamus, via and median eminence and hypothalamus-hypophysical portal system. Deficiency of gonadal steriods initiates such hypophysical stimuli.

It is unclear that, if any, is the relationship between the median eminence and the pars nervosa and the reproductive cycle. The supraoptic and paraventricular nuclei of *Agama bibroni* are replete with homogeneous secretory granules between July and September. These cells hypertrophy and degranulate during the whole time of reproduction. Neurosecretion is abundant in the neurohypophysis from September to January, but is less from March to June; it then increases again in July. In *Vipera aspis* neurosecretion it is abundant during hibernation, but during the period of reproduction it is generally reduced, especially in females. It is always absent from the pars nervosa *post partum*. Injection of extracts of the pars nervosa into various reptiles leads to increased body weight owing to reduced water loss. *Pitressin (vasopressin* fraction) or *pituitrin* (whole pituitary extract), injected into the alligator, elicits a reduction in the rate of glomerular filtration. *Heller* and *Pickering* (1961) showed that the neurohypophysis of *Testudo* and *Natrix* has a slow moving arginine *oxytocin (=vasotocin)* fraction and a fast moving *(oxytocin)* one; the antidiuretic action is mainly due to the slow moving component. Arginine vasotocin is probably present in the neurohypophysis of all non-mammalian vertebrates except elasmobranchs, and oxytocin also occurs in amphibians and turtles. *Arginine vasotocin* (slow moving), and a mixture of oxytocin and 8-isoleucine oxytocin, (=mesotocin; fast moving fraction) occur in the neurohypophses of *Natrix, Naja Chelonia, Lepidochelys,* and *Caretta*. Oxytocin and vasopressin elicit about equal amounts of diuresis in *Trachydosaurus rugosus*. Arginine vasotcin may well be the natural antidiuretic hormone in reptiles while arginine vasopressin is that in mammals.

In support of this conclusion *Butler* (1972) found that intravenous injection of arginine vasotocin into *Chrysemys picta belli* reduces the rates of glomerular filtration and urine flow and the net rate of sodium excretion. Previously *Munsick* (1966) are *Acher et al.* (1968, 1969) had demonstrated mesotocin and arginine vasotocin, identical with those of amphibians, in the neurohypophysis of *Vipera, Naja* and *Elaphe*. Knowledge of neurohypophysial hormones in reptiles is still rudimentary, even though these hormones certainly influence water balance and blood osmotic pressure by influencing kidney funtion. Details of the mechanism still need to be elucidated.

In mammals antidiuretic hormone (ADH), parathyroid hormone (PTH), and probably thyrocalcitonin (TCT) act by stimulating the synthesis of cyclic AMP in the membranes of renal tissue; this process is catalysed by adenylate cyclase situated on the inner surface of the cell membrane. *Vasotocin* (non-mammalian ADH) alone stimulates adenylate cyclase in kidney preparations of amphibians; PTH is active in birds and the alligator and stimulates preparations of rat kidney. The results suggest that, similar to the condition in mammals, ADH in amphibians and PTH in birds and reptiles are mediated via cyclic AMP. Hormones of the pars nervosa also influence muscular activity of the reptilian oviduct. Peritoneal injection of neurohypophysial extracts to gravid females of *Lacerta vivipara. Sceloporus undulatus,* and *Uta stansburiana* elicits premature oviposition. A similar, though delayed, response occurs in *Alligator* who weeks after administration of `post-pituitary extracts. Arginine vasotocin could well be responsible for oviducal contraction during oviposition in turtles, for it is consistently more potent than in osytocin on the oviduct of the lizard *Klauberina riversiana*. Though *Chiu* and *Lynn* (1971) recognize the pituitary hormones are concrned in the process of egg laying in snakes, they could not specify which ones. Perhaps these hormones originate in the pars nervosa.

The relationship of the thyroid to reproduction in reptiles is obscure. Reptiles are active after hibernation; their emergence is influenced by thyroid activity, and reproduction usually ensues soon thereafter. Among others *Miller* concluded that an increased level of thyroid function influences reproductive activity. However, cases, reported by *Chiu et al.* (1969), seem to show no correlation. The tropical Australian lizard *Carlia rhomboidalis* shows continuous spermatogenesis throughout the year, but thyroid activity may be correlated with yolk deposition in gravid females and with copulation and territory defence in males.

When thyroxine is injected to *Phrynosoma*. It results in atrophic ovaries and thyriodectomy of *Thamnophis* leads to the degeneration of specific cells in the adenohypophysis. Similar surgical treatment to *Agama agama* has a variable inhibitory effect on testicular function; some lizards show signs of spermatogenetic inhibition and all have reduced androgen production. Injection of thyroxine into normal animals, however, elicits more marked suppression of spermatogenesis, and in females both treatments led to atresia of all yolky follicles.

Thyroid hormones, therefore, probably have a direct effect on general bodily well-being and metabolism. Thus, they influence reproductive processes and perhaps the reproductive organs themselves. They may also act indirectly, by influencing the pituitary to elicit changes in endocrinal secretory activity. Finally reptiles show increased activity and enlargement of the interrenal glands (the cells and nuclei hypertrophy) during the final stages of the breeding season; these are, for example seen in *Xantuisa* and *Thamnophis sirtalis*, and at peak reproductive time in *Carlia rhomboidalis*. The adrenals of *Naja naja* increase in weight, and there is accompanying hypertrophy of cortical cells and nuclei during the spring breeding season, when the cells are actively secretory. After breeding the cells accumulate lipid and become smaller and again inactive. A similar relationship between summer maximum testicular size and function and the adrenals is recognized in *Calotes versicolor*.

2

Regeneration in Reptiles

Regeneration is the recovery of lost part. *Autotomy* is the regeneràtion of the body part given off willingly. It is a method of escape from enemies. Among the invertebrates, it is found in certain molluscs, crustaceans and echinoderms. Among the vertebrates, the tail appears to be the only organ that can be lost in this way. The capacity for caudal autotomy is developed to varying degrees in certain salamanders in the tuatara *(Sphenodon)* in many lizards in some amphisbaenians, in a few snakes, and in certain rodents. There is evidence that certain extinct reptiles could also shed their tails. It is interesting that autotomy seems to be unknown in fishes, despite the diversity of their structure and habits. In the great majority of cases, autotomy in reptiles involves the breaking or discarding of the tail *(urotomy;)* at one or more predetermined sites of weakness, which are known as *fracture planes* or *autotomy planes*. Autotomy is associated with special adaptations of the various caudal tissues, in particular, of the vertebrae. As a rule, each of these is more or less split at the site of the plane, so that the tail can break at any of these points along its length. Although the presence of these splits is usually fairly obvious in Recent and fossil material, the modifications of the soft tissues, such the muscles, blood vessels, and spinal cord, can only be recognized by fine dedection or by histological techniques.

Thus, *regeneration* may be defined as the ability to reproduce relatively complicated organs or structures after they have been lost through trauma or other causes. Although organs that can be autotomized usually have some power of regeneration, the two phenomena are not necessarily linked, and regeneration is much more widespread than autotomy. Examples of renewable organs among the vertebrates are the fin rays of certain teleosteam fishes, the limbs and tails of larval urodele amphibians (which are probably the best-regenerators among vertebrates), the mammalian liver, and the antlers of deer, which are physiologically shed and replaced each year. Among the reptiles, only the tail (in *Sphenodon* and most lizards) and, probably, the shell in Testudines have extensive powers of regeneration, although the new tail is not a perfect replica of the original. The limbs of certain lizards may show a limited power to regeneration under unusual circumstances, as may be tail in certain crocodilians.

The present combined study of autotomy and regeneration in reptiles begins with the morphological approach and then touches many areas of modern biology, from paleontrology to biochemistry. "It is an admirable feature of herpetologists that they are able to cross the boundaries between different aspects of their subject, which remains, perhaps more than many other branches of zoology, a single coherent discipline."

AUTOTOMY IN REPTILES

Fracture Planes of Sauria

Caudal vertyebrae of Sauria : In the majority of autotomous lizards, there is a fracture plane in each of the caudal vertebrae, except for the first few vertebrae. Generally, there are four to nine unmodified vertebrae, although there may be more in certain longtailed, limbless forms. These non-autotomous vertebrae may be called the *pygal series,* as opposed to the postpygal, autotomous series. However, in non-autotomous lizards, the distinction may be hard to draw. The *postpygal vertebrae,* unlike the *pygals,* are associated with autotomous modifications of the adjacent soft tissues. Autotomy through the pygal region would clearly be undesirable, because this region lies close to the cloaca and contains the hemipenes in the male. Moreover, the pygal vertebrae give origin to the caudifemoralis muscles that are inserted on to the femur and help to pull it backwards and downwards, thus playing an important part in locomotion. Some of the tiny, elongated vertebrae near the caudal tiip may also lack fracture planes.

The vertebral *fracture planes* have an important relationship with the paired transverse processes, which project laterally from the region in which the neural arch joins the centrum. These processes may perhaps represent fused caudal ribs.

Etheridge (1967) has shown that in lizards several different types of caudal vertebrae can be distinguished on the basis of the arrangement of the transverse processes. A vertebra may possess a simple, unmodified pair of these structures; the fracture plane, if present, may pass anterior to, through, or posterior to them. Single process vertebrae are commonly found throughout the whole tail in nonsaurian reptiles, including snakes, and are generally present in the non-autotomous pygal region of lizards. In certain lizards, some of the transverse processes appear to be *"doubled,"* so that there seem to be two of them on each side. *Etheridge* suggests that these are essentially anterior and posteror limbs of a single split process, rather than morphologically separate structures. These limbs may diverge distally to a varying extent or converge distally. Vertebrae with doubled processes are nearly always autotomous, and the fracture planes pass between the limbs of the processes, indeed, the autotomy plane may be the basic factor associated with the division of the processes. Some vertebrae, very often including those near the tail tip, may lack transverse processes altogether. The tails of most lizards contain more than one type of vertebra, and the different types of vertebrae are arranged in sequences that may vary in different groups and have some taxonomic significance.

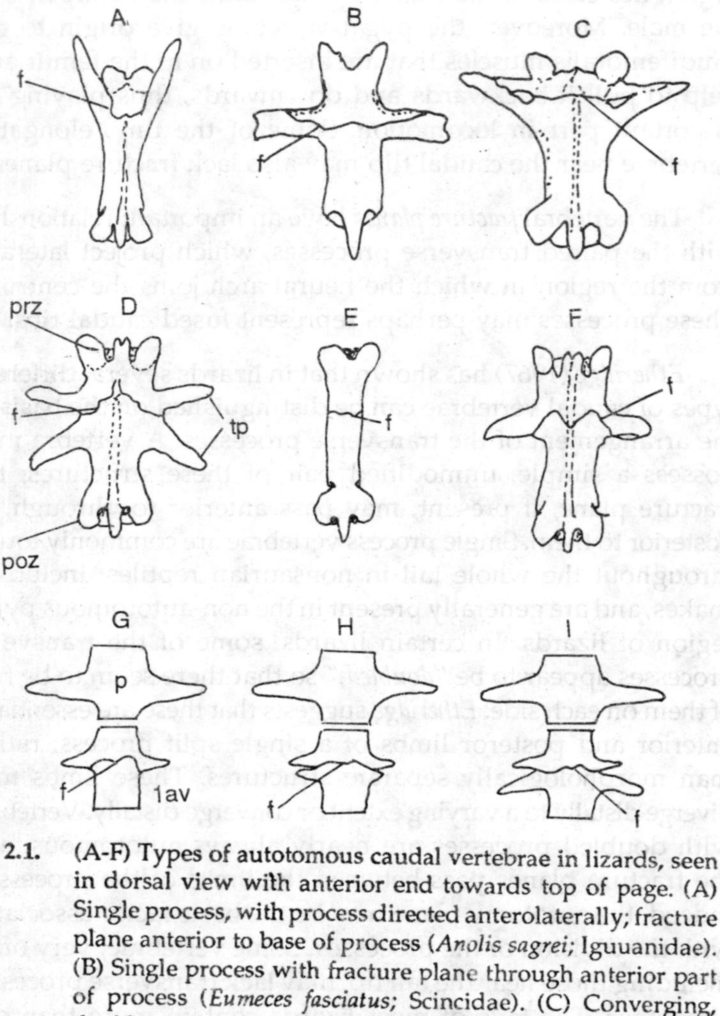

Fig. 2.1. (A-F) Types of autotomous caudal vertebrae in lizards, seen in dorsal view with anterior end towards top of page. (A) Single process, with process directed anterolaterally; fracture plane anterior to base of process (*Anolis sagrei*; Iguuanidae). (B) Single process with fracture plane through anterior part of process (*Eumeces fasciatus;* Scincidae). (C) Converging, double process, with fracture plane between the partly distinct processes (*Ophisaurus ventralis;* Anguidae). (D) Diverging process, with fracture plane between processes (*Dipsosaurus dorsalis;* Iguanidae). (E) No process (*Basiliscus vittatus;* Iguanidae). (F) Single process, with fracture plane through posterior part of base of process (*Sceloporus magister;* Iguanidae). (G-I) Types of caudal vertebrae in Lacertidae, dorsal view.

The last pygal (always a single process, non-autotomous) and the first post-pygal (autotomous) vertebrae are shown; the more posterior vertebrae may be different from the latter. (G) Pattern A. Post-pygal with single process and fracture plane passing essentially posterior to process, as in *Lacerta vivipara* and *L. viridis*. (H) Pattern B. Post-pygal with processes slightly divergent, posterior pair shorter, fracture plane between processes, as in *Algyroides* and some *Lacerta* species. (I) Pattern C. Post-pygal with divergent processes, posterior pair longer, fracture plane between processes, as in *Gallotia, Psammodromus,* most *Podarots,* a few *Lacerta,* f, Fracture plane; p, last pygal vertebra; poz, postzygapophysis; prz, prezygapophysis; tp, tr, transerve process; fav, first autotomous vertebra.

FRACTURE PLANES

Audit Vertebrae

As is true in lizards generally, the non-autotomous pygal vertebrae are of the single process type. There is also a terminal series without transverse processes. In some forms, all the other caudal vertebrae, including the most anterior postpygals, have single processes; this condition is found in some species of Lacerta (including *L. vivipara and L. viridis)* and among various other lacertid genera. Other forms show pattern B, with the most anterior post-pygals having slightly diverging processes with the posterior limbs smaller than the anterior ones, as in some species of Lacerta and in most of *Algyroides*. In some cases, this pattern B may be confined to a few vertebrae; patterns A and B may co-exist within a single population of a given species. Still other forms have pattern C and the most anterior posypygals have transverse processes with more widely divergent limbs, the posterior ones being the longer. This pattern is found in *Gallotia, Psammodromus, Podarcis, Acanthodactylus, Ophisops,* and some *Eremias.* All the postpygal vertebrae are said to process *fracture planes.* White's remark (1925) that they show no special peculiarities that determine the site of fracture of Lacerta vivipara is erroneous. In this species, the first postpygal vertebra, that is, the most anterior autotomous one, is the 7th or 8th. In other lacertid species, it may be any of the vertebrae between the 5th to the 9th inclusive, depending on interspecific or even on intranspecific variation.

The anatotomy of the *fracture plane* is most easily described in *Lacerta vivipara* or *L. viridis*. The *procoelous vertebra* is somewhat elongated and has a well-developed anterior neural spine and a lone posterior neural spine, which projects backwards between the postzygapophyses. Each *fracture plane* is represented by a split that divides the vertebra into two parts, the anterior part being slightly the shorter of the two. The split passes through the centrum, at which is edges are bordered by conspicuous bony ridges. The split extends dorsally through the neural arch on each side and through all or most of the anterior neural spine, sometimes stopping short at its tip. In general, the split runs posterior to the transverse process on each side, but it may pass obliquely through the posterior part of the base of the process.

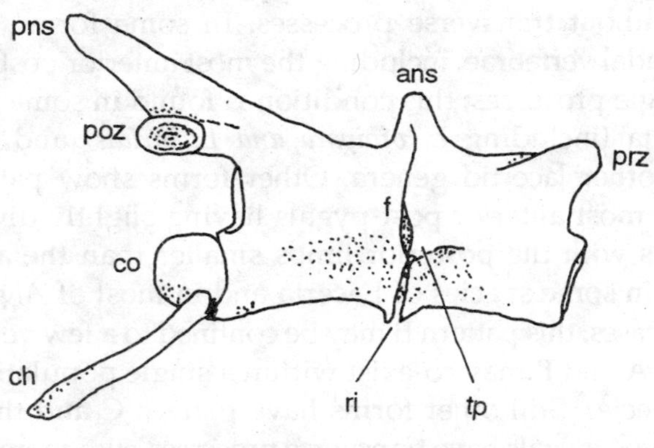

Fig. 2.2. *Lacerta viridis* (adult). Midcaudal vertebra seen from right side. Ans. Antenor neural spine; ch, chevron; co, vertebral condyle; f, fracture plane, pns, posterior neural spine; poz, posyzygapophysis; prz, prezygapophysis; ri, ridges bounding ventrolateral aspects of fracture plane; tp, transverse process.

In *Lacerta vivipara*, the centrum consists of a thin-walled, bony tube that is connected with each adjacent vertebra by a pad of intervertebral cartilage. From the walls of the tube, bony trabeculae project into the large marrow cavity; in the region of the split, this cavity is partitioned by transverse plates of bone that contain perforations. In some places, there is some connective tissue between the bony edges of the split, especially in the regions of the anterior neural spine and in the extremities of the ridges on the centrum. Some cartilage (or cartilage-like) cells can be seen embeded in or attached to the bone of the centrum adjacent to the split. However, it should be emphasized that the split is not occupied by a septum of cartilage or any other material, and much of the area of the split is almost devoid of tissue.

Development of Fracture Planes

As stated already that the fracture planes represent persistent lines of separation at the sites of the original intersegmental fissures; in non-autotomous vertebrae, these fissures become obliterated by the fusion of he pre-and postsclerotomites. This idea was originally suggested by *Albrecht* in a footnote to a paper on another subject and has been supported by some more recent workers, notably by *Werner* (1971a) in this study of vertebral development in geckos.

However, it has also been known, since the work of *Gegenbaur* (1862), that the fracture planes are not completed until the vertebrae become well ossified in early postnatal life. Thus, these planes appear to be *"secondary developments"* which somehow arise at or close to the original site at which the sclerotomites of adjacent segments had previously joined up.

According to *Bellairs* the development of the fracture plane is a more complicated process than had previously been envisaged. The split in the neural arch develops separately from that in the centrum. The split in the arch does indeed appear to represent a persistence of the intersclerotomic fissure and is recognizable from the earliest stages of chondrification.

The centrum arises as a continuous tube (perichordal tube) of cells that chondrify around the notochord. Although these

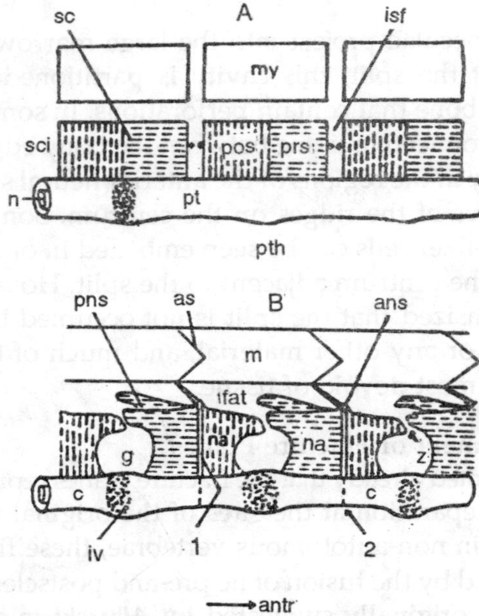

Fig. 2.3. Diagrams showing development of autotomous caudal vertebrae and their fracture planes in *Lacerta*. (A) Procartilage stage. Each sclerotome block consists of an anterior (pre sclerotomite) and a posterior (postsclerotomite) portion, more or less divided by a transient slit-like cavity, the sclerocoel. The sclerotome blocks are at first segmentally arranged, like the myotomes, and are separated by intersclerotomic fissures (width exaggerated here). The pre-and postsclerotomites of each block become separated around the level of the sclerocoel. In non-autotomous vertebrae, they recombine intersegmentally (suggested by small arrows), obliterating the intersclerotomic fissure. In autotomous vertebrae, the fissure partly persists as the fracture plane in the neural arch. Cells also migrate ventrally from the sclerotomes on each side to form a continuous perichordal tube around the notochord; this tube shows little, if any, sign of primary segmentation. The tube forms the vertebral centra, and perichordal thickenings develop along it opposite the sclerocoels; these perichordal thickenings (or rings) become the intervertebral pads of cartilage. (B) Later stage showing composition of definitive tail vertebra. The sclerotomites have separated and recombined so that each neural arch is derived from parts of two adjacent segments, which are separated by the fracture plane (1). Near the end

of embryonic life, fracture planes (2) also develop secondarily in the centra. The fat bands and muscles retain the primary segmental arrangement, though the latter interdigitate; the tissue between adjacent fat and muscle segments forms the connective tissue autotomy septum, which is bilaminar where it passes through the fat. The anterior neural spine is just beginning to develop. The chevrons are not shown ans. Anterior neural spine; as, connective tissue autotomy septum; c, centrum; fat, fat banding, position of spinal ganglion between vertebrae; isf, intersclerotomic fissure; iv, intervertebral pad; m, muscle; my, myotome; n, notochord; na, neural arch; pns, posterior neural spine; pos, postsclerotomite; prs, presclerotomite; pt, perichordal tube; pth, perichordal thickening; sc, sclerocoel; scl, sclerotome; z, zygapophysial joint; 1,2, fracture planes in neural arch and centrum respectively.

cells are of sclerotomic origin, the tube shows no obvious signs of primary somitic or sclerotomic segmentation; at least in *Lacerta*, the intersclerotomic fissures do not appear to extend ventrally into its substance. However, secondary segmentation is indicated by the development of perichordal thickenings, which in *Lacerta vivipara* give rise to solid intervertebral pads. In certain other lizards (such as *Auguis*) and in snakes, synovial cavities appear within the perichordal thickenings of both caudal and precaudal vertebrae, and the joints between the centra are of the diarthrodial type. The *Lacerta vivipara*, the split in the caudal centrum appears in late embryos and newborn young as a minute discontinuity in the ossifying shell of perichondral bone, which is bounded in front and behind by bony ridges, which become the ridges that demarcate the fracture plane of the adult. The split is at first most evident on the ventrolateral aspect of each side of the centrum, but later it spreads across the ventral midline and also extends dorsally through the bases of the transverse processes.

The margins of the split are related to a specialized region of the connective tissue autotomy septum, which appears close to the bone as an agregation of mesenchyme, the *"septal tissue."* This tissue has been interpreted as an invasive, Cascular sprout, but this interpretation is doubtful, and the precise significance of the septum is obscure. It is probable that it contains osteoblasts that build up the ridges of the centrum, and it is also possible that, by virtue of osteoclastic activity, it is involved in the absorption of bone to form the split between them. In so far as this split in the centrum develops late in ontogeny and is associated with ossification, rather than representing a failure of fusion across the original intersegmental boundary, it can be regarded as a "*secondary formation*."

Two further components of the vertebral fracture plane may also be recognized. The split in the anterior neural spine appears to arise secondarily in association with septal tissue, as does the split in the centrum. The fourth, and less clearly defined component of the fracture plane, is a small region of the centrum in the floor of the canal for the spinal cord at the level of the other splits; this region appears to remain unossified. All these different regions of the definitive vertebral fracture plane link up with each other during the first few months of postnatal life. Vascular channels called *central canals* pass vertically through the centra of caudal (and other) vertebrae in the vicinity of the fracture plane.

There is a mass of cartilage, which, in embryos of *Lacerta* and certain other forms, develops within the notochord at the same transverse level as the autotomy splits. Earlier descriptions imply that it persists as a cartilaginous septum between the two parts of an autotomous vertebra in the adult. This idea has been carried over into recent work. It may have some basis in *Sphenodon* and geckos in which the notochord, or remnants of it, and the notochordal cartilage, persist into maturity, but it does not apply to adult lizards in general. Usually the notochordal structures disappear in postnatal life and are replaced by endochondral bone and narrow spaces. However, an autotomy septum composed of connective tissue is a feature of the *perivertebral* soft tissues of the tail (see below). Hence, the use of the term *autotomy septum* in connection with osteological or fossil material is undesirable; the terms *fracture plane* or *autotomy plane* (or split) are preferable in a purely osteological context.

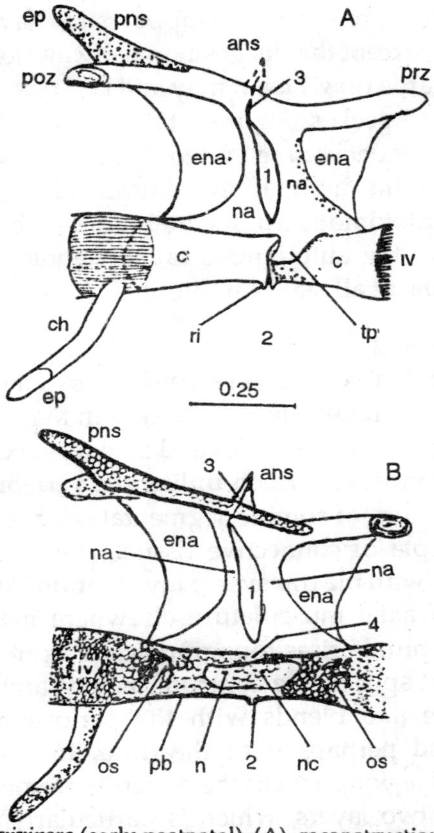

Fig. 2.4. *Lacerta vivipara* (early postnatal). (A), reconstruction of autotomous caudal vertebra seen from right-side. showing three components of fracture plane (1, 2, 3). The middle parts of the neural arch (na) ossify in cartilage; the cranial and caudal parts ossify as extensions into membrane. The anterikor neural spine has hardly begun to develop, but its future position is shown in (A) and (B) in broken lines. (B) Reconstruction of same vertebra sectioned longitudinally near the midline and seen from the right (the medial) aspect. The notochord and the our components of the definitive vertebral fracture plane (1-4) are shown ans, Anterior neural spine; c. centrum; ch. chevron; ena, extension of neural arch into membrane; ep, epiphysis; iv, intervertebral pad of cartilage; notchord; na, neural arch; nc, notochordal cartilage; os, ossification center in centrum; pb, perichondral shell of bone; pns, posterior neural spine; poz, postzygapophysis; prz. prezygapophysis; ri, ridges bordering fracture plane in centrum; tp. transverse process; 1. primary fracture plane in neural arch (persistent intersclerotomic fissure); 2. secondary fracture plane in centrum; 3, secondary fracture plane in developing anterior neural spine; 4. unossified area in floor of vertebral canal completing fracture plane in centrum.

The notochordal cartilage appears to have no relevance to autotomy, except that its position halfway along the centrum in young lizards (in which it may still be present) or in mature *Sphenodon* and geckos (Figure 1.11) makes it likely that it will be ruptured whenever the vertebra splits. Its non-relevance is supported by the fact that notochordal cartilages are present in the non-autotomous trunk vertebrae of embryo *Lacerta*, and are absent in the autotomous tail vertebrae of the limbless lizards *Anguis* at all stages in life.

Segmentation of the Tail

The vertebral column is surrounded by longitudinal bands of fatty tissue. These bands are absent in the non-autotomous pygal region, and the muscles and skin lie superficial to them. The fat and muscles, which unlike the vertebrae retain their primitive pattern of somitic segmentation, are subdivided by sheets or septa of connective tissue. The autotomy septum corresponds with the transverse myoseptum between adjacent segments of axial musculature elsewhere in the body. The autotomy septum arises topographically from the margins of the vertebral split in the centrum, neural arch, and anterior neural spine and blends with the periosteum around the vertebra, and perhaps with the dura. In late embryos, its perivertebral region contains the mesenchymatous septam tissue between its two layers, which is particularly evident in the vicinity of the central ridges. It continues the fracture plane out from the vertebra, through the soft tissue to the skin, and contains quite numerous pigment cells. At the deep surface of the muscles, the posterior layer of the septum blends with the muscle sheath, whereas the anterior layer continues peripherally as the intermuscular part of the septum and terminates at or in the dermis. The septum is perforated in the ventral midline by the caudal artery, vein, and paired lymphatic vessels. It does not cross the vertebral canal for the spinal cord.

The aututomy septa demarcate a series of autotomy segments throughout the tail, except for the basal, non-autotomous region. Each autotomy segment is essentially a

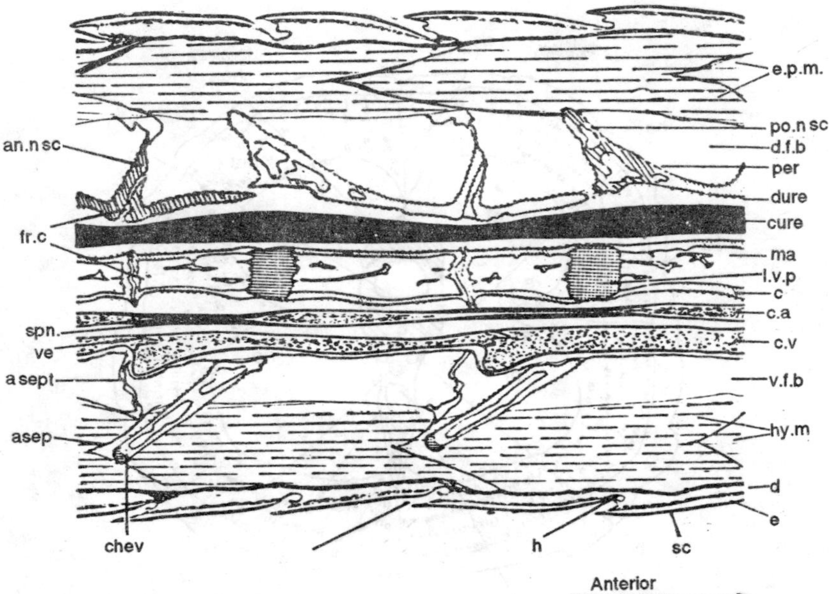

Fig. 2.5. *Lacerta vivipara.* Semidiagrammatic longitudinal section through midcaudal region of adult. The section is near the midline but shows the chevrons on one side. Arrows indicate where the skin breaks at autotomy An.n.sp, Anterior neural spine; a. sept, connective tissue autotomy septum separating muscles of adjacent segments; a. sep.b, bilaminar part of autotomy septum passing through fat; c, centrum; c.a, caudal artery; chev, chevron; c.v. caudal vein; d, dermis; d.f.b. dorsal fat band; dura, dura mater; e. epidermis; ep.m. epaxial muscle, fr.pl. fracture plane; h. hinge of scale; hy.m, hypaxial muscle; i.v.p, cartilaginous intervertebral pad; ma, marrow space in centrum; nc, spinal cord; per, periosteum, po.n.sp. posterior neural spine; sc, scale; sph. sphincter on caudal artery; val, valve in caudal vein; v.f.b. ventral fat band.

body segment containing a segment of skin, a muscle segment or myomere, a segment of fatty tissue, and the posterior part of one vertebra together with the anterior part of the next posterior vertebra. The submuscular, perivertebral fat layer arises in the embryo from the deeper portion of the myotomal region of the somite. It appears to develop directly

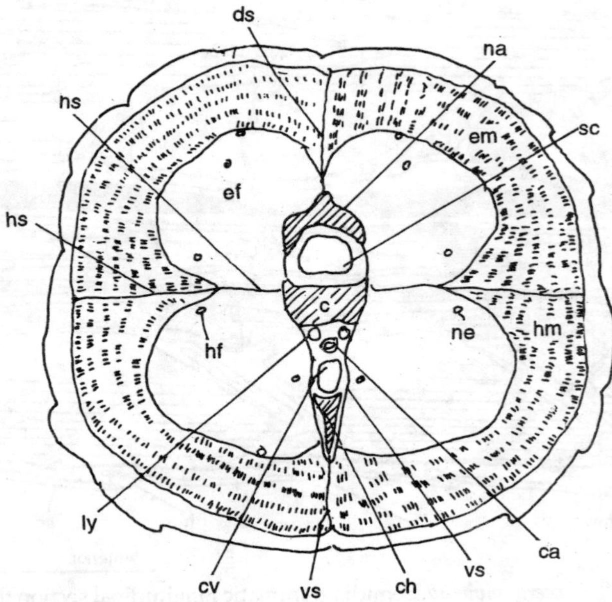

Fig. 2.6. Lacerta. Semidiagrammatic transverse section through autotomous part of tail, showing basic arrangement of muscles, fat, and longitudinal septa. For simplicity, the interdigitating processes of the muscles are not shown here, c, Centrum; ca, caudal artery; ch, chevron; cv, caudal vein; ds, dorsal median longitudinal septum; ef, epaxial fat band; em, epaxial muscle; hf, hypaxial fat band; hm, hypaxial muscle; hs, horizontal longitudinal septum; ly, caudal lymphatic; na, neural arch; ne, nerve; sc, spinal cord; vs, ventral median longitudinal septum.

in mesenchyme, rather than from the transformation of muscle cells; it is identifiable when the muscles and their septa begin to differentiate. This observation is of interest, because *Woodland* (1920), noting that the fat layer corresponds in position with the deeper caudal muscles of certain non-autotomous lizards (which have little or no fat layer), suggested that the fat arose from the degeneration of the deeper muscles. This may perhaps be true as a phylogenetic concept, but there is no evidence for it in the ontogeny of *Lacerta*. The fat layer is divided by the autotomy septum and by longitudinal septa (see below) into a series of discrete bands, each of which is completely enclosed by connective tissue. Each band would, therefore, seem to offer a useful system for the study of fat biology

The Muscles of Tail

The arrangement of the muscles in the autotomous part of the tail is complicated and not easy to describe. Each muscle segment is divided into four masses, which may here be termed *muscles*. On each side of the tail, there is one band of expaxial muscle and fat and similar hypaxial structures; these are separated by a horizontal longitudinal septum of connective issue. This septum is attached along is medial edge to the length of the vertebra in the region in which the centrum joins with the neural arch and also to the transverse process. Like the medial septum, it ends at the deep aspect of the dermis. Each epaxial and each hypaxial muscle and fat band is separated from its fellow of the opposite side by a part of the median longitudinal septum. The dorsal part of this, which separates the epaxial tissues on the two sides, arises from the dorsal midline of the vertebra including the anterior and posterior neural spines. The ventral median longitudinal septum, which separates the hypaxial structures, is distinctly bilaminar in the region of the fat bands, each layer arising from the centrum and intervertebral cartilage slightly to one side of the midline. The two layers of this septum enclose the caudal vessels and are attached to the two limbs of each chevron bone.

The real appearance of caudal muscles is more complicated, as each epaxial and each hypaxial muscle is prolonged into three (dorsal, middle and ventral) pointed processes in front and three behind. These processes interdigitate deeply with the corresponding processes of the muscles of adjacent segments and enclose recesses into which the tips of the latter fit. The anterior dorsal process of each hypaxial muscle, and the anterior ventral process of each hypaxial muscle are much smaller than the others. This arrangement is clearly evident in an end-on-view of the autotomized tail surfaces. Because of the length of the processes and the way in which they interdigitate, the muscles, seen in a true transverse section near the level of the fracture plane. The muscle processes are invested by the tissue of the autotomy septum, which separates them from the processes of adjacent segments. Thus, the septum has a complicated zigzag contour instead of being a flat transverse sheet, as it would be if the muscles were simple rectangular blocks. Superficially, the muscles are attached to the dermis. Their deep surfaces are mainly in contact with the fat bands, but are attached posteriorly to the vertebrae, whether directly or through the medium of septa. The main attachments in *Lacerta vivipara* are as follows:

The dorsal parts of the epaxial muscles (2) on each side are attached osteriorly to the posterior neural spine.

The posterior ventral epaxial process (6) and the posterior dorsal hypaxial process (8) are attached to the front of the transverse process in each side.

The posterior ventral hypaxial process (12) dis attached to the tip of the chevron.

Only the posterior processes of the muscles have firm bony attachments, at least in geckos and lacertids. Each anterior process extends into the recess between the posterior processes of the muscle in front and is attached only to the autotomy septum. Separation will, therefore, only take place at this anterior and weaker attachment.

Immediately after autotomy has occurred, four pairs of anterior muscle processes can be seen projecting conspicuously from the detached portion of the tail. The processes 1 and 11 do not project from the tail fragment.

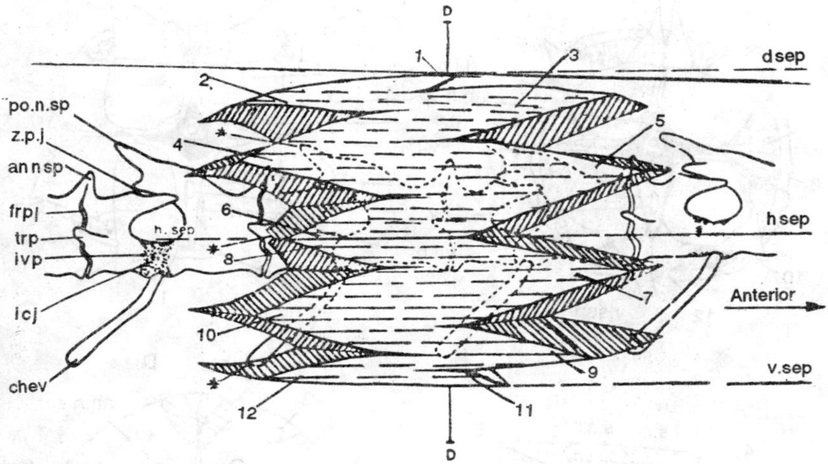

Fig. 2.7. *Lacerta vivipara*. Diagram showing relationships of muscles and vertebrae in mid-caudal region, as seen from the right side after removal of skin. The muscles of a single autotomy segment are shown. The recesses for the adjacent muscle processes are slightly deeper than is shown, giving appearance in transverse section. Asterisks show points of attachment of muscles to vertebrae. *Long broken lines:* outer surface of muscle, normally in contact with dermis; also, edges of longitudinal septa. *Short broken lines:* outline of vertebrae. *Slanting lines*: interdigitating surfaces of muscle processes; these are invested by the connective tissue autotomy seprum an.n.sp, Anterior neural spine; chev, chevron; d.sep. superior edge of dorsal, median longitudinal septum; fr.pl, fracture plane in verterba; h. sep, horizontal longitudinal septum; j.c. joint between centra; i.v.p. cartilaginous intervertebral pad; po.n.sp, posterior neural spine tr.p, transverse process; v.sep, infrior edge of ventral median longitudinal septum; z.r.j. zygapophysial joint; 1,3,5, anterior processes of epaxial muscle; 2,4,6, posterior processes *tt* epaxial muscle; 7,9,11, anterior processes of hypaxial muscle; 8,10,12, posterior processes *tt* hypaxial muscle.

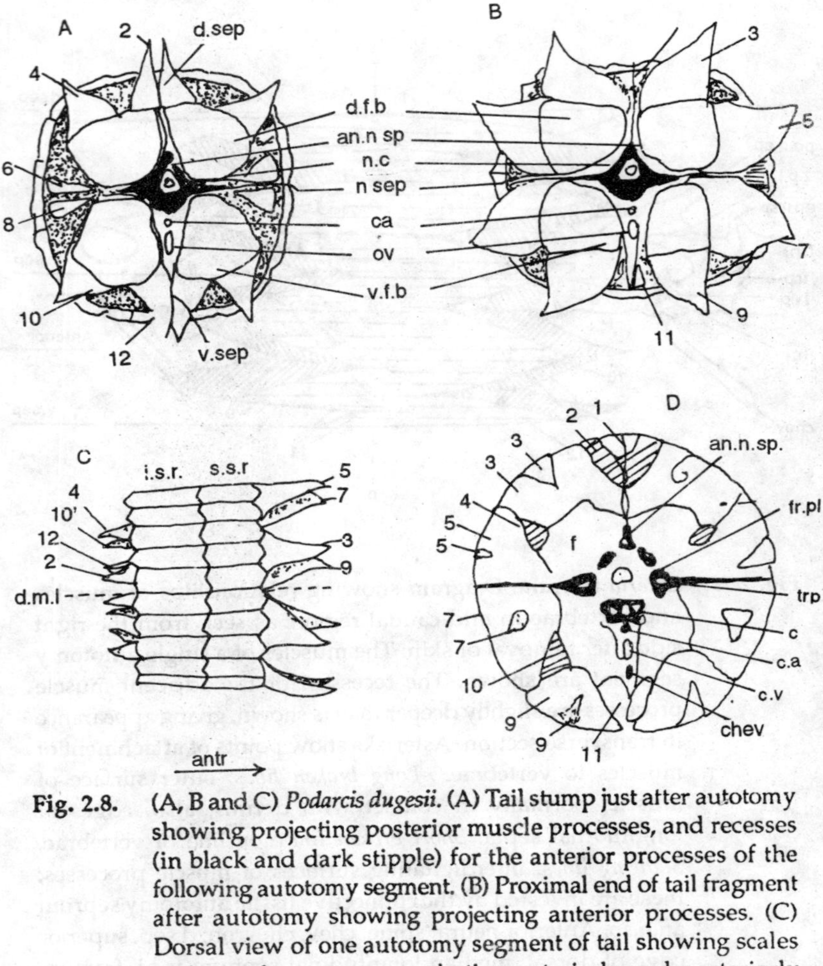

Fig. 2.8. (A, B and C) *Podarcis dugesii.* (A) Tail stump just after autotomy showing projecting posterior muscle processes, and recesses (in black and dark stipple) for the anterior processes of the following autotomy segment. (B) Proximal end of tail fragment after autotomy showing projecting anterior processes. (C) Dorsal view of one autotomy segment of tail showing scales and muscle processes projecting anteriorly and posteriorly after autotomy. (D) *Lacerta vivipara*. Transverse section through middle of tail near level of fracture plane. Sections of interdigitating muscle processes from three adjacent segments are seen: those from the middle segment being unshaded, those from the segment in front in slanting lines, and those from the segment behind in stipple. Lymphatic vessels not shown an.n.sp, Anterior neural spine; c, centrum; c.a, caudal artery; chev, chevron, c.v, caudal vein; d.f.b. dorsal fat band; d.m.l, dorsal midline; d.sep, dorsal median longitudinal septum;

f, fat band; fr.pl, fracture plane; h.sep, horizontal longitudinal septum; l.s.r, long scale row; n.c. spinal cords; s.s.r, short scale row; tr.p, transverse process; v.t.b, ventral fat band; v.sep, central median longitudinal septum; 1,3,5, anterior processes of epaxial muscle; 2,4,6, posterior processes of epaxial muscle; 7,9,11, anterior processes of hypaxal muscle; 8,10,12, posterior processes of hypaxial muscle.

The broken surface of the stump shows four pairs of recesses that, before autotomy, accommodated the anterior processes of the caudal fragment. These recesses lie beneath the skin and between the posterior processes of the muscles in front. The latter (except for 6 and 8) project slightly beyond the broken skin of the stump, but are less prominent that the anterior processes attached to the tail fragment. Soon after autotomy, these posterior processes appear to shrink and may become more or less hidden by the skin around the raw surface of the stump. After the tail has broken off, the autotomy septum adheres to the posterior processes of the stump; this has certain implications for the regenerative process.

The high glycogen content of the caudal muscles suggests that these are physiological of the quick-contracting type, as would be appropriate in view of the their role in autotomy.

Mechanism of Autotomy in Lizards

The tail of a lizard can be wrenched off by the sheer mechanical force of an outside agent, without any participation of the caudal muscles. Such *"passive rupture"* may occur in a dead lizard; if the species is *autotomous* the break is generally located at one of the fracture planes. It is likely that there is an element of passive rupture in many autotomies and that this is produced by the sheer tension on the tail resulting from the attempts of lizard to escape. The force necessary to produce breakage has been studied in *Podarcis sicula* by holding the lizard and attaching various weights to its tail. The breaking force is, in fact, quite considerable, ranging between 500 and 780 gm in animals kept in the light at 21° to 25°C. It is uncertain how far these experiments reproduce the conditions of natural autotomy, in which active contraction of the caudal muscles is most likely involved.

The fact that the muscles of a lizard may be capable potentially of producing autotomy on their own is indicated by observations that the tail may occasionally be shed without having actually be touched. This "*spontaneous autotomy*" has been observed in *Lacerta dugesii* and *L. vivipara* and in the gecko *Gehyra variegata*, although *Woodland* (1920) does not report it in his classical study of *Hemidactylus*. It is likely that, in some species, interference with the tail is a necessary prerequisite for autotomy. The readiness with which the tail is shed certainly varies in different forms and also depends on a variety of circumstances and factors, including temperature. *Sheppard* and *Bellairs* (1972) filmed 24 autotomies in six specimens of *Lacerta dugesii*. Whenever, the tail is grasped by forceps the fracture usually occurs through the plane in front of the autotomy segment that was being held; it never takes place more than three segments anterior to this. Whenever, the tail of a lizard is grasped, the regions in front of and behind the forceps usually flex strongly; in many cases the forceps seem to be used as a fulcrum for leverage. A similar effect has been observed by *Woodland* (1920), who believed that a gecko was unable to autotomize unless the tail was fixed in some way. In nature, a fixed point might be provided by the jaws or paw of a predator, and it is possible that even in apparently spontaneous autotomous, in which the tail is not touched, some irregularity or feature of the surroundings might serve as a fixed point for leverage.

The caudal muscles have two distinct functions; normal flexion of the tail at the intervertebral joints and the production of autotomy. The first function involves distribution of the strain along a number of vertebrae, wheras in the second the strain is concentrated at a single fractue plane. Flexion of the tail in a single direction often may be sufficient by itself to produce autotomy; if the tail is held, the contraction of the muscles of several segments may occur in such a way that the stress is concentrated at one fracture plane. So far as could be observed in the film, the skin generally splits first on the stretched, convex aspect of the tail, and then the muscles on this aspect separate from their weak anterior attachments. The split then appears to travel right through the remaining caudal tissues. In some instances, a second flexion occurs in the opposite direction before the tail autotomized completely. This has been ascribed to contractoin of the muscles of adjacent segments, which might complete, by momentum, the rupture of the still intact muscles on the side toward which the first flexion had been directed. The flexion that produced autotomy is mainly in the lateral plane, but sometimes this movement is combined with slight dorsoventral flexion to produce a rotatory effect. In a few instances, the lizard's tail seems to split straight across, virtually without bending. "*Passive*" rupture may be an important element in this type of autotomy, which seems to be caused mainly by the attempt of the animal to pull away.

In *Hemidactylus, Gekko,* and *Tachydromus*, rather different conclusions about autotomy have been reported. Two flexions of the tail, one each in the opposed directions are supposed to be necessary. Also, it has been claimed that the flexion is limited to the muscles of a single autotomy segment. Possibly, those muscles which produce, and which are on the same side as the initial flexion, become detached by their own force, perhaps splitting the vertebra at the same time. The process is then repeated by the contralateral muscles, which produce the second flexion in the opposite direction. It seems possible that the fat layer may play some part in the actual mechanism of *autotomy*. It may for example, provide an easily compressible medium around the vertebrae which offers less resistance to the sudden bending and subsequent tearing apart of the two parts of the vartebra than would a firm investment of deep muscles. It is also conceivable that, by displacing the muscles from the longitudinal axis of the vertebral column, the fat layer increases their mechanical advantage, and, thus, enhances the strength of flexion. These possible functions of the fat layer are particularly suggested by the appearances of a horizontal section.

Neuropsychological Factors

The neurological basis of autotomy seems to have received little attention. However, *Slotopolsky* (1929) found that *autotomy* could be provoked by electrical stimulation of the tails of lacertid lizards that had been decapitated or in which the spinal cord had been transected in front of the lumbar region. Autotomy could also be elicted in a detached fragment of tail. However, it was more difficult to provoke under these circumstances than in intact animals. Although no localized *"autotomy center"* has been noted in the central nervous system, there appears to be an especially excitable region in the spinal cord. In *Lacerta,* this region begins shortly posterior to the middle of the back; in *Anguis,* it is confined to the tail. Stimulation of this region produced a strong twisting of the tail. No special type of nerve fiber associated with autotomy yet has been described. *Mahendra* (1950) suggested that an enigmatic, filamentous structure called *Reissner's fiber* might be involved in autotomy. This seems unlikely, but the matter does not appear to have been critically investigated. The fiber, which is present in most vertebrates, is secreted by the subcommissural organ, which lies in the roof of the brain ventral to the pineal complex and runs down the whole length of the central canal of the spinal cord.

There are two neurological components governing autotomy. A reflex one, which is based on reflex arcs through the caudal nerves, and a cerebral or "psychic" one, which is evoked by the sensations of fear or restraint. This would agree with the general impression that lizards autotomize more readily when freshly caught in the field than when they have become more or less take in captivity.

Fracture Planes in Different Groups

In most Scincidae, the first four to ten caudal vertebrae (pygals) lack fracture planes. In most genera, the plane passes through the single transverse process, dividing it unequally into a narrow anterior and a wider posterior part, the two parts may be separated by a slight gap. In a few forms, there may be an approach to the converging process type, as is seen in *Chalcides*. In *Egernia cunninghami*, the posterior part of the tranverse process is forked distablly, giving the impression of a diverging process vertebra, but the fracture plane passes anterior to the bifurcation. In certain skins (e.g., some species of *Acontias* and *Mabuya*), the fracture plane lies wholly anterior to the transverse process, whereas in others, such as *Trachydosaurus*, the plane is posterior to the process. The split tends to disappear in mature individuals of *T. rugosus* and *Tiliaua scincoides* and is apparently absent in mature *Egernia depressa* and *Corucia zebrata*.

In Cordylidae there are three types of caudal sequence. In *Gerrhosaurus* and also in *Tetradactylus*, the postpygal vertebrae consist of an anterior series of diverging process vertebrae and a terminal series of no process vertebrae, all with fracture planes. *Cordylus* resembles most skinks in having a long series of anterior postpygals with single process that are traversed by fracture planes and a short, terminal, no-process series, which is also autotomous. The very long tail of *Chamaesaura*, containing over 80 vertebrae, shows a unique sequence. There are about 15 non-autotomous pygals, followed by a series of postpygals with expanded transverse processes, each traversed by a fracture plane. These are followed in turn by a series of autotomous vertebrae with progressively diverging processes; the latter gradually disappear towards the caudal tip.

In most Teiidae, there is the usual non-autotomous pygal series with single processes. This is followed posterioly by a diverging process series and in turn by a single process series (the process corresponding with their anterior or posterior process of the last series, so that the fracture plane may be either in front of it or behind it). The caudal tip is formed of a no-process series. The divergent process series is absent in *Dicrodon* and *Teius*. All the postpygal vertebrae of the teiids examined by Etheridge possess fracture planes. However, in *Tupinambis teguixin,* only the 15th to the 19th caudal vertebrae have persistent fracture planes in the adult, although in the young some of the posterior vertebrae are also autotomous. Thus, some teiids may show a reduction of autotomy with age, as described later in some iguanids, but individual variation may also occur.

In Xantusiidel the postpygal vertebrae often consist of an anterior, single-process series with the process traversed by the fracture plane, then a divergent process series, then a single process series with the process anterior to the fracture plane, and finally a no-process series. There is some individual variation. All the caudal vertebrae, except the five or six pygals, have fracture planes.

In most geckos there are some four to eight non-autotomous pygal vertebrae of the single process type. All or most of the remaining caudal vertebrae are also of the single-process type, with the fracture plane essentially posterior to the process. However, all, or almost all, vertebrae of postpygal series of some forms lack transverse processes.

The autotomous adaptations have been described for *Hemidactylus flaviviridis, H. turcius; Gekko gecko,* and *Tarentola mauritanica.* There appear to be no significant differences. Each postpygal vertebra is divided into approximately two halves by the fracture plane. *Woodland* mentions a *"hyaline septum"* occupying the vertebral fracture plane; he does not describe its relationship with the notochordal cartilage and seems to use the term *"septum"* in a different sense to that of *Gadow.* However, this may again be responsible for reference to an "autotomy septum" in osteological material. The accounts of *Mahendra* and *Werner* make it clear the no such hyaline septum exists, though its presence might possibly be simulated by artifacts of preparation. However, as previously stated, the notochord in the amphicoelous vertebrae of most geckos may persist throughout life, at least in some places, and its substance

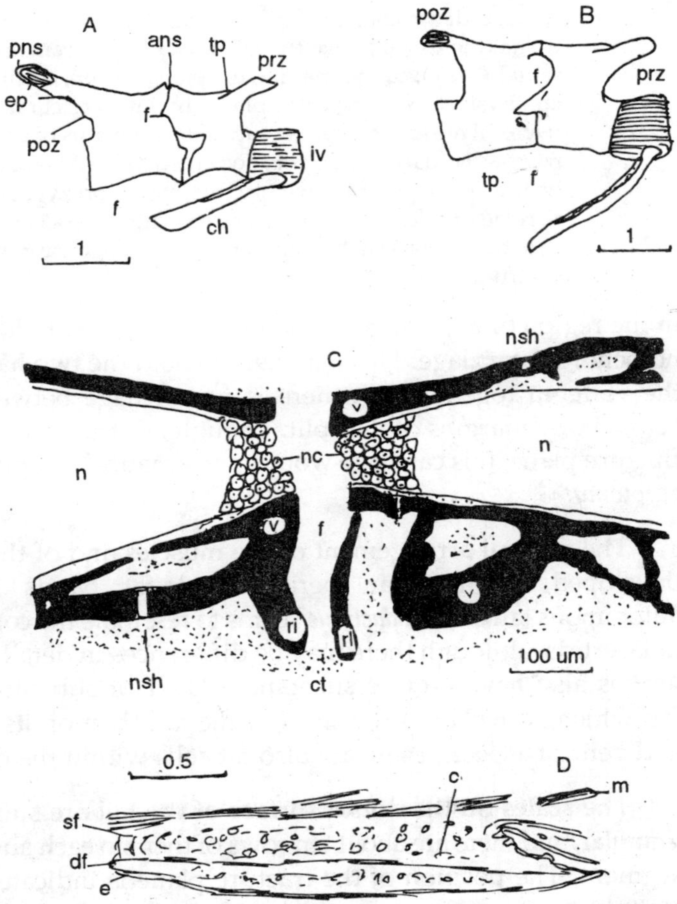

Fig. 2.9. (A) Oedura lesueurii (Gekkonidae). Autotomous 10th caudal vertebra from right side with preceding chevron. (B) Pygopus lepidopodus. Autotomous 20th caudal vertebra from right side. (C) *Oedura lesueurii*. Longitudinal section through (adult) autotomous caudal vertebra (centrum only), showing bony fracture plane (somewhat widened) and a rupture of persistent notochordal cartilage following postmortem preparation. (D) *Phelsuma abbotti* (Gekkonidae). Longitudinal section through large ventral scale showing dermal and subcutaneous fat.

cont....

[(A-C) Scales of A,B, and D are in mm. ans, Anterior neural spine; c, dense collagenous layer; ch, chevron; ct, connective tissue (the tissue above the label may be the remains of the septal tissue); df, dermal fat; e, epidermis; ep, cartilaginous epiphysis; f, bony fracture plane; iv, intervertebral pad; m, hypaxial muscle; n, space occupied by remnants of notochord; nc, notochordal cartilage; nsh, notochordal sheath; pns, posterior neural spine; poz, postzygapophysis; prz, prezygapophysis; ri, ridges bordering fracture plane on ventral aspect of centrum; sf, subcutaneous fat; tp, transverse process; v, bone of vertebral centrum.

in the region of each fracture plane is chondrified to form the notochordal cartilage. This may help to hold the two halves of the vertebra together, but there is no cartilage between the actual bony margins of the split. Though not traversed by the fracture plane, this cartilage would presumably be reptured at *autotomy*.

The general arrangement of the muscles and of the well-developed submuscular. perivertebral fat bands in the autotomous tail of *Hemidactylus* seems to resemble the condition in lacertids, although there may be differences in detail. Many geckos also have a quite substantial layer of sub-cutaneous fat, which is thicker at the sides of the tail than on its dorsal and ventral aspects; there are also fat cells within the dermis.

The scales on the dorsal surface of the tail are small and granular, and there are about ten rows of them to each autotomy segment. The position of the fracture plane is indicated by a straight transverse row on either side of the line of cleavage elsewhere, the dorsal scales have a less regular arrangement. The scales at the sides of the tail are bigger, and at the midventral line, there are only two large rectangular scales to each autotomy segment. As in lacertids, the existence of some kind of *structural* fracture plane, presumably continuous with the connective tissue autotomy septum, has been mentioned or implied in *Hemidactylus* and *Tarentola;* the plane passes through the subcutaneous fat and the skin, or at least through its dermal component.

In a number of geckos, at least when adult, the fracture planes are limited to a few vertebrae at the front of the postpygal region, and at autotomy, all the rest of the tail is shed. Thus, in some species of *Stenodactylus*, only two vertebrae are extensively split, the splits in a few successive vertebrae becoming increasingly restricted. In some forms that have a more extensive series of fracture planes, the tail only breaks readily at the base. The Australian spiny knob-tailed gecko, *Nephrurus asper*, is the only gecko reported to have no fracture planes at all. Its tail is very short, and there is virtually nothing for a predator to grasp.

A form of defense, akin to autotomy, is practiced by the Australian gecko *Gehyra mutilata*, which also autotomizes. The species allows large areas of skin to be ripped from its body when it is seized; bleeding is minimal and the skin is soon regenerated. This also occurs in *Teratoscincus* and in *Geckolevis*.

The anatomy of the fracture planes of the snake-like Pygopodidae seems to be basically similar to that in geckos, with the planes or splits passing posterior to the single transverse processes and beginning between caudal vertebrae three and seven; some of the more anterior postpygals may be only partly divided. In *Pygopus lepidopodus*, *Delma fraseri*, and *Lialis burtonis*, the split in the neural arch of a typical pygal vertebra is very wide and forms a considerable gap between the two halves of the arch. *Pletholax gracilis* is the only species that is virtually non-autotomous, the bony fracture plane being represented only by a small aperture in the neural arch on each side.

Iguanids

Iguanids show much variation in the pattern of autotomy, and a number of forms lack fracture planes. A single genus (e.g., *Anolis*) may contain both autotomous and non-autotomous species. Four different types of caudal sequence occur within the family.

In the first type, all the caudal vertebrae are of the single process type, however, the single processes differ in size and orientation in different parts of the tail. Most forms with this type of sequence are autotomous, the fracture planes beginning at any vertebra between the 5th and 15th and passing posterior to the bases of the transverse processes, as in *Sceloporus*. The following genera with this type of seqence lack any trace of fracture planes: *Crotaphytus, Hoplocercus, Leiosaurus, Phrynosoma* and *Uracentron*.

Iguanids with the second type of sequence have a short series of single-process vertebrae (of which only the last may have a fracture plane) that are followed by a long-no process series, which may or may not be autotomic. This condition is found in some species of *Anolis* (including *A. carolinensis*), in which only about the ninth to the 17th caudal vertebrae are autotomous, as well as in *Basiliscus* among other genera. Completely non-autotomous forms with this type of sequences are *Phenacosaurus, Chamaeleolis, Corythophanes, Laemanctus, Enyalius, Polychrus, Urostrophus,* and some species of *Anolis*.

In certain of *Anolis* and related genera, the caudal vertebrae show a third type of sequence. They are all of the single-process type, though the orientation of the process changes abruptly at one point in the series. The fracture planes, when present, pass anterior to the bases of the transverse processes, as in the species of *Anolis* with this sequence. Fracture planes are absent in *Chamaelinorops, Anisolepsis* and *Aptycholaemus*.

Iguanids showing the fourth type of sequence have a distinct pygal series with single processes, then a series with divergent processes, then another single-process series, and finally a no-process series, as have some teiids. This sequence is found in most of the big iguanas. Fracture planes passing between the transverse processes of the second series and either anterior or posterior to the single processes are found in *Iguana iguana, Ctenosaura, Cyclura, Enyaliosaurus, Sauromalus,* and *Dipsosaurus,* these planes are absent in *Iguana delicatissima,*

Conolophus, Amblyrhynchus, and *Brachylophus,* all non-autotomous forms.

Although *Anolis carolinensis* has been extensively used in regeneration studies, the anatomy of its fracture planes has been only briefly described. As previously noted, only about eight caudal vertebrae are completely split. *Cox* states that autotomy can occur through the posterior part of the tail, which lacks fracture planes.

The arrangement of the caudal muscles of *Anolis* seems to resemble that in *Lacerta,* except that ten, instead of eight, muscle processes project from the front of the caudal fragment; probably the extra pair corresponds either to the small dorsal or to the ventral processes of lacertids. The fat bands in *Anolis* are relatively small. The fracture plane does not pass through the skin, but the posterior borders of the scales just in front of each plane are aligned in such a way as to give the tail a segmented appearance.

In Agamids

There is a anterior, single process series of caudal vertebrae in agamids, followed by a long, no-process series. No intravertebral fracture planes have been reported in any form except in *Uromastys,* in which traces of an obliterated plane have reputedly been observed. In quite a number of agamids, including some species of *Agama,* in *Gonocephaulus subcristatus,* and in *Diporiphora bilineata,* breakage of the tail is fairly common. The tail generally breaks between adjacent vertebrae, but sometimes through the slender but non-autotomous centrum itself. The position of the break may be difficult to judge in view of the fact that parts of a vertebra adjacent to a break may undergo absorption or *"ablation"*. It is uncertain whether the facility for tail-break in agamids is a positive adaptation associated with any automous anatomical feature; *Arnoid* (1984) suggests that it is a secondary acquisition of certain species after fracture planes had been lost in ancestral agamids.

The soft tissues of the tail have been described in *Calotes versicolar*, in *Uromastyx hardwickii* and in *Agama agilis*. The muscles are differently arranged from those in autotomous lizards, and the submuscular fat bands are absent or poorly developed. However, in view of the previous observations, anatomical study of both the caudal skeleton and soft tissues of agamid species, which have the facility for caudal fracture and regeneration, would seem desirable. The specialized prehensile tail of chameleons contains neither fracture planes nor fat bands.

In Anguids

Most anguids and *Anniella* have a series of pygal single-process vertebrae that are followed by a long series of vertebrae with converging processes. The level of the first fracture plane is quite far back in limbless Anguidae, involving the 17th caudal vertebra in *Anguis fragili.* Each fracture plane usually passes between the party distinct anterior and posterior limbs of the process, in some forms, it lies entirely anterior to the process. Fracture planes, passing anteriorly through the bases of the transverse processes, were present in the Upper Eocene anguid. *Placosaurus.*

The caudal centra in the genus *Ophisaurus* are undivided by fracture planes but are so thin that they break more readily than they disarticulate.

The anatomy of the fracture planes of *Anguis fragilis* has been described. The vertebral split lies much nearer the front than the back of the vertebra, at least in the midcaudal region. It traverses the centrum and neural arch, including the short anterior neural spine, but stops at the anterior border of the gap between the two convergent limbs of the transverse process. At autotomy, the base of the anterior limb is retained on the stump, whereas the longer, remaming part of the process is attached to the separated fragment. In the embryo, the fracture plane is situated relatively further back along the vertebra than in the adult, and its final anterior position may, to some extent, be due to relative growth phenomena.

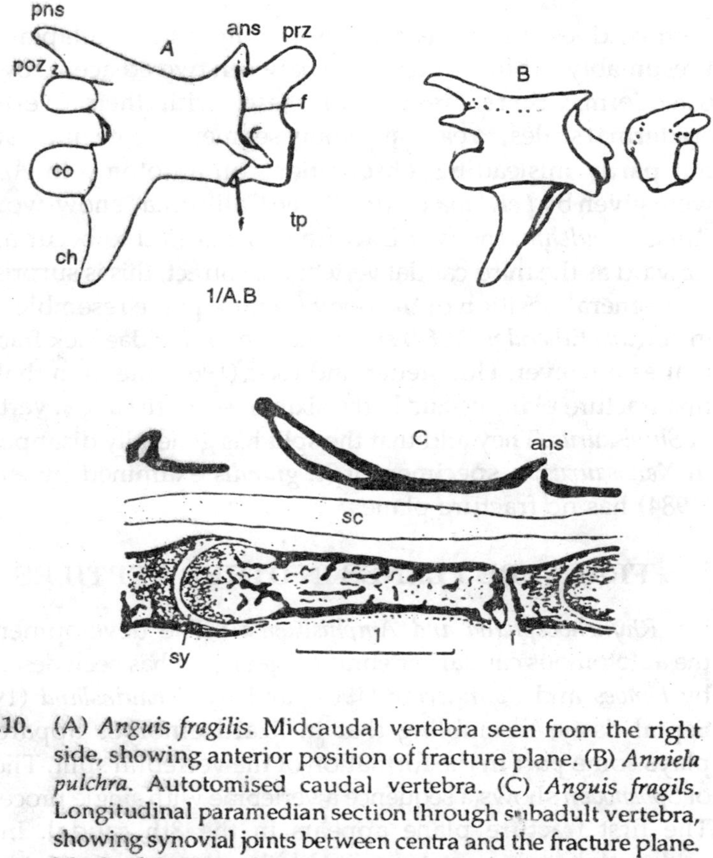

Fig. 2.10. (A) *Anguis fragilis*. Midcaudal vertebra seen from the right side, showing anterior position of fracture plane. (B) *Anniela pulchra*. Autotomised caudal vertebra. (C) *Anguis fragils*. Longitudinal paramedian section through subadult vertebra, showing synovial joints between centra and the fracture plane. Much of the articular region is cartilaginous. The chevron is not shown. Scales in mm. ans, Anterior neural spine; ch, chevron; co, condyle; f, fracture plane; pns, posterior nèural spine; poz, prz, pre- and postzygapophyes; sc, spinal cord; sy, synovial cavity; tp, transverse process.

The arrangement of the muscles appears, on the whole, simdilar to that in Lacerta, but these are partly attached to both the anterior and posterior convergent limbs of the transverse process, so that the latter is fractured at autotomy. The submuscular fat bands are less well-developed than in *Lacerta* but, as in this genus, the spinal cord is constricted opposite each fracture plane. The skin, with its well-developed layer of dermis, does not appear to show autotomous adaptations. Presumably rupture takes place between two adjacent rows of osteoderms; there are two of these, with their overlying epidermal scales, to each autotomy segment. Some interesting, but partly misleading, observations on autotomy in *Anguis* were given by *Leighton* (1903). In the California "show-worm", *Anniella pulchra*, the first fracture plane is said to occur as far forward as the third caudal vertebra; if correct, this is surprising. The general position of the bony fracture plane resembles that in *Anguis*. *Etheridge* (1967) states that Xenosauridae lack fracture planes however, Hoffstetter and *Gasc* (1969) mention that the first fracture plane occurs in the sixth or seventh caudal vertebra of *Shinisaurus*. They add that the split has generally disappeared in *Xenosaurus;* a specimen of *X. grandis* examined by *Arnold* (1984) has no fracture planes.

FRACTURE PLANES IN OTHER REPTILES

Rhynchocephalia and Amphisbaenia : The development of the autotomous caudal vertebrae of *Sphenodon* has been described by *Howes* and *Swinnerton* (1901) and by *Schauinsland* (1906). Septal tissue resembling that in lizard embryos apparently plays some part in the formation of the vertebral split. The tail of *Sphenodon* shows a sequence a vertebrae with single processes. The first fracture plane appears in the 8th caudal. In the midcaudal region, at least, it divides the vertebra into two portions of nearly equal length; it passes through the small transverse process but nearer its anterior than its posterior aspect. As in many lizards, the split in the centrum is bordered laterally and ventrally by bony ridges. The caudal muscles and their interdigitating processes seem basically similar to those in lizards, although the processes seem more slender. The supposed differences noted by *Ali* (1941) mainly involve the projection of an extra pair of muscle processes from the front of the autotomized fragment, a condition which perhaps resembles that in *Anolis*. The difference may be illusory, because *Lacerta* shows a pair of muscle processes in a corresponding position, but these are small and do not project from the surface. It seems desirable that in future studies of autotomy, the same worker should examine several different species of reptiles; some of the alleged variations among different; forms may be due mainly to differences of methodology and illustration.

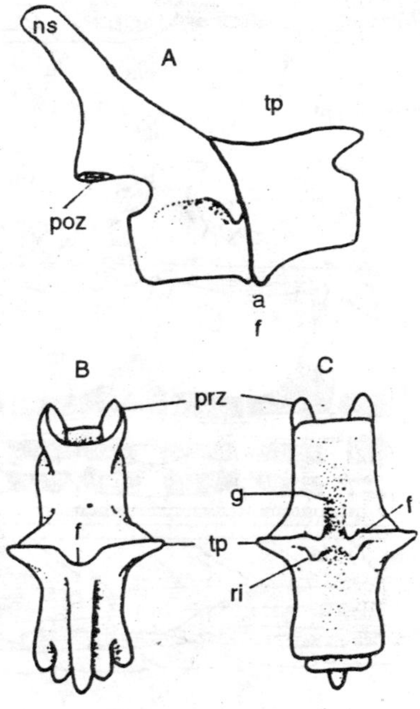

Fig. 2.11. *Sphenodon punctatus.* Midcaudal vertebra seen in (A) right lateral. (B) in dorsal, and (C) in ventral view, f, Fracture plane, g, groove on ventral aspect of centrum, roofing over caudal artery; ns, neural spine (= posterior neural spine of *Lacerta)*; poz, postzygpophysis; prz, prezygapophysis; ri, ridges bordering fracture plane on ventral aspect of centrum; tp, transverse process.

Sphenodon has a fairly well-developed submuscular fat layer around the vertebrae, which, according to *Ali* (1941), is not divided tranversely by autotomy septa. The arrangement of the longitudinal septa is somewhat different from that in

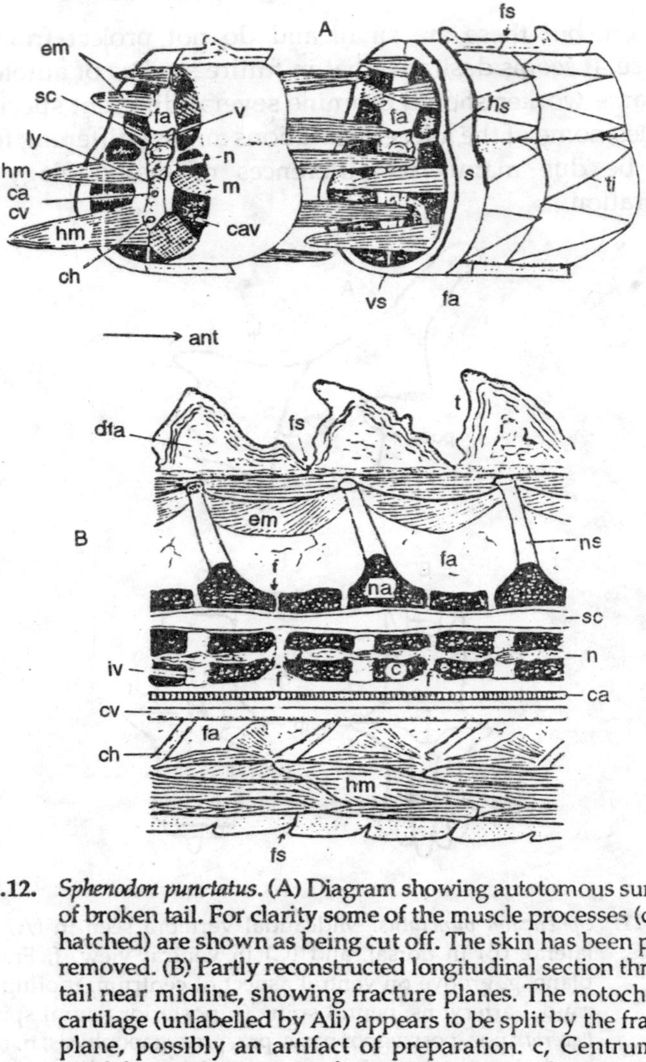

Fig. 2.12. *Sphenodon punctatus*. (A) Diagram showing autotomous surfaces of broken tail. For clarity some of the muscle processes (cross-hatched) are shown as being cut off. The skin has been partly removed. (B) Partly reconstructed longitudinal section through tail near midline, showing fracture planes. The notochordal cartilage (unlabelled by Ali) appears to be split by the fracture plane, possibly an artifact of preparation. c, Centrum; ca, caudal artery; cav, cavity for muscle process; ch, chevron; cv, caudal vein, dfa, dermal fat deposit, em, epaxial muscle process; f, fracture plane in vertebra; fa, submuscular fat layer, fs, fracture plane in skin; hm, hypaxial muscle process; hs, horizontal longitudinal septum; iv, intervertebral pad; ly, caudal lymphatic (not shown in B); m, cut muscle procession, notochord; na, neural arch; ns, neural spine; s, skin; sc, spinal cord; t, skin tubercle; ti, tendinous inscription; v, vertebra; vs, ventral median longitudinal septum.

Lacerta, because a median dorsal seprum is lacking, and there are two pairs of thin lateral septa above and below the well-marked horizontal septum. There is no subcutaneous fat layer, but the dermis contains fat deposits beneath the dorsal and lateral skin tubercles. The fracture planes are evident on the surface of the skin. Each autotomy segment possesses a large tubercle mid-dorsally and is demarcated from adjacent segments by an annular groove. The skin apparently shows a line of cleavage related to each fracture plane. The caudal artery is devoid of sphincters. It has been seen that the tail of a dead *Sphenodon* was difficult to break, and suggested that its autotomous adaptations are more primitive than in lizards. However, *Dawbin* (1962) states that, in the wild, a high proportion of specimens show loss of portions of the tail.

A single fracture plane occurs in one, but not in the other, subspecies of the South American *Amphisbaena angustifrons;* this situation is paralleled by that in the closely related species pair. *A. occidentalis* and *A. townsendi* of Peru. The presence of the amphisbaenid plane is indicated in most individuals of the appropriate form by a slight external constriction of the tail at a species-specific site; some species may have a vertebral fracture plane but almost no surface constriction. These observations suggest that the ontogenetic or phylogenetic changes that are required to produce a split in the vetebra or a weakness in the skin are relatively minor ones; however, this may not apply to the muscular adaptations to autotomy.

In Snakes

The tails of the several species of snakes including *Pilocercus elapoides* are liable to break off when seized. In all but the first few caudal vertebrae of *Pilocercus* each expanded transverse process is grooved in such a way as to suggest an intravertbral plane of weakness. Comparable, although shallower, grooves are found in the sibynophine, *Scaphiodontophis venustissimus.* However, *Etheridge* (1767) and *Arnold* (1984) believed that in these snakes the breakage usually occurs between adjacent vertebrae.

In Early Fossil Reptiles

The most ancient reptiles in which caudal fracture planes have been observed are the Lower Permian cotylosaurs *Captorhinus* and *Labidosaurus*. The splits begin at about the 7th caudal vertebra, at which level the transverse processes diappear. Each split divides the vertebra into approximately equal portions and seems to involve only the lateral and ventral aspects of the centrum. Fracture planes were also present in the Lower Permian mesosaurs (Figure 2.16) an enigmatic group of small aquatic reptiles with very doubtful affinities. One would not expect autotomy in these forms, because their long, laterally compressed tails were presumably important in swimming. Among the lepidosaurs, fracture planes seem to have been absent in many eosuchians, including the Triassic *Prolacerta* and *Macronemus*, forms which have been regarded, perhaps erroneously, as being close to the origin of lizards. They were also absent in the Upper Permian reptiles *Weigeltisaurus* and probably in the allied *Daedalosaurus*. These animals were gliders with rib-supported "wings" like the modern lizard *Draco*, and they probably used their tails as rudders when airborne

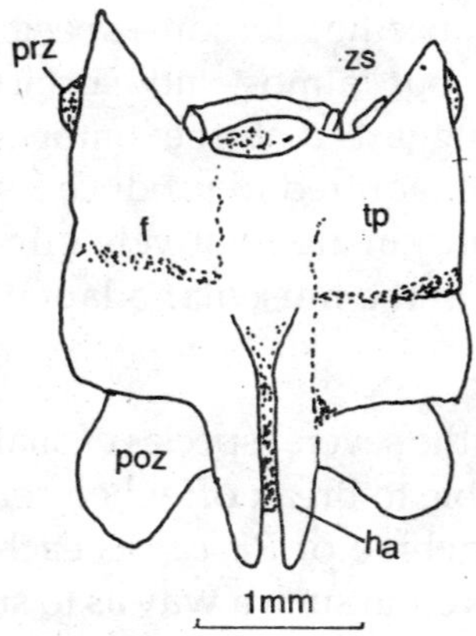

Fig. 2.13. *Plocercus elapoides laticollaris* (Serpentes). Ventral view of caudal vertebra showing groove in transverse process which appears to be a fracture plane. f, Fracture plane; ha, haemapophysis; poz, prz, post-and prezygapophyses; tp, transverse processes; zs, zygosphene.

However, the small, superficially lizard-like eosuchian *Gephyrosaurus* from the Lower Jurassic had fracture planes. In the anterior autotomous vertebrae, the splits pass through the transverse processes and there are central canals near the level of the fracture planes, as in some modern lizards. Further posteriorly, the transverse processes are absent, and the splits do not reach the dorsal surfaces of the neural arches. The fracture planes become almost obliterated near the caudal tip, suggesting that ontogenetic closure is occurring in posteoanterior sequence, as described in certain Iguanidae.

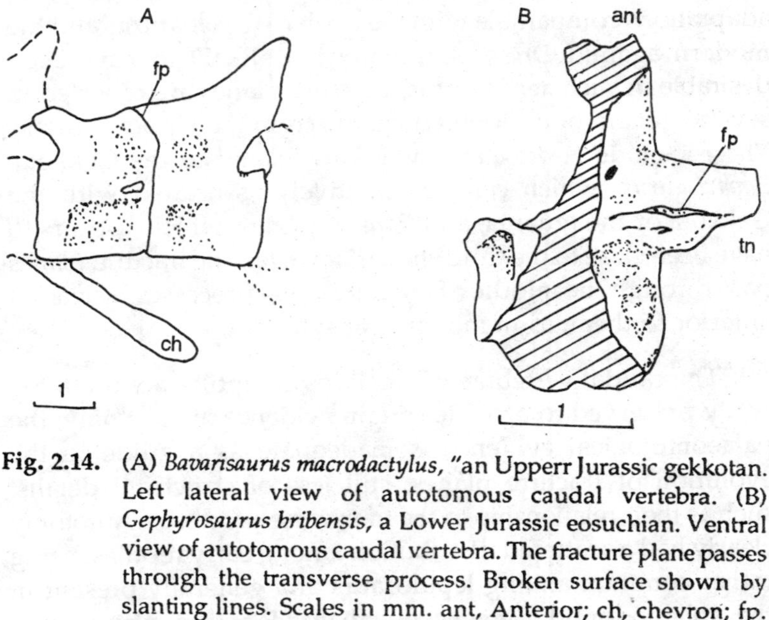

Fig. 2.14. (A) *Bavarisaurus macrodactylus*, "an Upperr Jurassic gekkotan. Left lateral view of autotomous caudal vertebra. (B) *Gephyrosaurus bribensis*, a Lower Jurassic eosuchian. Ventral view of autotomous caudal vertebra. The fracture plane passes through the transverse process. Broken surface shown by slanting lines. Scales in mm. ant, Anterior; ch, chevron; fp. fracture plane; tr, transverse process.

Fracture planes have also been described in *Tanystropheus*, a curious, longnecked Triassic reptile regarded by *Wild* (1973) as a "true, highly specialized lacertilian." However, it may have been an eosuchian and has been classified as a protorosaur, a group which may perpaps be referred to the Eosuchia. The fracture planes of *Tanystropheus* occur throughout a series of about 16 midcaudal, vertebrae, which lack transverse processes, the planes are restricted to the ventral parts of the centra. The processes may show ontogenetic obliteration. This reptile may have been terrestrial when young and partly marine in later life. The hind legs seem to have been used for swimming, and the function of the moderately long, slender tail is problematic.

Among the earliest supposed members were the Permo-Triassic *Paliguana* and *Saurosternon*. The latter form, at least, seems to have lacked fracture planes, as did the Upper Triassic *Kuehneosaurus*. *Kuehneosaurus* was a gliding reptile showing adaptations comparable with the earlier *Weigeltisaurus* and the modern agamid *Draco*. Autotomy is unlikely to have been desirable in such aerial forms. Fracture planes are reported in several genera of undoubted lizards from the Upper Jurassic. These include *Ardeosaurus*, which is related to the Gekkota, *Bavarisaurus*, which can be tentatively associated with the Gekkota or the Iguania, and *Paramacelloaus*, which appears to have been close to the Cordylidae. However, the fracture planes pass through the middle of the transverse processes instead of anterior to them, as in modern cordylids.

The caudal vertebrae of small fossil reptiles are often too badly preserved to provide certain evidence of autotomy; the palaeontological evidence is inadequate as a guide to the evolution of fracture planes, still less of structural details, such as their relationship to the transverse processes. Autotomy clearly had a scattered distribution among early reptiles, being rather common among lepidosaurs and generally present in the sphenodontid rhynchocephalians. Fracture planes may have constituted a common primitive rhynchocephalian and saurian feature. In that case they must have been secondarily lost in forms such as the kuehneosaurs and in the more recent agamids, chameleons, and platynotans, perhaps in response to specific functional needs. On the other hand, the absence of fracture planes in the Paliguanidae, perhaps the earliest known lizards, suggests that they may have been independently evolved in various later saurian groups and in some Amphisbaenia.

They must almost certainly have evolved independently in the very few snakes that have been thought to possess them.

Significance of Autotomy

The adaptive value of autotomy as an antipredator device can hardly be doubted by anyone who has tried to catch lizards. Its value is apparently confirmed by the recovery of tails from the stomachs of predators and the high incidence of broken or regenerated tails in wild populations. There is also some experimental evidence. *Congdon et al.* (1974) exposed geckos (*Coleonyx variegatus*) with intact tails and others rendered tailless, to attacks by the night snake *Hypsiglena torquata ochrorhynca*. In the course of 30 trials, 37% of the initially intact geckos escaped while losing portions of their tails whereas all the initially tailless ones were captured. Comparable results have been obtained with *Hemidactylus turcicus*.

The violent writhing of an autotomized tail, perhaps a reflex triggered off when the spinal cord is ruptured, has often been noticed; it both distracts the predator and offers him an easily obtained meal. The thrashing autotomized tails of the skkink *Scincella lateralis* are highly effective in attracting attacks from a cat; such movements also delay the swallowing of the fragment by a snake (*Lampropeltis triangulum*). In both cases, the lizard had a better chance to escape. The writhing tail fragments also develop much higher concentration of lactate than do intact tails, suggesting a high capacity for anaerobic metabolism.

In *Coleonyx variegatus*, the distracting effect of a potentially autotomous tail is enhanced by waving it in the air as a response to threat. The tail may also be of a different colour from the rest of the animal. In some lizards, such as the skink *Eumeces fasciatus*, the colour contrast is only striking in the young. However, *Clark* and *Hall* (1970) have suggested that the juvenile blue tail colour has a different function; it may act as a visual signal that inhibits adult males from attacking young hatched in the same territory. Not all autotomies are due to attacks by enemies. In some species, tails may be broken off as a result of intraspecific (e.g., territorial) fighting, as reported in *Hemidactylus flaviviridis*, *Sceloporus olivaceus*, *S. magister*, and *Anolis carolinensis*.

Autotomy has such an obvious survival value that it is interesting to inquire why the ability to perform has been lost, reduced, or never gained in such groups as agamids, some iguanids, chameleons, a few geckos and skinks, and all platynotids. A common view is that autotomy does not occur in lizards in which the tail performs any specialized and more or less indispensable function. It has been elaborated by *Vitt et al.* (1977), who classified the tails of lizards broadly into *"actively functional"* and *"passively functional"* categories, the latter being useful mainly by virtue of their power of autotomy.

In *chameleons*, the prehensile tails essential for their arboreal life-style, and the powerful, heavy tails of monitor lizards (*Varanus*), which may be employed both for swimming and for defenc, clearly fall into the actively functional category, as do the tails of the extinct aquatic mosasaurs. Predictably, their tails lacked fracture planes. Other examples of nonautotomy may be found among the Iguanidae, especially the larger species, and include the Agamidae. Thus, the laterally compressed tail of the marine iguana, *Amblyrhynchus*, which is essential in swimming, is apparently nonautotomous and seems to have only limited powers of regeneration. In this lizard, loss of any substantial part of the appendage would probably be incompatible with survival except in those areas of its habitat where food can be obtained by the edge of the sea. The tails of such forms as *Iguana* and *Basiiliscus* are also used for swimming and, by the latter, as a counterpoise in bipedal locomotion. If only one-third of the long, heavy tail is removed, the basilisk is unable to run bipedally. Some of the larger iguanids also use their tails as weapons. Fracture planes tend to be absent, reduced, or prone to undergo ontogenetic obliteration in such lizards. Examples are some of the smaller iguanids that practice bipedal locomotion, such as *Crotaphytus*, *Anolis*, and other forms in which the tail is an adjunct to climbing or perching.

Some samples of *Anolis carolinensis* show a low percentage of caudal regenerates. Similar considerations probably apply in the case of the bigger agamids, such as *Hydrosaurus, Physignathus, Chlamydosaurus* and *Uromastyx*. In each case, the tail is used either for swimming, as a weapon, in bipebal running, or for more than one of these purposes. The absence of fracture planes in the smaller members of the Agamidae may perhaps be correlated with such factors as the prevalence of bipedal running, the use of the tail in intraspecific territorial fighting, or its use as a rudder in the case of the gliding lizards (*Draco*). However, in some agamids, as previously noted, the tail, especially its more distal part, may be broken off at fragile intervertebral sites.

The tail may play some part in the gliding flight of the lacertid *Holaspis guentheri*; it is interesting that tail regenerates were not observed in wild specimens, despite the conspicuous colour of the appendage and, apparently, the presence of fracture planes.

Large members, family Teiidal such as the caiman lizard, *Dracaena*, which uses its tail for swimming; *Callopistes*, a little-known, monitor-like predator with bipedal tendencies, and *Tupinambis*, the tegu lizards. All these forms possess vertebral fracture planes. *Dracaena* may be an exception to the general principle that actively functional tails are not autotomous. However, it is unknown how readily autotomy occurs in these big lizards, at least in the adult; in *Tupinambis* some of the vertebrae may lose their feature planes with age. All these descriptions are of skeletons, and it is possible that the muscular and other adaptations to autotomy are less complete, also that the *"psychic factor"* postulated by *Slotopolsky* (1929) is somewhat in abeyance. The large regenerates sometimes observed in *Tupinambis*, as in *Iguana iguana*, which also shows ontogenetic reduction of the bony fracture planes, may reflect autotomy occasioned when the specimens were young. It seems possible that mere size may be a factor that promotes the reduction of autotomy. The tail becomes an effective weapon only in a lizard of substantial dimensions, and increase in size may be correlated, ontogenetically or phylogenetically, with changes in behaviour. A big lizard may find it advantageous to turn and bite or to lash with its tail, rather than to shed it. Such actively defensive behaviour is perhaps foreshadowed in the medium-sized (maximum total length over 40 cm) alligator lizard. *Gerrhonotus multicarinatus,* which may turn on an enemy or use its semi-prehensile tail in various defensive fashions. Although autotomous, this lizard is less prone to shed its tail than are smaller forms, such as *Coleonyx variegatus* and *Eumeces skiltonianus.* A similar trend is exemplified by other species of *Coleonyx. C. brevis* and *C. variegatus* wave their tails as a distracting threat gesture and have a high incidence of autotomy. *C. reticulatus,* a larger and more aggressive lizard, has a prehensible tail that is not used in display; its incidence of autotomy is considerably lower.

In the members of the family Gekkonidae autotomy is restricted to the basal region of the tail. Some of the geckos, such as *Stenodactylus,* are ground-living desert dwellers with slender tails. *Werner* has shown that certain species autotomize reluctantly; the tail may have an active function as a stabilizer, being applied to the ground when the lizard stands with its long legs extended and its body elevated from the substratum. In other geckos with restricted autotomy, the tail is expanded posterior to a basal constriction, a condition to some extent comparable with that in certain autotomous salamanders. The tail is often modified; it is very fat in the barking gecko *phyllurus,* knob-ended in *Nephrurs laevis,* flattened and leaf-like in *Phyllurus,* and furnished with cutaneous expansions in the bark geckos, *Uroplatus.* The shape to the tail in geckos, such as *Phyllurus,* would seem to inhibit autotomy except at the basal constriction; more distal breakage through the broad part of the tail might result in a dangerously large wound. Moreover, in *Phyllurus* and *Uroplatus,* the tail appears to form part of a general cryptic pattern and, hence, may fall into the "actively functional category." However, *Phullurus* autotomize quite readily and it would appear that anatomical restriction of fracture planes is not invariably associated with a reduced incidence of autotomy. *Arnold* (1984) suggests that basal autotomy may have the function of providing a predator with a large enough fragment to distract it; this could be particularly useful in the case of slow-moving species of geckos. However, the significance of basal autotomy is not always clear; it contrasts with the more economical type of autotomy practiced by lacertids and by geckos such as *Hemidactylus*.

Several others species of geckos of the genus *Diplodactylus* possess sub-cutaneous caudal glands producing a viscous exudate, which can be squirted at an aggressor. These geckos have fracture planes all along the tail, but in most species the incidence of autotomy is low, probably in correlation with the actively defensive function of the tail. However, *Diplodactylus elderi*, which is a ready autotomizer, is an exception

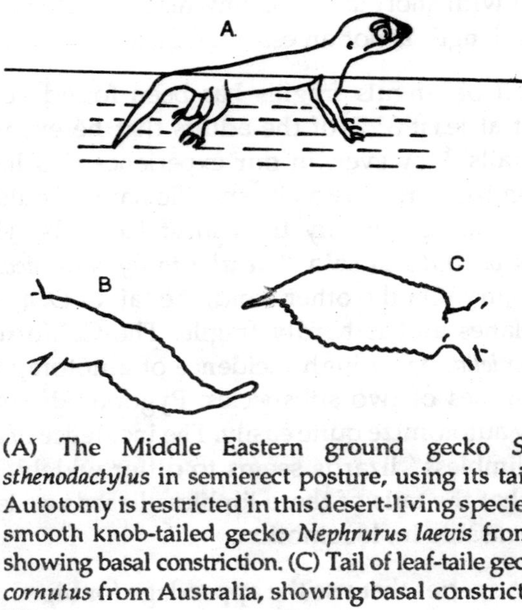

Fig. 2.15. (A) The Middle Eastern ground gecko *Stenodactylus sthenodactylus* in semierect posture, using its tail as a prop. Autotomy is restricted in this desert-living species. (B) Tail of smooth knob-tailed gecko *Nephrurus laevis* from Australia, showing basal constriction. (C) Tail of leaf-taile gecko *Phyllurus cornutus* from Australia, showing basal constriction. (B) and (C).

The case of *Lygodactylus klugei* is perhaps comparable. The tail of this gecko has a scansorial pad, which helps it to cling to trees; nevertheless, it has a high rate of autotomy.

The various types of limbless or nearly limbless lizards are also difficult to fit into a clear-cut scheme of active or passive tail function. Most of these reptiles are autotomous; however, one would expect the tail to be an important instrument of propulsion. Actually, there seems to be little critical information on the extent to which limbless locomotion, either above or below ground, is dependent upon the integrity of the tail. The intact, long, and quite prehensile tail of the slow-worm, *Anguis fragilis*, was long ago claimed to play a significant part in its movements by a dedicated amateur herpetologist. Yet *Anguis* moves well after loss of a substantial part of the appendage. Like some other members of its family, this species is a poor regenerator; the not infrequent discovery of individuals in the wild with short tails and tiny regenerates suggests that normal tail length is not an essential factor in survival.

The tail of Anguis fragilis has been found very fragile found that at least 35% of the adults that he examined had damaged tails. However, in our experience this lizard does not autotomize as readily as its specific name would suggest; it sheds its tailless readily than most lacertids. The tail of *Ophisaurus ventralis* is said *"literally to fly into pieces"* at the slightest injury. On the other hand, the tail of *O. apodus* lacks fracture planes and is hardly fragile. The California "slow-worm" *Anniella*, has a high incidence of autotomy (59% and 69% in samples of two subspecies. Pygopodids, in general, apparently autotomize quite easily. The incidence of autotomy in other "limbless" lizards seems to differ widely. One may conclude that the tails of these limbless lizards assist, but are not indispensable in, locomotion.

Certain species that use the appendage for bipedal running may have a lower incidence of autotomy. The tail is also important in undulation and swimming; one would expect that even predominantly terrestrial lizards may sometimes enter or find themselves in water. The tail may function in certain types of defensive and sexual dislay and may also act as a status symbol. In the hierarchical communities of the chuckwalla (*Sauromalus obesus*), a fairly large iguanid, "tyrant" males lose their dominant position after injury to the tail. "A male that loses 30 to 50% of his tail appears to sustain psychological as well as physical damage and thus loses social status." It is perhaps significant that this lizards shows ontogenetic reduction of its fracture planes. However, comparable observations have been made on the fully autotomous *Uta stansburiana;* it is possible that the loss of the tail may have its deleterious effect by reducing overall size of the lizard.

Fat and Automony

Most autotomous lizards store subcutaneous fat in considerable amounts in submuscular, perivertebral bands of caudal fat. There may also be some subcutaneous fat, as in many geckos. The fat layers are reduced or lacking in the majority of non-autotomous forms, *Heloderma suspectum* and *Trachydosaurus rugosus* being notable exceptions. Fat storage seems to have become the *main* function of the stumpy tails of these lizards; this storage might thus be regarded as actively functional. In most other lizards, fat storage appears to be a supplementary function, although, as previously mentioned, the fat may conceivably play some part in the mechanism of autotomy. In *Lacerta*, the incidence of autotomy and the seasonal accumulation of caudal fat may be positively correlated.

The amount of fat stored in the tail and elsewhere in the body seems to vary substantially. In the gecko *Coleonyx variegatus*, caudal fat is said to represent only a small part of the total body fat. On the other hand, in *C. brevis* "caudal energy reserves represented 60% of the total reserves of tailed females and were one-third greater than the total energy reserves of tailless females." In *Lacerta vivipara* (examined during the summer), the total fat forms about 39% of the total body weight, 17.3% of this being in the tail and 21.7% elsewhere, mainly in the abnorminal fat bodies. However, in some skinks, the tail may be the only organ of fat storage.

Fat is utilized when food is short. *Lacerta vivipara*, emerging from hibernation in the spring, may have lost as such as one-half their previous summer fat; 25% of the loss is derived from the tail. The main value of fat storage may be maintenance of activity in times of drought, hence, its importance in lizards such as *Heloderma*, which live in arid or semiarid conditions. *Scincella lateralis*, with tails amputated near the base, survive starvation an average of 24 days, whereas those with intact tails survive for 35 days. Fat reserves are also important for reproduction in some species, and females of *Coleonyx brevis* normally draw on both abdominal and caudal fat for the formation of yolk in their eggs. If the tails is lost, only about half the normal caloric energy may be available for vitellogenesis. Tailless reproducing females must eat more to make up for the energy loss, and the eggs that they lay are reduced in mass (dry weight) and energy content. The nutritional handicap of tail loss may occur at a critical period, as in females during reproduction; the effect may be temporary because fat is redeposited as the tail regenerates. Another more curious form of compensation for autotomy has been observed. Some lizards, notably *Scincella lateralis*, have the habit of eating their autotomized tails if they can and, hence, are able to recover the lost fat.

Thus, we see that the tails of most lizards have various roles, of which autotomy on the one hand and locomotion and nutrition on the other are the most significant. The relative importance of these partly incompatible roles shows all gradations among different lizards. In some forms, such as the gecko *Lygodactylus klugei*, they appear to be of almost equal weight; moreover, a strong trend in favour of autotomy could be balanced by a high regenerative capacity. This, to some extent, invalidates the distinction between actively and passively functional tails, although it remains a useful concept. In a large number of species, the integrity of the tail is probably not essential under normal circumstances; the tail is indispensable only in non-autotomous lizards. The relative costs and benefits may vary among different species, perhaps even among different members (e.g., sexes) of the same species, and are related to such factors as the extent of fat storage and the energy required for regeneration and other purposes. Autotomy can be regarded at a compromise adaptation, often involving sacrifice, which a lizard cannot easily afford but which is preferable to the alternative of certain death.

Temperature and Autonomy

The activities of ectothermic animals are, in general, enhanced as the temperature rises toward the maximum tolerable limit. As a rule, autotomy seems to conform with this principle and can generally be elicited with greater ease in warm than in cold lizards. As the body temperature of the lizards rises from 22° to 30°C, the force required to cause breakage of the tail in the iguanid *Uta stansburiana* diminishes progressively from 15.1 to 6.9 ounces (428-194 g, the ounces are presumed avoirdupois for the purpose of conversion; *Brattstrom*, 1965). However, if the temperature is elevated still further, the required force rises further, reaching 11 ounces (312 g) at 42°C. The temperature range (30°-38°) at which the required forces are lowest (6.9-8.1 ounces or 194-230 g) may correspond to the optimal activity range of the species. Comparable observations on the effects of the initial temperature rise have been made on *Podarcis sicula* and on the reluctantly autotomizing gecko, *Stenodactylus*.

However, there is evidence that the relationship between autotomy and temperature may be more complicated in some species. The Australian gecko, *Gehyra variegata*, autotomizes more readily when it is in a cold torpor at night (ambient temperatures around 4°C) than when it body temperature is suitable for activity (20°-30°C). Moreover, cold individuals often shed almost the whole of their tails spontaneously, whereas at 20° to 30°C, autotomy is very rare unless the tail is actually grasped, and then only a small part of it is shed. Autotomy frequently takes place at low (4°C) and high (19° and 29°C) body temperatures, but seldom between 8° and 15°C. The group at 4°C loses some 80% of the tail, whereas those at 15° to 29°C only lose mean percentages of 21 to 52.

Regeneration

As stated already that the regeneration is the recovery of lost port.

The process of caudal regeneration in lizards is, in many ways, comparable with that seen in urodele amphibians. It falls into three principal phases: Wound healing, blastema formation and differentiation and growth. Generally speaking, an autotomized tail undergoes no elongation during wound healing and early blastema formation (the so called "latent period", see below); it then elongates rapidly for several weeks. Growth finally slows down to almost nothing after three months or more except in so far as it may form part of the general body growth of an immature animal. The wound that results from natural autotomy in a lizard, such as *Hemidactylus* or *Lacerta*, causes relatively little damage to the local tissues. In theory, at least, the only structures that are actually torn are the spinal cord and meninges, the peripheral nerves, the caudal blood vessels and lymphatics, the skin or at least its epidermal layer, and perhaps small regions of the above, such as the tip of the anterior neural spine with its adjacent periosteum. The connective tissue of the autotomy septum will probably also be damaged, but most of it will remain adherent to the tissues of the stump. The integrity of the muscles and fat bands is more or less preserved by this septum, whereas the vertebra, as previously described, is traversed by a preformed split. However, in *Anolis* the bony adaptations to autotomy seem less complete, and the muscles do not separate quite cleanly; fragments of the anterior muscular processes of the broken-off tail are left adhering to the stump.

Bleeding is reduced owing to the action of the caudal arterial sphincters and venous valves, at least in forms in which these are present. However, the stump is covered by extravasated blood, which soon clots, plugs the ruptured spinal cord, and fills up the recesses between the protruding muscular processes of the stump. Within minutes of autotomy, the skin around the raw surface contracts and covers these processes; the central part of the surface becomes hidden by a scab, consisting of clotted blood and necrotic tissue, through which the autotomized surface of the vertebra may protrude. Various tissue reactions proceed beneath the scab, including the appearance of phagocytic white cells, lymphocytes, and eosinophil erythrocytes. The epidermis proliferates around the margins of the raw surface and, as autoradiography shows, begins to incorporate tritiated thymidine. The substance is taken up by the replicating DNA of cell nuclei; in general, its uptake is an indicator of impending mitosis. The broken end of the spinal cord retreats from the scab and becomes closed by the proliferation of cells derived from its ependymal lining. These processes occur within one to two days after autotomy. Slightly later, the epidermis migrates (centripetally) across the autotomized surface beneath the scab until it reaches the margins of the autotomized half-vertebra. Then follows the remarkable process described by *Werner* (1967a) in geckos, whereby the distal one-half of the half-vertebra is cut through by the action of osteoclasts and is shed or "ablated," together with the scab. This vertebral ablation apparently occurs rather rapidly a few days autotomy, the neural arch being shed separately from and before the fragment of the centrum. Only about one-fourth of the originally autotomized vertebra will therefore remain in the distal end of the tail stump. Vertebral ablation also takes place in *Scincella lateralis*. The tips of the posterior muscle processes of the stump may also become incorporated into the scab and subsequently be ablated at least in *Anolis;* the proximal parts of the muscle fibers later undergo repair.

After vertebral ablation and the shedding of the scab, the

new epidermis completely covers the wound surface so that healing is now effected. The multilayered epidermis becomes thickened at the margins of the autotomy site and at the tip of the young regenerate; there it forms an apical cap of cells. One or more papillae of the epidermis may extend into the residual necrotic tissue beneath the cap, and the epidermal cells may engulf dead erythrocytes and connective tissue debris. Blister-like protuberances also form on the deep surface of the epidermis and perhaps release substances into the underlying tissues. Removal of the apical cap and of the adjacent skin in lizards inhibits tail regeneration as does replacement; of the cap by a graft of mature skin. This apical cap is comparable with that seen in amphibian tail and limb regenerates, and perhaps also with the apical crest or ridge, which appears at the tip of the limb of amphibian and amniote embryos. The importance of this cap and of its interaction with nerve fibers in the regeneration of the amphibian limb is well-known; naked contact must occur between the epidermis and blastema without the interposition of dermal tissue.

Some four to eight days after autotomy in lizards, the ependymal lining of the tip of the spinal cord elongates and grows posteriorly to form the ependymal tube. The tip of this becomes dilated to form a sac and approaches the epidermis of the apical cap. Large numbers of mesenchyme-like and fibroblast-like cells, together with many pigment cells, appear around the ependymal tube and its terminal sac they form what may be broadly called a *blastema*, a mass of dedifferentiated cells deriving from pre-existing tissue and capable of poliferation and radifferentiation. According to Cox (1969a), the blastema of the regenerating tail of *Anolis* differs somewhat from that of regenerating amphibian limbs on which classical accounts have been based. The cells beneath the apical cap of the lizard hardly proliferate at all, and the main areas of proliferation, with the highest tritiated thymidine labeling indices, lie somewhat more anteriorly in the stump; here, they contain aggregations of strongly basophilic cells, which will give rise to myoblasts. Blastema cells also aggregate around . the ependymal tube and soon differentiate into procartilage. Blood vessels appear within the blastema and presumably soon become connected with the vessels of the stump, although this process does not seem to have been described. Observations on *Hemidaztylus* are, to some extent, conflicting, because mitotic activity occurs in the cells beneath the apical cap. Our observations are inconclusive on this point

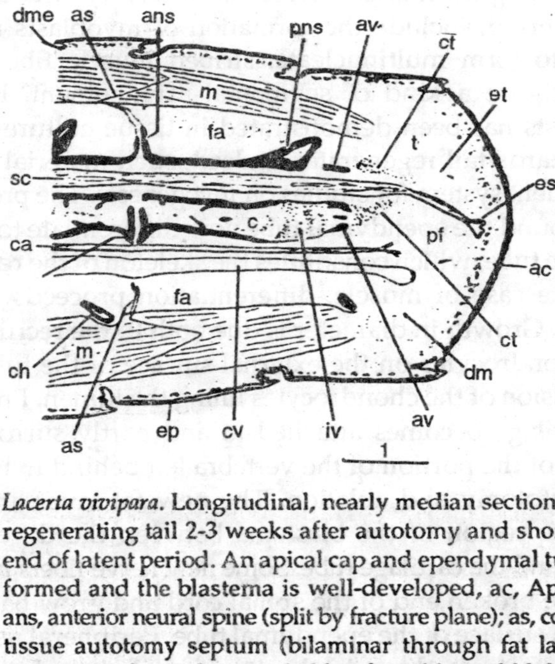

Fig. 2.16. *Lacerta vivipara.* Longitudinal, nearly median section through regenerating tail 2-3 weeks after autotomy and shortly after end of latent period. An apical cap and ependymal tube have formed and the blastema is well-developed, ac, Apical cap; ans, anterior neural spine (split by fracture plane); as, connective tissue autotomy septum (bilaminar through fat layer); av, autotomised vertebra, probably after ablationl bl, blastema; c, centrum; ca, caudal artery; ch, chevron; ct, developing cartilage tube; cv, caudal vein; dm, developing muscle (pro-muscle aggregate); dme, dermal melanophore; ep, epidermis; es, ependymal sac: et, ependymal tube; fa, fat layer; iv, intervertebral pad; m, original muscle; pns, posterior neural spine; sc, spinal cord.

Despite the intense cellular activity that has been described, the regenerate as a whole does not begin to elongate perceptiibly for some ten to 15 days after autotomy. This interval, or latent period, appears to be relatively constant in most of the species studied. At the end of the period, the regenerate becomes externally visible as a smooth, darkly pigmented cone, a millimeter or so in length. During the next few weeks, it elongates very rapidly; the blastema proliferates and differentiates to form the new tissues.

Histological changes, which occur around the end of the latent period, include the formation of myoblasts and their fusion to form multinucleate, striped muscle fibers, which soon acquire a kind of segmental arrangement. Fusion of myoblasts has been demonstrated in tissue cultures derived from lizards tail regenerates; indeed, such material provides an excellent system for studies on myogenesis. The procartilage cells around the ependymal tube now differentiate to form the cartilage tube, which constitutes the skeleton of the regenerate; as in the case of muscle, differentiation proceeds proximo-distally. Growth in diameter is due both to the recruitment of new chondrocytes on the external surface of the tube and to cell division of the chondrocytes lining the lumen. Proximally, the cartilage becomes attached to and partly surrounds the surface of the portion of the vertebra left behind in the stump after autotomy and ablation. The new fat layer arises later from the region of the blastema between the differentiaing muscle and the cartilage tube. Some new nerve fibers regenerate from the broken end of the spinal cord and grow back on the external surface of the ependymal tube. Peripheral nerves also regenerate from those in the stump and extend posteriorly through the blastema. The regeneration of the ependyma and nerves is describes in greater detail below.

The development of the scales begins between two and three weeks after autotomy in *Sphaerodactylus*, at which time the regenerate is about 2 mm long. The process has been described in a number of species and is of considerable interest. Before scale formation begins, the epidermis is several cells layers in thickness and the outermost layer is becoming keratinized. The basal layer of the epidermis gives rise to a series of ridges, which grow into the underlying blastema, the tissue of which will become the dermis. These ridges become slanted so that their deep extremities point anteriorly, away from the tip of the regenerate. Subsequently, each ridge splits longitudinally to give rise to the greater part of the inner surface of one scale and the outer surface of the scale behind it. The tissue at the bottom of the split will become the hinge region, and the overlapping pattern of the normal scales in lizards, such as lacertids, will be reproduced. Keratin appears as a sheet, down the middle of the ridge, and when the split takes place, a thicker layer of this substance remains on the outer rather than on the inner scale surface. As the scales differentiate in a proximo-distal sequence, all stages of the process can be seen in a fairly advanced *Lacerta* regenerate. Eventually, the scales may encroach on the region of the original apical cap, although in *Anguis* tha latter appears to keratinize as a single, smooth conical scale

The development of the new scales on the regenerating tail follows a different pattern from that in the embryo. In the embryo, the scales arise as a series of elevations, which soon become asymmetrical and are slanted posteriorly. No epidermal ridges projecting into the dermis are involved in their formation, and no substantial keratinization occurs until the scales have almost reached their definitive form. The scales of the lizard tail regenerate show intriguing resemblances to developing mammalian hairs. The form and pattern of the scales on the regenerating talls of adult geckos can be influenced by temperature variations and by transplanting pieces of ectopic tissue into the adjacent part of the original tall.

Nerves and Ependyma in Regeneration

The nerve supply of the original tail in *Chalcides ocellatus* seems to resemble that in *Sphaerodactylus* and, perhaps, also that in *Lacerta*, although further comparative studies are required to conform this. In all three forms, each spinal nerve, after leaving the vicinity of its ganglion, passes back through three successive autotomy septa and supplies three successive autotomy segments. Each segment is innervated by three nerves on each side: the nerve from the segment immediately in front of it and the nerves from the two successive anterior segments. Each nerve traverses a part of its own segment, but apparently does not supply it. The situation is further complicated by the fact that each nerve breaks up into several branches.

Towards the end of the latent period after autotomy, these branches, which are derived from the last three pairs of spinal nerves within the stump, grow backwards through the blastema, although only ahbout half of them approach the tip of the regenerate. No new dorsal root ganglia are regenerated, and the cells in the last three ganglia in the stump increase in volume, the peripheral sensory field, innervated by their fibers is expanded.

It is well-known that the spinal cord undergoes partial regeneration after autotomy; the ependyma, some of the menings, and some glial cells regenerate, together with a considerable number of descending nerve fibers, the majority of which are unmyelinated. Reissner's fiber also regenerates, although not to the. tip of the ependymal tube. In early regenerates, the nerve fibers end in the blastema near the apical cap; later they become surrounded and, eventually perhaps, are even enclosed distally by the closure of the cartilage tube. They have no distal connections, and their function remains enigmatic. No new cell bodies are formed. The functional nerve supply of the regenerate is probably derived entirely from the peripheral nerves of the stump. Most of the descending fibers of the spinal cord are closely associated with the ependymal tube, but a few pass within the regenerated. Apparently, the ability of *Anolis* to regenerate nerve fibers in the spinal cord is confined to the tall; transection of the spinal cord in the thoracic region produces little if any regeneration of the ependyma and nerves.

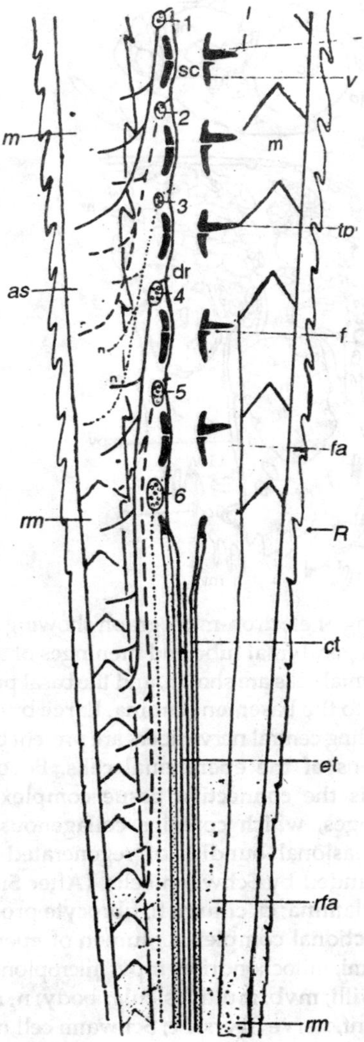

Fig. 2.17. On left, diagram showing nerve supply to original tail and regenerate, and on right, arrangement of caudal muscles as seen in horizontal longituudinal section of tail of an autotomous lizard. On the left, only alternate muscle segments are stippled. For clarity the first two original segments are shown as being supplied by one and two spinal nerves respectively, instead of by three, as is actually the case. The nerve distribution is greatly simplified and it is uncertain how many branches from the last three nerves supply each segment of the regenerate. The segments of the latter are shown much more regular than they really are. Successive nerves are shown in solid black, broken lines, and connected dots. Ventral roots are not shown. Transverse processes and fat are shown on the right side only as, Connective tissue autotomy septum (for clarity shown as unilaminar throughout); ct, cartilage tube; dr, dorsal nerve root; et, ependymal tube; f, fracture plane; fa, original fat; m, original muscle; n, nerve; R, approximate level of regeneration; rfa, regenerated fat; rm, regenerated muscles; sc, spinal cord; tp, transverse process; v, vertebra; 1-6, spinal ganglia.

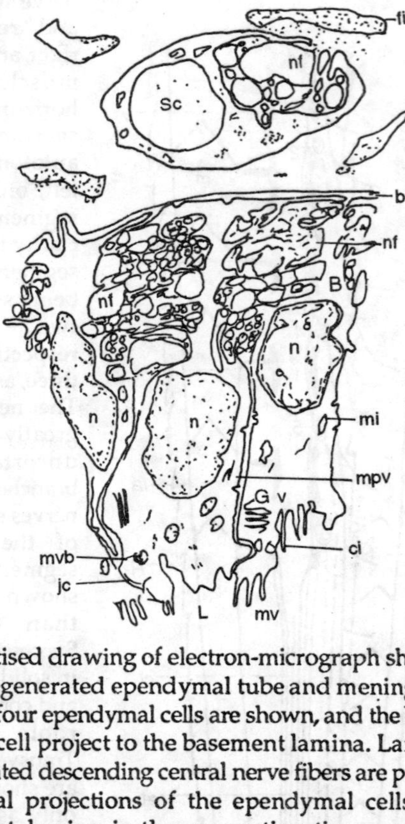

Fig. 2.18. Schematised drawing of electron-micrograph showing portion of the regenerated ependymal tube and meninges of a lizard. Parts of four ependymal cells are shown, and the basal processes of each cell project to the basement lamina. Large bundles of regenerated descending central nerve fibers are present between the basal projections of the ependymal cells. Beyond the basement lamina is the connective tissue complex of the regenerated meninges, which contains collagenous fibers, fibrocytes, and occasional bundles of regenerated central nerve fibers surrounded by Schwann cells. (After Simpson, 1970) bl. Basement lamina; ci, cilium; fi, fibrocyte process; G. Golgi body; jc, junctional complex, L, lumen of ependymal tube; m, meninges, mi, mitochondrion; mpv, micropionocytotic vesicle; mv, microvilli; mvb, multivesicular body; n, nucleus of ependymal cell; nf, nerve fibers; Sc, Schwann cell nucleus.

Ultrastructural studies have shown that ependymal cells send out processes during regeneration, and these overlap at the interface with the meninges, forming large channels. The regenerating nerve fibers pass into these channels so that the ependyma appears to act as a guide for the fibers. This process is nerve penetration can be followed in a proximodistal sequence of material taken from successive regions of the ependymal tube. More recently, the effect of nerve fibers on limb regeneration has been critically studied by Singer and his colleagues, who have shown that regenerative capacity is highly dependent on the concentration of fibers near the site of injury. Thus, the limbs of the frog Rana and of lizards which do not normally regenerate after amputation, can be induced to do so by experimentally augmenting their nerve supply. This "trophic" function of the nervous system in promoting regeneration is ascribed to some chemical substance produced by the nerve, which perhaps interacts with the cells of the apical cap and stimulates cells division in the underlying blastemal tissues. However, the precise mechanism is not yet understood.

Kamrin and *Singer* (1955) applied this principle to the tail of lizards/Destruction of the spinal cord in *Anolis carolinensis* for about 1 cm proximal to the site of amputation induced no significant regeneration, a finding that was attributed to the reduction of nerve fibers at the injury site. However, *Simpson* (1964), working on *Scincella lateralis,* showed that damage to the spinal cord does not prevent regeneration after amputation, as long as significant portion of the ependymal lining of the cord survives. In further experiments, cartilage tubes containing ependyma from mature tail regenerates were implanted as autografts within the dorsal muscles of the tails of the same animals and produced small but normally structured regenerates at the sites of implantation. No regeneration occurs at control sites to which grafts of cartilage tube without ependyma have been implanted. No viable nerve fibers were included with these implants, as the fibers inevitably degenerate whenever, they are separated from their cell bodies after amputation of the mature regenerate. The possible role of the peripheral nerves, some of which were probably left intact in Simpson's experiments, has been studied by removing the three pairs of

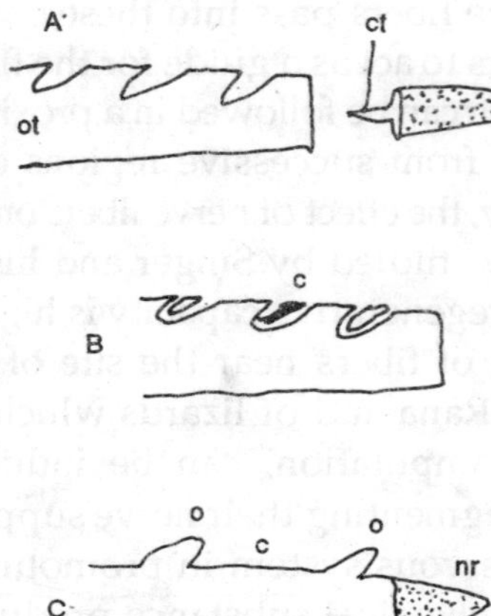

Fig. 2.19. Diagrams illustrating Simpson's experiments (1964) to demonstrate induction of regenerative outgrowths by ependyma. (A) Three channels are cut in the dorsal aspect of the original tail of *Scincella,* anterior to a previous (old) regenerate. The latter is amputated and portions of cartilage tube containing ependymal tube are dissected from it. (B) Portions of cartilage tube containing ependyma are implanted in the front and rear channels. A portion of cartilage tube *without* ependyma is implanted in the middle channel as a control (black). (C) Regenerative outgrowths appear only at the two sites at which ependyma was included in the implants. c. Control; ct, cartilage tube with ependymal tube in old regenerate; nr, new regenerate; o, regenerative outgrowth; or, old regenerate; ot, original tail.

dorsal root ganglia proximal to the site of injury and disrupting the accompanying motor tracts but leaving the spinal cord with its ependyma intact. This drastically reduces the number of peripheral nerve fibers at the autotomy surface from the normal 4.5 fibers per unit area of 100 mm^2 to 0.34 fibers per 100 mm^2 : normal, although somewhat delayed regeneration takes place.

Thus, we see that the ependyma, rather than the nerve fibers, is the critical agent in inducing normal regeneration of the tail. The saurian tail, therefore, affords a striking contrast to the amphibian limb, regeneration of which is definitely correlated with the quantity of nervous tissue present at the wound site. It is possible that the amphibian tail agrees with the saurian tail in this relationship, but the idea has not yet been critically investigated. The ependyma also has a distinct and somewhat specific function in inducing the formation of the cartilage tube. A few experiments involving damage to the spinal cord show that an accessory, dorsally directed tail develops at the autotomy site in addition to the normally directed regenerate. The ependymal tube enters the accessory regenerate, which has a cartilage tube, whereas the normally directed regenerate lacks this structure. These findings seem to indicate that, whereas the ependyma may initiate the development of two regenerates from closely adjacent sites, the development of the cartilage only occurs in the continued presence of ependymal tissue. Insertion of autografts of ependyma (taken from the regenerating tail) into the muscles of the caudal stump also produces cartilage tubes. This demonstrates the importance of ependyma in the induction of cartilage.

These results seem to conflict with the observations of *Weiss* (1930, 1939;474), which have seldom been quoted by later investigators. He transplanted blastema from the tails of *Lacerta* to the shoulder region after amputation of the forelimb. The younger transplants generally failed to develop or remained only as small nodules; these observations have been confirmed in comparable experiments on *Eublepharis*. However, in some cases, the older *Lacerta* transplants gave rise to substantial regenerates that obtained a good nerve supply from the shoulder region. These resembled normal tail regenerates in many ways and seem genuinely different from the superficially tail-like regenerates that sometimes form after loss of a limb. Thus, Weiss noted that the arrangement of the scales and muscles was typically caudiform and that some, although not all generates, included cartilage tubes but no spinal cord. However, these experiments are not conclusive, because it is probable that the blastemas had been exposed to the influence of the ependyma for a long-enough period before transplantation to allow induction of cartilage to take place. The results of an apparently similar experiment have been illustrated. Despite the problems raised by these and other observations, the major inductive role of ependyma, at least under a range of experimental conditions, has been well-established. As in the case of the induction by nerves, the production of some form of "trophic" substance, which influences the blastema and/or the apical cap, has been hypothesized. Indeed ultrastructural changes, suggestive of secretory activity, have been shown to occur in the ependyma just proximal to the autotomy site in the tail of *Anolis*. Both ependyma and nerves arise from the neural tube of the embryo, and "it is not unreasonable to assume that the ependyma may be acting in the same way as the nerves [do] in other regenerating systems."

Nevertheless, it would be wrong to discount completely the role of nervous tissue in reptilian tail regeneration. After removal of the spinal cord and ependyma, *Anolis* shows a very limited regeneration of the autotomized tail as long as 50 to 75% of the peripheral nerve fibers remain intact. Possibly, sympathetic trunks have an as yet unconsidered role, as these regenerate on either side of the caudal artery at the same time as nerve fibers are growing back from the spinal cord. The critical question as to whether a hyperinnervated tail can regenerate in the absence of ependyma remains to the answered; the technical difficulties for doing so are substantial.

Mention must be made of other experiments of *Whimster* (1978) in which accessory tail regenerates were induced by transecting and deviating the spinal cord or by implanting pieces of it. Unfortunately, his conclusions that nerve fibers are all-important seems to have been reached without appreciation of the work of *Simpson*, of *Cox* or of *Bryant* and *Wozny*. Many of his results can be interpreted on the assumption that the ependyma and not the nerve fibers is the effective agent.

Biochemistry and Histochemistry of Regeneration

Biochemical and histochemical observations & are made on three species: *Hemidactylus flaviviridis, Mabuya striata,* and *M. carinata.* They studied the distribution and fluctuating concentrations of enzymes and other substances during the various phases of regeneration, the associated metabolic changes that occur in organs such as the liver, and the role of thyroid and other hormones. Some comparative studies on the non-autotomous lizard *Calotes versicolor* have also been made.

The Mature Regenerate

In some lizards, as previously mentioned, the size of the mature regenerate may approach or even exceed that of the original; this may occur in certain lacertids, skinks, and geckos, and in *Xantusia vigilis.* In *Lacerta dugesii,* regenerates may measure about 90% of the portion lost by 12 weeks after autotomy, whereas regenerates of *Eumeces skiltonianus* average 125% of the volume of the original tail. However, in many other lizards, the regenerate never appears to approach the size of the original. The volume of the regenerate is only about 30% that of the original in *Eumeces gilberti* and the anguid *Gerrhonotus multicarinatus. Anguis fragilis* never seems to grow more than a small conical regenerate, at most 2 cm in length; it seems unlikely that this species iis capable of replacing more than 20% of the portion of tail lost. *Anniella pulchra* and the non-autotomous *Ophisaurus opodus* also appear to be poor regenerators, If *O. apodus* loses a part of its tail, the wound heals into a blunt, conical stump and this way also be true for the autotomous *O. ventralis.* However, *Bustard* (1970), perhaps referring to other species, stated that regenerates in this family are shorter than the originals and show differences in scalation. In many cases regenerates can be distinguished by differences in the colour, size, and pattern of the scales and by such features as the absence or the irregularity of tubercles and spines. The distinctive early appearance of the regenerate is striking in geckos, such as *Hemidactylus,* and in *Sphenodon,* in which the new scales show no signs of segmental arrangement.

It has been suggested that the scales pattern on the regenerated tail of certain lizards, such as *Ophisaurus gracilis,* shows a reversion to a more primitive type, as seen in related species. Such as a reversion has even been claimed in the case of *Srhenodon* on the basis of apparent resemblances between its regenerated scales and the fossilized impression of original caudal scales of the Jurassic rhynchocephalian *Homoeosaurus.* However, there is no real evidence for this idea of recapitulation. Although the gross form of regenerated scales on tail regenerates may often differ from the original form; specialized structures, such as the sensory hairs of geckos, often regenerate, although their distribution may be atypical. Such regeneration is apparently dependent upon the regeneration of nerve fibers. After autotomy, the scansorial pad on the ventral aspect of the tail tip of the gecko *Lygodactylus* is regenerated and regains functional efficiency. In the gecko *Terratoscincus,* the dorsum of the tail is furnished with broad overlapping scales, which can be rubbed together to produce a rustling sound. Here the regenerate completely reproduces the original dorsal scale pattern, which is necessary for stridulation, although the scales on the ventral tail surface may differ from those of the original. The regenerated dermal ossifications of *Anguis* are smaller and less regular than the original ones.

The caudal squirting glands of certain species of *Diplodactylus* are regenerated after autotomy. However, the new glandular chambers are not segmentally arranged, as are those in the original tail and lack the normal exist slits for the secretion. The defensive efficiency of the squirting apparatus of regenerated tails is unknown.

Unlike the originals, the regenerated muscles have no attachment to the skeleton, represented by the cartilage tube. *Woodland* (1920) believed that the muscles of *Hemidactylus* show no evidence of segmentation, but this appears to be incorrect, as indicated by *Mufti* and *Munir* (1973). In geckos and other lizards, the new muscles arise as a series of myomeres and develop processes that interdigitate with the muscles of

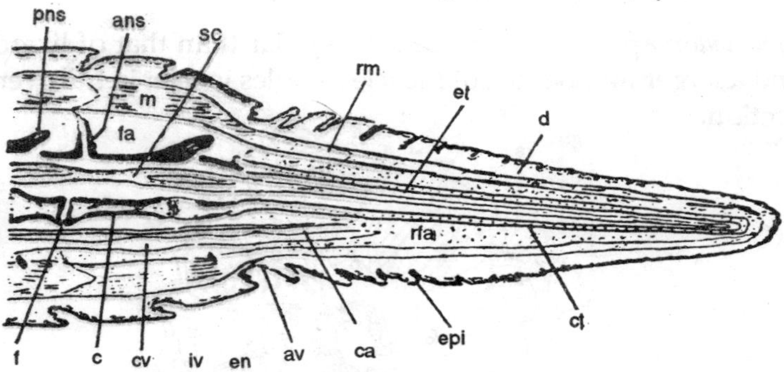

Fig. 2.20. *Lacerta vivipara.* Longxudinal section near the midline through an advanced, but not completely mature, regenerate. The cartilage tube appears to be closed posteriorly in this section, but it is uncertain whether it really has no posterior opening. Scale in mm. ans, Anterior neural spine (split by fracture plane); av, autotomised vertebra; c, centrum; ca, caudal artery; ch, chevron; ct, cartilage tube; cv, caudal vein; d, dermis; epi, epidermal ingrowth forming new scale; et, ependymal tube; f, fracture plane in centrum fa. fat (label on epaxial fat band); iv, intervertebra pad; m, original muscle; pns, posterior neural spine rfa, regenerated fat interspersed with pigment cells; rm, regenerated muscle; sc, spinal cord.

adjoining segments in a fashion that is, to some extent, comparable with that of the original tail. The adjacent segments of muscle are separated by thin inscriptions of connective tissue, which are less evident than the connective tissue autotomy septa of the original. However, the arrangement of the muscles is not exactly the same as that of the original ones. They are less regular, and there are no clearly defined horizontal and median longitudinal septa separating them into four quadrants. Instead, radial septa partition the muscles into many longitudinal bundles; in lacertids, a transverse section may pass as many as 14 bundles of a single segment. However, the muscles of one segment often interdigitate with fibers of an adjacent segment. The first muscle segment of the regenerate is attached of the last segment of the original caudal stump.

The segmentation of the regenerated tail muscles of

Sphenodon appears to be still less regular than that of lizards, and a larger number of cut muscle bundles is seen in transverse section.

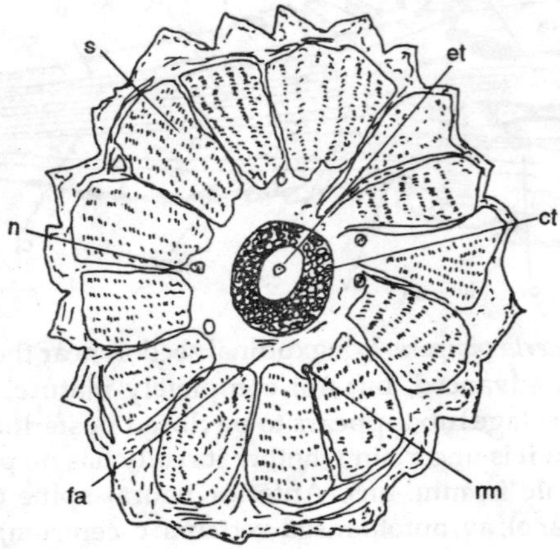

Fig. 2.21. *Podarcis muralis*. Transverse section through mature regenerated tail. ct, Cartilage tube (probably calcified); et, ependymal tube; fa, fat; n, never; rm, regenerated muscle; s, radial longitudinal septum.

In lizards, the regenerated layer of submuscular fat is undivided by septa, and thus forms a continuous, unsegmented investment for the cartilage tube; in *Anguis*, little if any new fat is formed. The subcutaneous fat layer is regenerated in some geckos. The regenerated caudal artered and vein lack sphincters and valves, respectively.

The cartilage tube may become calcified or perhaps even ossified in mature regenerates of some species and may, therefore, show up in X-rays. In *Hemidactylus*, the tip of the cartilage tube eventually becomes closed, although the chondrocytes in this region retain the ability to divide until the regenerate reaches its full length. Regenerates of *Lacerta* often show the ependymal tube issuing from the unclosed tip of the cartilage tube, but in such cases, the regenerate may not be completely mature. In addition to the tissue regenerated from the spinal cord, the cartilage tube also contains connective tissue and blood vessels, a few of which may pass through holes in its walls.

Woodland (1920) graphically described the regenerated tail of lizards as a "jerry-built" structure; in many respects, its functional value falls short of that of the original, even in forms in which it may eventually approximate to the original length. However, it does facilitate locomotion in balancing or swimming. The regenerated muscles, despite their irregularity of skeletal attachment, provide mobility, although flexibility is reduced, probably to complete rigidity in cases of extensive calcification.

In one respect, as a site for fat storage, a regenerated tail may become more efficiency than an original. In *Dipsosaurus*, the volume of the cartilage tube appears to be considerably less than that of the original vertebrae; thus, there is greater room for the deposition of fat in a regenerate of equivalent size. This finding conforms with observations that the mature regenerated tails of some lizards may have a higher total energy content than the original ones

Abnormal Regeneration

Lizards with bifid tails, that is, with a regenerate growing from an original tail or joined to another regenerate, are sometimes seen in nature, and the same condition is known in *Sphenodon*. There are many variations in the size and position of the extra tail; the plane of bifurcation may be horizontal, vertical, or oblizue, the first peing the post common. Trifid tails have also been reported, and tails with even more than three branches are conceivable. Such abnormalities may result from incomplete autotomies or ruptures and from other injuries, either to original tails or to previous regenerates.

Similar examples can be produced experimentally by partly breaking the tail or damaging its vertebrae, by wounding the tail or by implanting grafts of ependyma. Ensuing regenerates do not always contain cartilage tubes, and conversely, a single

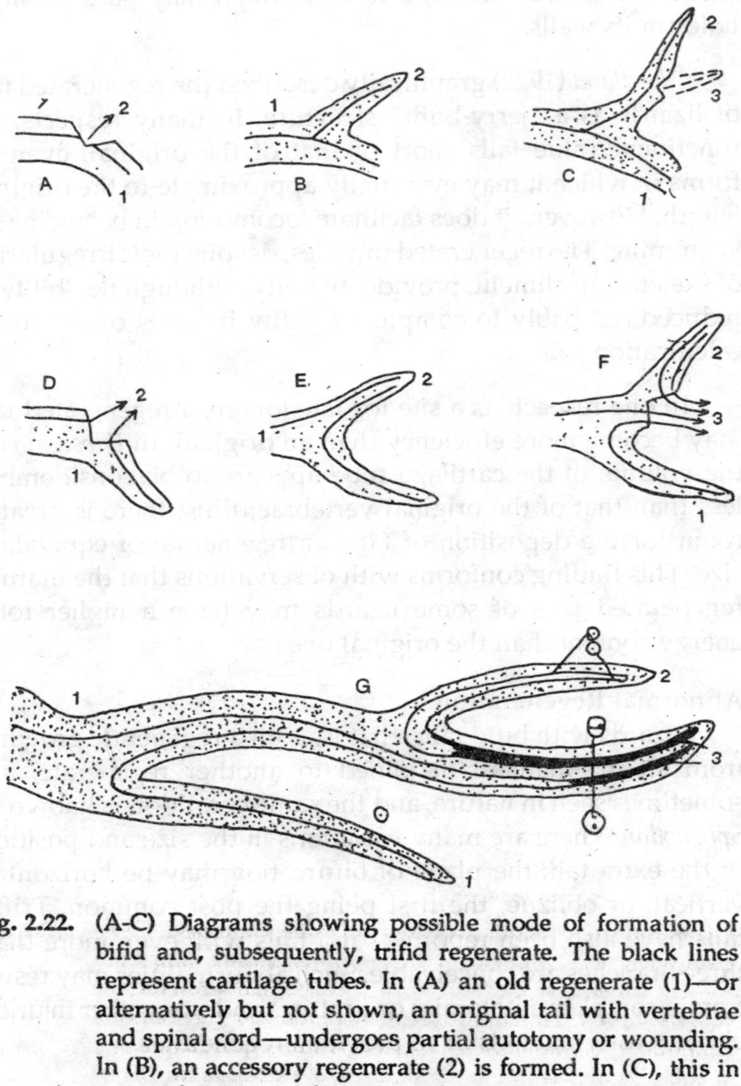

Fig. 2.22. (A-C) Diagrams showing possible mode of formation of bifid and, subsequently, trifid regenerate. The black lines represent cartilage tubes. In (A) an old regenerate (1)—or alternatively but not shown, an original tail with vertebrae and spinal cord—undergoes partial autotomy or wounding. In (B), an accessory regenerate (2) is formed. In (C), this in turn undergoes partial rupture and another accessory regenerate (3) appears. (D-G) Diagrams showing possible mode of formation of trifid tail with *seven* cartilage tubes (black lines), as found by a Graper (1909) in a specimen of *Lacerta agilis*. This could follow partial rupture of an old regenerate (1) in

(D), though in the actual specimen the condition arose from an original tail. The possiobility of growth (shown by arrows) from the proximal ends of the tubes is implied. Some of the cartilage tubes are fused, as shown by the solid black segments. Transverse sections of the cartilage tubes show in (1) a single tuube containing ependyma; in (2), two separate tubes each containing ependyma; in (3) (upper), partly fused tubes each devoid of ependyma, and (lower), cartilage tubes fused into one containing two tubes of ependyma.

tail regenerate may contain more than a single cartilage tube. The three tail regenerates of Graper's specimen of *Lacerta agilis* taken in the wild contained no less than seven cartilage tubes between them, although some had apparently undergone partial fusion. This account is also of interest for its implication that cartilage and ependymal tubes can regenerate after repture from both their proximal and their distal ends.

Woodland (1920) and *Das* (1933) suggested that the vertebrae of an original tail or the cartilage tube of an old regenerate must be damaged if the new regenerate is to contain a cartilage tube; *Tornier* (1897) obtained extra tails containing two such tubes by damaging adjacent vertebrae. However, such injuries are likely to involve damage to the spinal cord and its ependyma, thus evoking regeneration. It is doubtful whether injury to bone or cartilage only would be followed by the formation of new cartilage, because implants of cartilage without ependyma fail to evoke regeneration of any kind. In the light of these findings, the observations of *Brindley* (1898) and *Graper* (1909) that a cartilage tube may (exceptionally) lack an ependymal tube, at least over some part of its length, are difficult to explain. Possibly ependyma is not always necessary for the maintenance of cartilage growth; conceivably, an ependymal tube might undergo degernation within the cartilage tube of a mature regenerate.

Small extra tails, which lack cartilage tubes (and presumably ependymal tubes), like some of those seen in wild specimens, were produced by relatively superficial injuries confined to the muscles and overlying tissue. This suggests that some regeneration can occur in the absence of ependyma, under certain circumstances. Still more superficial injuries, involving the removal of scales from small areas of an original or a regenerated tail, are followed by the development of new, atypical scales and not by the growth of a new regenerate.

Some degree of regeneration at sites outside the normal region of autotomy in lizards can be provoked by various experimental procedures such as cauterization, and deviation of the sciatic nerve or of the cpinal cord to the subcutaneous tissue. Moreover, normal regenerates will sometimes form in *Hemidactylus* after amputation through the pygal region, which does not show autotomous adaptations of the vertebrae and muscles.

In the experiments cited above, the regenerates were produced from the caudal base or from adjacent areas, such as the dorsal pelvic region. Therefore, the results conform with the concept of *"territories of regeneration."* According to this, tail-like (or limb-like) regenerates can only be produced from tissues in or near the normal site of the appendage, that is, within its "territory." The growth of tail blastemata transplanted to the shoulder, as in the experiments of *Weiss*, does not conflict with this concept because the regenerates remain true to their origins. No experiments on reptiles have determined whether regeneration can be provoked from regions such as the midflank, which are distant from all appendages; in amphibians (adults, at least) the results of such experiments are negative.

The tail generally regenerates after amputation through an intervertebral site, that is, through the region of the intervertebral and between fracture planes, at least in geckos and lacertids. The latent period may be slightly prolonged, perhaps because of the greater damage to the tissues. However, this is by no means certain, and *Anolis* shows no significant correlation between regenerative performance and the position of the amputation in relation to the fracture plane.

However, *Jamison* (1964) has reported that after intervertebral amputation of the tails of a number of genera (*Scincella, Eumeces, Gerrhonotus, Sceloporus* and *Anolis*), regeneration was significantly retarded as compared with specimens that had undergone normal autotomy. In view of the above findings, these results seem to require confirmation.

(a) Rates of Regeneration

The nonautotomous lizards such as *Crotaphytus* and *Conolophus*, appear to have little or no power of regeneration. If the tail is broken by force or amputated, it heals over without elongation, as in a snake. However, some regeneration has been observed in the completely non-autotomous iguanids *Brachylophus* and *Amblyrhynchus*. Regenerates of up to about 3 cm have been observed in the latter, but the caudal crest is not reproduced.

In some lizards the regenerates is every limited. Regenerates up to 75 mm long and containing cartilage tubes have been observed in some species of *Agama*; regenerates of 93 to 112 mm have been observed in the big Australian agamid *Physignathus lesueuri*. One captive *Physignathus* grew a regenerate of 93 mm in about two years and nine months. An interesting specimen of *Calotes cristatellus* had a trifid tail, with two of the three branches (the longest being 47 mm) and most of the third being regenerates and lacking vertebrae

It has been suggested that in non-autotomous forms extensive regeneration with cartilage tube formation only occurs after tail loss involving dammage to an actual vertebra, and not as the consequence of the more common intervertebral separation. This is an interesting point, as the spinal cord and its ependyma would be reptured in either type of tail-break. However, loss or "ablation" of part of a vertebra occurs after intervertebral tail-break in a way comparable with that described in lizards with fracture planes. It seems possible that the condition of such regenerating vertebrae was interpreted as due to the initial injury, rather than to the osteoclastic activity which may have occurred later.

It is uncertain whether any autotomous lizards are completely unable to regenerate and, conversely, whether, as in the gecko *Lygodactylus klugei*, ease of autotomy is always correlated with speed and completeness of regeneration; the tail of the latter species with its scansorial pad has both actively and passively functional characteristics. Many geckos, lacertids, and skinks certainly shed their tails freely and are good regenerators. *Anguis fragilis*, which, sometimes at least, is a rather reluctant autotomizer, grows only a small regenerate and has one of the slowest regeneration rates known among lizards. The fastest regeneration was shown by a specimen kept around 27°C, which produced a regenerate of 5 mm after 14 weeks. The fastest rate achieved during this period was about 0.07 mm a day. Another captive specimen, kept indoors and not hibernating, grew a regenerate 8 mm long in three years. Regenerates of 1.5 cm or more probably take several years to produce in the wild, allowing for periods of zero growth during hibernation.

Although the histological appearance of regeneration in *Anguis* are essentially similar to those in *Lacerta*, one has the impression that differentiation of tissues predominates over proliferation; although the volume of blastema remains small, the blastema relatively quickly produces new tissues, such as muscle and cartilage. The anguis *Gerrhonotus multicarinatus* appears to give a much better regenerative performance than does *Anguis*, but regenerates are smaller than original tails. The long-bodied Indian sand skink, *Ophiomorus streeti*, may have an even lower regeneration rate than *Anguis*. Regenerates are some 2 to 3 mm long after three months, 5 to 6 mm long after six months, and 8 to 10 mm after a year. The average growth rate for a regenerate 2.5 mm long after 12 weeks (minus three weeks latent period as estimated for *Anguis*), would thus be about 0.04 mm a day.

At the other end of the scale, the fastest regeneration rates recorded for lizards appear to be those indicated by the graphs of *Baranowitz* et al. These rates were bassed upon a mathematical analysis of regeneration in *Lacerta lepida*, a big lizard a maximum total length of around 80 cm, the ten specimens used by *Baranowitz et al.* averaged about 15 cm in snout-vent length and about 26 cm in length of the intact tail. Their Fig. 2.2A shows that the tail of one specimen elongated from about 38 to 76 mm (38 mm) between the 25th and 33rd days after amputation with a razor of 70% of the intact original tail. The regeneration rate during this eight-day period (which was probably the period of maximum growth in this individual) was therefore about 4.7 mm a day; slightly highér figures were attained, at least translently, by certain other specimens. A mean figure for five individuals was about 3.2 mm a day. Although the higher figures may represent unusual performances, which could not be matched in the wild, they do have genuine biological significance as indicating the best achievement of which a species is known to be capable.

These rates for *Lacerta lepida* are significantly higher than any previously known. They were obtained from lizards probably kept under optimal conditions at around 34°C, with 12 hours of light and 12 hours of darkness in a rigorously controlled experiment. These rates may be compared with the rate for the elongation of the antlers of deer, which at more than 1 cm a day, probably represents the fastest rate of postnatal growth in the animal kingdom. *Lacerta dugesii* shows a maximum regeneration rate during the fifth week after autotomy of 2.6 mm a day. However, the average growth rate of regenerates of ten individuals was only about 1.3 mm a day during the same period. Integumentary markers show that the terminal region of the regenerate grew faster than the proximal region, and also faster than the regenerate as a whole.

The figures cited in the table represent a *"mixed bag"* of observations, mostly made in laboratories under different circumstances, and do not indicate a *"normal value"* for any species. Rates in the wild would probably vary immensely. Indeed, it is doubtful if such a concept as a norm is valid at all. Very considerable variation in the extent and rate of tail

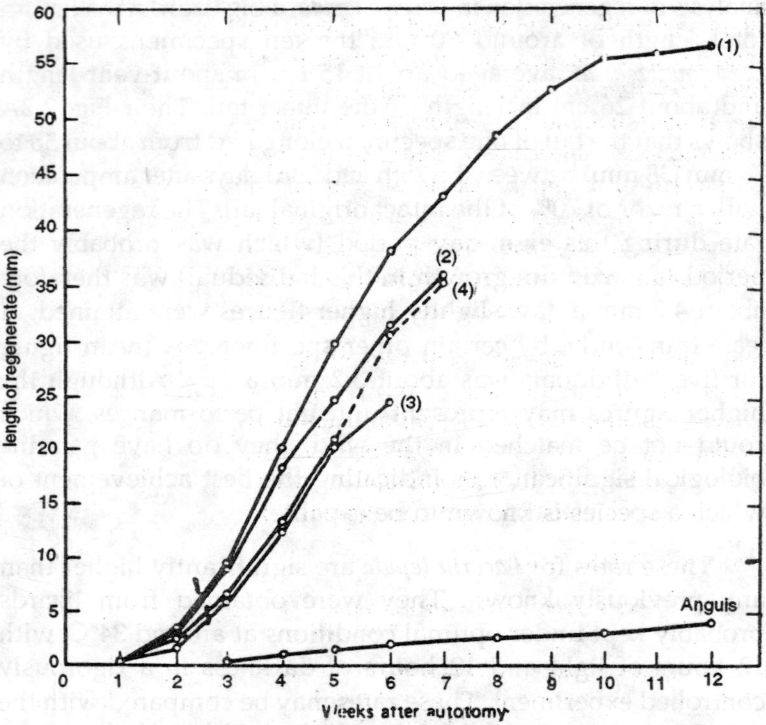

Fig. 2.23. Graph showing growth in length of tail regenerates of *Podarcis dugesii*, and of a single individual of *Anguis fragilis*. (1) Average growth in ten lizards (*P. dugesii*) after first autotomy of complete original tail. Growth during the seven to 12-week period was only observed in six specimens. (2) Average growth in 14 lizards after second autotomy, proximal to a mature regenerate. (3) Average growth in eight lizards after third autotomy, induced six weeks after the second autotomy, the growth of which is included in (2). (4) Average growth in 19 lizards after the third or fourth autotomies; six of these lizards were observed only for five weeks; and the broken line shows the average further growth for the remaining 13. *Anguis*, fastest growth observed in specimen kept at 27°C.

replacement can occur even among individuals kept under a particular experimental regime. Such variability is found among groups of different ages and sexes, and among groups maintained under different environmental conditions. Moreover,

TABLE 2.1

Rates of Caudal Regeneration for Various Lizards

Species	*Daily Regeneration Rates, MM* [a]	*Total Length of Lizard, cm (Ref. No.)*	*Author*
Lacerta agilis	1.36 (S)	19 (5)	Hooker (1912)
L. lepida	4.7 (M) and 3.2 (Av)	40 + (1)	Baranowitz et al. (1979)
Podarcis dugussi	2.6 (S) (M) and 1.3 (Av)	20 (3)	Bryant and Bellairs (1967a)
Mabuya heathi	0.75-1.6	?	Viu (1981)
Eumeces fasciatus	0.6 (S)	22 (4)	Fitch (1954)
E. skiltonianus	0.52 (Av)	22 (4)	Vitt et al. (1977)
E. gilberti	0.57 (Av)	30 (4)	Vitt et al. (1977)
Ophiomorus streeti	0.04 (AV) first 12 wks Aa	Up to 20 (3)	Rathor (1971)
Xantusia vigilis	0.18 (Av)	10 (4)	Zweifel and Lowe (1966)
Coleonyx brevis	0.38 (S)	11 (4)	Mulaik (1935)
C. variegatus	0.7 (Aw)	14 (4)	Vitt et al. (1977)
Hemidactylus garnoti	0.7 (S)	13 (2)	Cagle (1946)
H. turcicus (juvenile)	0.3 (Av)	?	Werner (1967a)
Sphaerodactylus argus	0.47 9Av)	4 (3)	Hughes and New (1959)
Anolis carolinensis	1.5 (Av) (M)	18 (4)	Maderson and Licht (1968)
Anguis fragilis	0.07 (S) (M)	35-40 (5)	Bryant and Bellaris (1967a)
Gerrhonotus multicarinatus	0.82 (Av)	40 (4)	Vitt et al. (1977)

[a] Aa, After autonomy, Av, average of a sample; M, maximu,; S, single specimen.

the performance of any lizard may differ after first and second amputations, so that no individual can be classified as an inherently *"good"* or *"bad"* regenerator. Howver, it would appear that the whole course of a single regenerative performance by an individual lizard is, to some extent, predictable, given adequate data obtained from earlier stages of regeneration by the application of the Gompertz Curve.

(b) Amount of Tail Lost and Size of Lizard

In *Anolis* and lacertids, the greater the amount of tail removed, the faster was the regeneration rate. The length of the regenerate decreases as the breakages site shifts toward the caudal tip. The amount of tail eventually replaced differs little; this is also suggested by observations on skinks (*Eumeces fasciatus*) in the wild. The correlation between a relatively short stump and a relatively long regenerate has also been reported for geckos of unstated species. However, in these experiments, the overall length of the stump plus regenerate was often reduced in those animals that have the shortest stumps.

The combined lengths of stump and regenerate may represent a more biological significant figure for records of experiments of this type than the amount of original tail replaced the former can be compared directly with the length of the original intact tail, which is likely to relate to the optimum functional length for the individual lizard concerned. An alternative measurement is the percentage of the volume of the original tail, which is regenerated. The length of a regenerate also might be expressed as a percentage of the snoutvent length of the animal.

The size of a lizard may have some bearing on the amount of tissue lost during au.otomy, and perhaps on the rate of regeneration. If a large lizard and a small one with similar body/tail proportions each lose the same linear proportion of their tail, the former will have lost a greater absolute length (and a greater volume) of tail than will the latter. The larger lizard will need to regenerate relatively faster than the smaller one in order to restore the optimal functional tail length in the same period. This trend is perhaps indicated in the high regeneration rates reported for the big *Lacerta lepida* and the medium-sized *Lacerta dugesii.* However, it might be wrong to say that these forms show a better regenerative capacity than a tiny *Sphaerodactylus* gecko just because their new tails elongate faster.

Baranowitz et al. (1979) have suggested that the diameter of the tail at the site of injury, which is probably related to the number of cells involved, may have some bearing on the rate of regeneration. This factor deserves further consideration, particularly as this area increases as the square rather than the cube of linear dimension.

(c) Number of Amputations

The average rate of regeneration in a group of lizards tends to drop for the second or subsequent autotomies or amputations through the original stump although the overall quantity of tail lost necessarily increases. However, individual exceptions occur. Whereas 16 to 44% of the length of the original tail of *Anolis* removed has been replaced seven weeks after the first amputation, only 7 to 21% of the total original (i.e., intact) tail is replaced after the second amputation. Consequently, successive losses of the tail may reduce (at least for a time) the ability of the animal to grow a new tail, which approches the length of the oiginal intact appendage.

(d) Photoperiodic Effect

Temperature appears to have a clear-cut effect upon regeneration. During the phase of rapid growth, the average regeneration rate of *Anolis carolinensis* kept at 32°C is 1.5 mm a day. At 21°C, a temperature well below the thermal preference of the species, the latent period increases from about eight to 30 days, and the corresponding regeneration rate decreases to about 0.15 mm a day. Moreover, the average length of the regenerates after six months is very much less in the lizards kept at 21°C than in those kept at 32°C after six weeks. Regeneration rates in *Hemidactylus* and *Mabuya* are considerably higher in the hot season than in the cold one. In *Mabuya*, regenerates reach their full size after 50 days in the hot season, whereas similar growth takes 90 days in the cold season.

These findings suggest that the normal thermal preferences of different species and their rates of regeneration may be correlated. Thus, fossorial, cryptic, and nocturnal forms tend to regenerate more slowly than thermophilic forms. It remains to be seen whether slowness of regeneration is necessarily correlated with inability to replace more than a small proportion of the lost appendage as it seems to be in *Auguis*.

A surprising relationship exists between regenerative performance, that is rate and length of tail regenerated in seven weeks, and the photoperiod. Adult *Anolis carolinensis* kept at 32°C with eight hours of light out of 24 hours show the best performance, those kept at 32°C with 16 hours of light give the worst performance, and those kept under fluctuating conditions approximating to nature are intermediate.

However, rather different results have been obtained by *Turner* and *Tipton* (1972). In the same species, appetite, growth of the whole body, and the rate of tail regeneration were stimulated by exposing the lizards to 18 hours of light per day at a temperature of 32°C. Thus, blastema formation starts a day or two earlier than in animals kept with a six hour light priod at the same temperature, and the growth phase tapers off much earlier. However, after four to five weeks, the tail length does not differ significantly between the two groups of experimental animals.

(e) Age, Sex and Skin-Shedding

Young *Sphenodon* regenerate their tails more rapidly than do older individuals and regenerative capacity may be greater in young than in old lizards. However, if part of the tail is lost before maturity, it may be difficult between the later phases of regeneration and the normal growth of the animal. Moreover, the size factor may be significant in intraspecific as well as interspecific comparisons. Thus, graphs for adult *Anolis* seem to indicate that juveniles have lower regeneration rates than do adults, but this may reflect only their smaller size.

There is evidence for sexual differences in regeneration rate. Adult males of *Coleonyx variegatus, Eumeces skiltonianus,* and *Hemidactylus flaviviridis* regenerate more rapidly than do females. *Mabuya heathi and Lygodactylus klugei* show the opposite trend. The figures cited in the previous table are irrespective of sex. Pond found that the relative length of the mature regenerate in Dipsosaurus dorsalis is greater in males than in females. It has been suggested that the higher regeneration rates in the males of some species are due to the fact that females, while building up eggs in their ovaries, can spare less energy for regeneration. The differential may be correlated with the effects of male sex hormones.

(f) Nutrition and Drugs

Regeneration in *Anolis carolinensis* is unaffected by nutritional state, at least as long as the body weight remains constant.

Supplementation of the diet of regeneration *Lacerta viridis* accelerates the formation of the cartilage tube and in some cases its calcification. It also changes the histology of the junction between the cranial end of the tube and the vertebral fragment remaining in the stump of the original tail.

Oral administration of thalidomide to *Hemidactylus flaviviridis* before amputation of the tail does not obviously affect regeneration. Nitrogen mustard and certain antimetabolites and other substances produce inhibitory or delaying effects upon regeneration.

THE EFFECT OF HORMONES

Hormones also influence tail regeneration in lizards. Removal of the pars distalis of the pituitary (adenohypophysis) several days before or at the time of amputation of the tail in *Anolis carolinensis* delays the formation of the blastema; after 21 days, the length of the regenerate is 15% that of controls. Inhibition particularly affects the differentiation of mukscle. If the parts distails is removed after the regenerate has begun to elongate, the subsequent rate of regeneration soon becomes much reduced. The effects are proportional to the amount of pars distalis tissue removed. However, administration of combined adenohypophysical hormones restores regeneration to normal levels. The effect is probably produced by growth hormone, thyrotropic hormone, and probably prolactin. Subcutaneous injection of ovine prolactin leads to an increase in the weight, although not in the length or the water content of the regenerates. There is evidence that this harmone may have complex effects on growth in a variety of vertebrates, and may also exert an indirect effect through the thyroid. After hypophysectomy, *Hemidactylus flaviviridis* shows similar responses, as well as a reduction in the level of circulating posphatase, an enzyme associated with phagocytic activity, such as that seen during the early, wound-healing phases of regeneration.

Administration of thyroxin to *Anolis* accelerates blastema formation and reduces the latent period; it also causes active growth to taper off two or three days earlier than usual. Conversely, the latent period is protracted by two weeks after thyroidectomy, and the elongation of the regenerate is much inhibited.

Adenohypophyseal and thyroid functions appear to be associated; if thyroxin is administered to regenerating, hypophysectomized *Anolis*, The regenerates differentiate and grow normally. Apparently, the pituitary exerts its influence on regeneration, partly at least, through the thyroid. It has been suggested that thyroxin, to some extent, controls the outgrowth of the ependymal sac and tube into the young blastema; thus, variation in the timing of this event could be responsible for the variability of regenerate phenomena.

A rather different interpretation of the significance of thyroxin has been advanced by *Magon* (1975a), who studied regenerating *Hemidactylus flaviviridis* at different seasons. The secretory activity of the gland has been assessed on the basis of the height of the follicular cells; apparently, an optimum rather than a particularly high concentration of thyroxin is necessary for regeneration. Normal *Hemidactylus* has a more active thyroid during the winter than during the (warmer) summer. Amputation of the tail in winter is followed by a reduction in thyroid activity, whereas amputation during the summer enhances such activity; in both cases the optimum concentration of thyroxin is achieved.

Later, in studies on *Mabuiya striata*, *Magon* (1977a) showed that thyroid activity fluctuates during different phases of regeneration. It decreases during wound healing, increases during blastema formation, and declines again during the phases of differentiation and growth. Normal lizards tend to have greater thyroid activity than regenerating ones, at least during some phases of the process.

A functional relationship between the hormones of the thyroid and the male gonads during tail regeneration has been demonstrated in *Hemidactylus flaviviridis*. The delay in regeneration, which normally follows thyriodectomy, is considerably more marked in males than in females, and the male gonads show degenerative changes. In males, the delay is due both to the deficiency of sex hormone and to the effect of thyriodectomy on general metabolism, whereas only the latter factor is involved in females. Administration of testosterone to geckos with the thyroid intact enhances the rate of regeneration in both sexes, but especially in females, possibly because the males wound have produced the optimal concentration of androgens under normal circumstances. Injection to testosterone also increases the regeneration rate of previously thyroidectomized lizards

Both the adendhypophysis and the neurohypophysis are

in part under hypothalamic control, and there is evidence that neurohypophysial hormones may also be involved in regeneration. Regenerating *Hemidactylus flaviviridis* shows histological changes, characterized by degranulation, which occurs in the cells of the supraoptic and paraventricular centers of the hypothalamus. They interpreted these changes as indicating that neurosecretory activity in these centers was much increased during the phases of wound healing, blastema formation, and differentiation; the activity falls off some 30 days after amputation. This activity was probably associated with the conservation of water and electrolytes. In these experiments, many of the lizards were kept under different conditions of illumination. Neurosecretion was particularly marked in lizards exposed to constant red light, whereas in those exposed to other colours of light and to constant darkness, such activity was less marked than in the controls.

Regeneration in Embryos

Regenerative capacity in many animals tends to be greater in youth than maturity. This is true in amphibians, and possibly for reptiles as well, although we lack clear-cut evidence for the latter group. The question as to whether regenerative phenomena occur in reptilian embryos is therefore of interest. *Marucci* (1915a) briefly reported that the tails of embryos of *Podarcis muralis* fail to regenerate after amputation, but he did not give details of his experimental technique. *Moffat* and *Bellairs* (1964) amputated the tails of many embryos of *Lacerta vivipara*, which were explanted from the mother and maintained in a simple form of culture. The amputated tails healed very completely without regeneration, unless the operation was performed at a very late stage of embryonic life, quite soon before the embryos hatched from their explanted eggs.

Bryant and *Bellairs* (1967b, 1970) showed that amputation in earlier stages of embryonic life was almost invariably followed by the development of constrictions around the tail, which became actively extruded through the fused amnio-allantoic membrane after the operation. The constriction, derived from

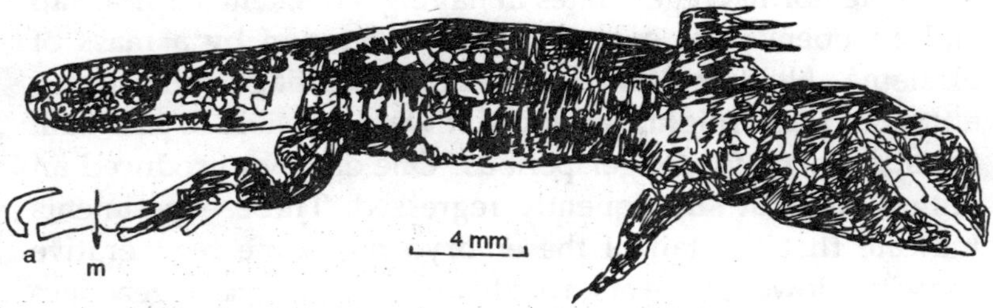

Fig. 2.24. *Lacerta vivipara*. The baby lizard had just hatched in culture. The tail and right hind limb had been amputated 19 days previously, and there are no signs of regeneration. However, the appearance may be due to the development of constriction bands causing autoamputation of the extremities proximal to the experimental lesions, as shown in inset. Inset: a, site of experimental amputation; m, constricting amnio-allantoic membrane; ts, tail stump; stippled region has sloughed off.

the contractile membrane, produced the effect of amputation by ligature, somewhat proximal to the site of the experimental amputation. Auto-amputation by this means produces a different type of lesion from that caused by a blade; there is no raw, exposed surface, and the original epidermis remains unbroken throughout the process, so that no apical cap is formed. The distal part of the tail undergoes avascular necrosis and falls off into the allantoic cavity. Such amputation is perhaps comparable to that produced in adult lizards by suturing the scales just proximal to the site of amputation. The epidermis, although severed, had its edges brought into apposition, and no regeneration occurs. A probably similar type of amputation of limbs and digits has been observed a human fetuses and newborn, presumably after rupture of the amnion by fetal movements or by some other kind of intrauterine trauma. The fact that very late lizard embryos regenerated in the normal fashion can probably be ascribed to the greater resistance of the keratinized scales and differentiated tissues to the action of the ligature. These procedures almost always result in small curved outgrowths from the caudal stump. These outgrowths resemble normal regenerates in having a thickened apical cap and an ependymal sac or tube surrounded by a mass of blastema. Neither new muscle nor a cartilage tube forms, although these might develop if it were possible to maintain the specimens for longer periods. One embryo produced an outgrowth that subsequently regressed. These experiments indicate that the tail of the embryo has some regenerative capacity. However, the outgrowths are unlike normal regenerates in that they do not arise from the entire stump area but only from a small region of the stump opposite the spinal cord; it is unlikely that they could ever replace the tissue that has been removed. The question as to whether embryos can regenerate tails in anything like the normal fashion remains unanswered.

Regeneration of Limbs in Lizards

Regeneration in the repitilian limbs has also been studied. Simple amputation of the limbs of young and adult lizards, at least of lacertids and of *Anolis,* is normally followed by healing and not by regeneration. The process of healing of the fore limbs of *Anolis,* amputated between wrist and elbow, has been studied by *Barber* (1944), who compared it with tail regeneration. Healing of the hind limbs after amputation through the thigh has been described in *Lacerta vivipara* and *L. dugesii* and in *Anolis carolinensis.*

The initial stages of healing are, in many ways, similar to those which follow amputation of the tail. A scab is formed and new epidermis grows across the wound. Like the apical cap of the regenerating tail, it becomes many layers thicker than normal. The tip of the cut bone or bones may protrude into the scab, and small portions of it may be cut off by osteoclasts and sloughed, like the fragment of autotomized vertebra in the tail. The cut muscles dedifferentiate extensively (unlike those in the tail after normal autotomy) and may contribute to the mass of blastema-like cells that develop around the damaged bone. However, they soon acquire a more fibroblast-like appearance than those seen in the blastema of amphibian limb regenerates. Nerve fibers regenerate into the blastema, but they are considerably less numerous than those in the regenerating limb of the newt. Extensive formation of cartilage, probably derived from the blastema cells, occurs around the injured bone by encasing its tip and plugging its cut end which may undergo partial osteoclastic resorption. This cartilage may eventually become partially ossified. The profuse formation of cartilage is apparently a characteristic sequel to bone injury in amphibians and reptiles. Later changes involve the conservation of much of the blastema into connective tissue and the formation of new scales over the former wound surface; in *Lacerta*, scale formation begins about three months after the operation.

Amputation of the hind limbs of cultured embryos of *Lacerta vivipara* is also followed by healing with extensive cartilage formation. Similar healing without regeneration was reported by *Marcucci* (1915a) in embryos of the oviparous *Podarcis muralis*. However, it is possible that the results of these experiments, like those on the embryonic tail, were falsified by autoamputation due to the development of constricting bands from the ruptured embryonic membranes. The results of amputating the digits in lacertids are very similar to those of limb amputation. The failure to regenerate the lost tissue or any semblance of the claw is of some practical interest in view of the use of toe clipping as a means of individual identification in ecological studies.

Despite these findings, it is well-known that the limbs of lizards, at least of certain species, are not entirely devoid of regenerative capacity. Indeed, this also appears to be true of birds and mammals, at least in embryonic and juvenile life stages. The finger tips of human children can even regenerate after trauma if the wound surface is not sutured, but covered by a nonstick dressing.

In wild species the lizards show partial regeneration of the forelimbs or hindlimbs. These regenerates do not look like normal limbs. They are usually curved or tapering, almost tail-like structures covered with irregular scales, they show no evidence of digit formation. Such appendages fall for short of the virtually perfect limbs regenerated by newts. However, they may be of some functional value as they increase the length of the stump and may serve as levers or props. In the two specimens of *Lacerta vivipara* described by *Poyntz* and *Bellairs* (1965), the original injury, perhaps caused by the bite of a predator, had apparently occurred just above the knee. The regenerated bone extended from the lower end of the femur and was somewhat expanded; its tip was attached to an irregular mass of cartilage. The sectioned regenerate showed comparatively little formation of new muscle. *Brindley* (1898) described a specimen of *Podarcis muralis*, which had a tail-like structure arising from the dorsal aspect of an otherwise normal hind limb, and he found evidence of a previous injury near the site.

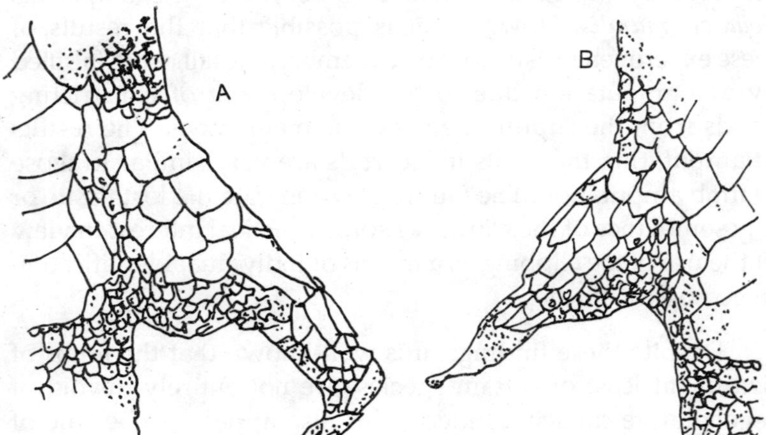

Fig. 2.25. (A,B). Natural limb regenerates of *Lacerta vivipara* in ventral view.

Similar types of partial regeneration, which never properly produce the morphology of the lost tissues, have been provoked in lizards by a variety of experimental procedures. Thus, the

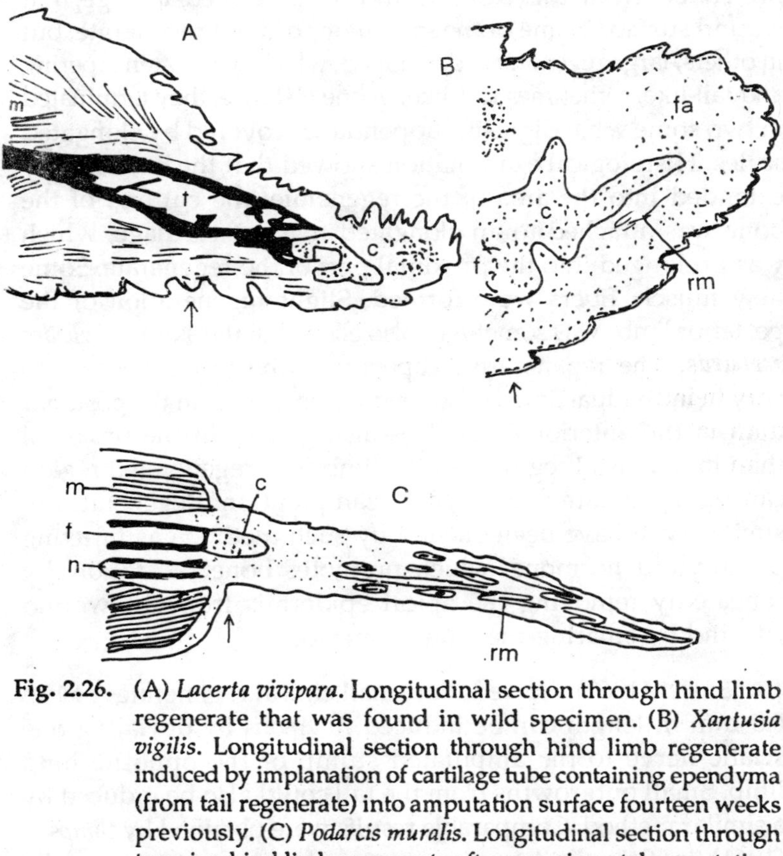

Fig. 2.26. (A) *Lacerta vivipara*. Longitudinal section through hind limb regenerate that was found in wild specimen. (B) *Xantusia vigilis*. Longitudinal section through hind limb regenerate induced by implanation of cartilage tube containing ependyma (from tail regenerate) into amputation surface fourteen weeks previously. (C) *Podarcis muralis*. Longitudinal section through tapering hind limb regenerate after experimental amputation. Arrows in (A), (B) and (C) show approximate levels of amputation. c, Cartilage of regenerate; f, femur; fa, fat and connective tissue; m, original muscle; n, nerve; rm, regenerated muscle.

formation of small tail-like structures was induced by the infliction of extensive longitudinal wounds in the thigh of *Podarcis muralis*. The front limbs and hind limbs in this species were amputated by *Marcucci* (1930 and earlier) by means of transverse and oblique cuts at different levels. Shortly after operation, he removed the tip of the humerus or femur, which protruded from the wound, and so produced a large, flat wound surface. Some of the specimens did not regenerate, but in others, large regenerates developed, which were often tapering and tail-like, sometimes rod-like; in one instance, they terminated in two some what digit-like appendages covered by elongated scales. Histological examination showed that the bone hardly extended into the area of the regenerate; the cut end of the bone was attached to an elongated mass of cartilage, which was prolonged into the proximal part of the regenerate. Some new muscle fibers were formed. Slight regeneration of the posterior limbs was sometimes also elicited in the skink, *Chalcides ocellatus.* The regenerative capacity of the limbs appears to vary in individual lizards and tends to be greater in the posterior than in the anterior limbs. It is also greater in the proximal than in the distal segment of the limb. The regenerated region can be regenerated again after reamputation. Regenerates of similar type have been elicited by such methods as injecting parathyroid hormone, which promotes bone resorption, by repeatedly removing the apical epidermis from the wound site and by enlaring the wound surface.

Singer (1961) showed that small cinical regenerates of 1 to 6.5 mm in length can be induced in *Anolis* by deviating one sciatic nerve to the amputated stump of the opposite hind limb. Small outgrowths from the tail could also be induced by a similar method. Comparable results were obtained by *Simpson* (1961) in *Scincella lateralis* as a consequence of cutting the spinal cord and deviating its cut end into the stump of the amputated thigh. Similar observations have been made by *Kudokotsev* (1961). The histological structure of the small regenerates obtained by these methods is variable, but muscle, cartilage, and bone are often present.

It is uncertain whether limb regeneration in lizards can be elicited by implants of *disconnected* nervous tissue. However, this has apparently been achieved in the pouch-young of the opossum (*Didelphis marsupialis*). The formation of seemingly recognizable, foot-like structures has been induced by implanting grafts of cerebal hemisphere into the subsequently amputated limbs of opossums one or two days after birth. Fore limb regeneration has also been evoked in chick embryos by implants of neural tube.

It would appear that failure of lizards limbs to regenerate, except incompletely and under unusual circumstances, reflects the fact that these limbs, like those of the frog *Rana* and of the mammals, contain too little nervous tissue. This is, of course, suggested by some of the experiments described above, but *Singer* and his collaborators have provided a quantitative basis for the hypothesis. They have shown that a critical factor in limb regeneration is the actual amount of axoplasm present in a given area of tissue. This factor is usually expressed by the number of nerve fibers per unit area of 100 μm^2 in a cross section of the limb, which corresponds with the area of the wound surface. It is not necessarily related to the size of the limb (as in different species), but is apparently quite constant throughout the thickness and length of a limb, including individual digits of any given species. The diameter of the individual fibers may also be significant, because a relatively few large neurons (as in the clawed toad, *Xenopus laevis*) may have an equivalent trophic effect to that of a larger number of small ones (as in the urodele *Notophthalmus viridescens*). In fact, both species regenerate, *Xenopus* only partly. In the forms discussed below, however, the diamter of the neurons is assumed to be roughly similar (although the actual diameter of the neurones of *Anolis* is not indicated in the papers cited). Thus, the number of fibers in each species gives a measure of the quantity of axoplasm.

Adult *Notophthalmus* have a mean number of 24.5 nerve fibers per 100 μm^2 of forelimb tissue (excluding bone). Reduction of their number by experimental denervation below a threshold value of 10.8 will keep regeneration from occurring. The forelimbs of *Anolis carolinensis* have about 12.3 fibers per 100 μm^2; the corresponding figure for the mouse (*Mus*) is four, about the same as that for the tail of *Anolis*. Thus, there seems to have been an actual reduction during evolution of the concentration of nerve fibers in the limb, although the central nervous mechanisms associated with limb function are much more complex in mammals than in urodele amphibians.

By urodele standards, the innervation of the forelimb of *Anolis* is slightly above the threshold for potential regeneration. In spite of this, the limb of the lizard normally fails to regenerate, suggesting that additional factors may be involved and perhaps that lizard tissue may be less responsive than newt tissue to the trophic action of nerve fibers. It is possible that this responsiveness may be increased by the techniques used by *Marcucci* (1930) and others; also perhaps, a lacerating bite from a predator in the wild might have a comparable effect of enlarging the wound surface. This wound partly explain why a regenerative response may occasionally occur in the absence of hyperinnervation.

Limited limb regeneration has been induced in *Scincella lateralis* by means of ependymal implants and the regenerates are "as good if not better than the regenerates induced by nerve augmentation." More extensive experiments using ependyma were made on *Xantusia vigils*, which was kept at 32°C on a daily cycle of eight hours light and 16 hours of darkness, a regime shown to be optimal for tail regeneration in *Anolis*. Nine days after amputation of the left hindlimb through the midthigh, when the wound scab had fallen off, a 3 mm segment of cartilage tube containing ependyma was implanted a few millimeters proximal to the wound surface; the implants were taken from the lizard's own mature tail regenerates. In nine out of 11 specimens treated in this way, cones of blastema were visible within ten days after the operation. The regenerates grew comparatively fast during the first, second, and third weeks and, eventually, attained lengths of 0.5 to 4.23 mm after six weeks. In specimens maintained beyond this time, no further growth occurred. In most cases, the regenerate was broad and rather flat, but the two longest regenerates were think and tail-like. No digits were formed. Virtually no regeneration occurred in control groups, including specimens into which cartilage tubes without ependyma were inserted.

A histological study showed that the wound surface of these regenerates was covered by a thickened apical cap about two weeks after amputation; an extensive region of apparently dedifferentiated cells was present beneath the epidermis. By three to four weeks, cartilage, connective tissue, some fat, and a little muscle had appeared. After five or six weeks, the regenerates seemed mature, with differentiated scales and dermis.

The condition of the implanted tissue is of interest. In most cases, the ependymal tube had grown distally out of its cartilaginous investment, and its tip was dilated to form a vesicle, as in early tail regenerates. However, after about four weeks, the ependymal tube degenerated it did not induce the formation of a new cartilage tube, as it did in *Simpson's* (1964) experiments on the tail. The new cartilage of the regenerate, which appeared to have arisen (like the other new tissues) from the blastema was irregular in shape and attached either to the implanted cartilage tube to the cut end of the femur, or to both. The small quantity of new muscles present in these regenerates was perhaps related to the paucity of ingrowing nerve fibers; a similar sparsity of muscle was noted in a limb regenerate of *Lacerta* grown in the wild and in the hyperinnervated limb regenerates of *Anolis*.

Therefore, it appears that the regenerative capacity of the saurian limb is only vestigial, and falls short of that of the tail, which is itself imperfect. There are apparently no reports of the formation of a new limb joint, of recognizable bones of the lower leg, forearm or extremities, or of an integrated muscular system. Even toes do not appear to regenerate.

Regeneration of Tail and Jaws in Crocodilia

Caudal regeneration has been reported for *Caiman crocodilus* and *Melanosuchus niger*. The skeleton of the regenerates consisted of an elongated rod of calcified cartilage (21.5 cm long) in *Melanosuchus*. The scutes covering the regenerates were abnormal. Such regeneration does not seem to have been noted in other crocodilians, although injuries to the tail often occur in the wild.

Fig. 2.27. *Melanosuchus niger*. Last two vertebrae (v) and 21.5 cm tail regenerate (r).

A substantial part of the maxilla was regenerated in a captive *Crocodylus palustris*, which was bitten by a pool-mate. A Piece of bone and adjacent tissue measuring some 70 x 40 mm (20 mm thick), together with the third maxillary tooth, was lost, and other tooth sockets were injured. The damage was almost repaired after 45 months, and the missing tooth was replaced by a tooth of greater size. In this, as in other instances of limited regeneration (or extensive healing) in captive crocodilians mentioned by Brazaitis, the animals received medication.

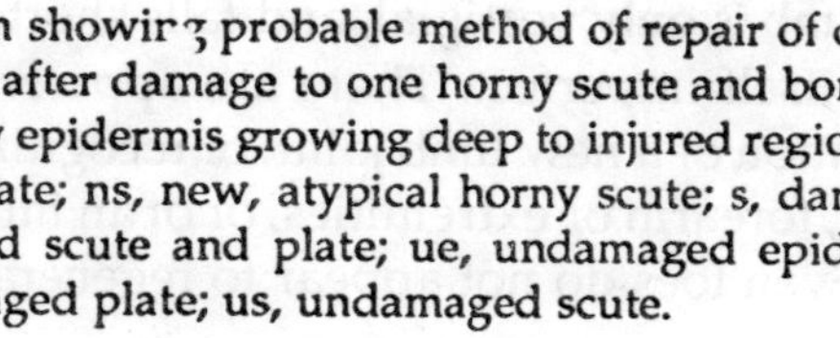

Fig. 2.28. Diagram showing probable method of repair of carapace of tortoise after damage to one horny scute and bony plate(s). ne, New epidermis growing deep to injured region; np, new bony plate; ns, new, atypical horny scute; s, damaged and sloughed scute and plate; ue, undamaged epidermis; up, undamaged plate; us, undamaged scute.

3

Development

Reptiles are oviparous i.e., lay large *polylecithal* eggs. containing all of the nutrients needed to produce viable offspring. These eggs usually are deposited in subterranean nests or in other sheltered sites, and embryonic development then proceeds under physical conditions dictated largely by the location of the site, the type of nest, and prevailing weather.

Many species of turtiles, including all cheloniids, chelydrids, and dermochelyids and most pelomedusids and emydids, produce eggs with flexible shells. However, morphology of flexible eggshells varies considerably among species. Eggs of sea turtiles of the family Cheloniidae have a shell comprised of a relatively thick, fibrous membrane overlain by a calcareous layer formed of numerous small; often poorly defined, shell units. Eggshells of chelydrids and emydids are similar, except that shell units are larger and more clearly defined and the shell membrane accounts for a smaller proportion of the total thickness of the eggshell. Large gaps between adjacent shell units (or between groups of shell units) provide for the exchange or respiratory gases and water between embryos and the nest environment. The crystalline material in the eggshell usually is aragonite, but traces of calcite may form of the nucleation sites for shell units Calcite may also comprise a portion of the calcareous layer of eggshells produced by animals reared in captivity which raises the possibility that characteristics of eggshells may be modified by the diet on which females are maintained. Other species of turtles, including carettochelids, chelids, dermatemyids, kinosternids, testudinids, trionychids, and some pelomedusids and emydids, lay eggs having rigid shells. The shell of such eggs is comprised of a relatively thin shell membrane (or two) overlain by a calcareous layer.. The calcareous layer is formed of crystals of aragonite arranged into discrete shell units that about tightly, thereby causing the layer to be rigid and non-compliant. The crystalline layer is interrupted occasionally by distinct pores that provide for an exchange of respiratory gases and water between the contained embryo and the nest environment. These pores are more concentrated at the equator of elliptical eggs than at the poles.

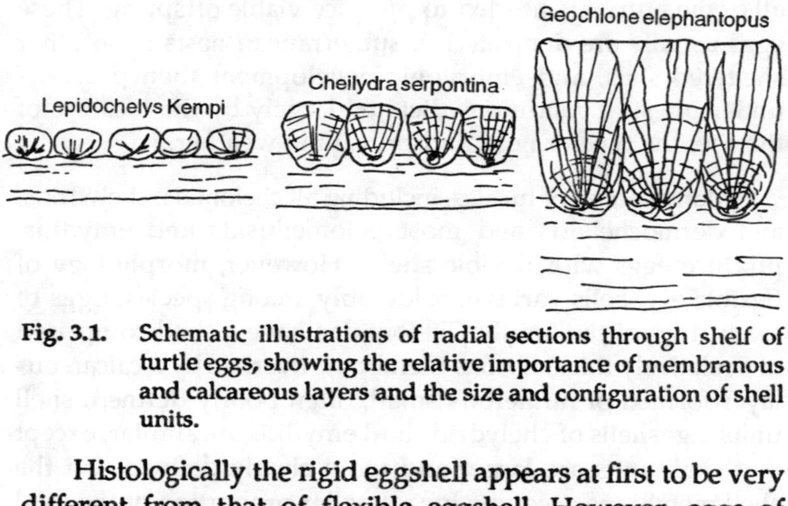

Fig. 3.1. Schematic illustrations of radial sections through shelf of turtle eggs, showing the relative importance of membranous and calcareous layers and the size and configuration of shell units.

Histologically the rigid eggshell appears at first to be very different from that of flexible eggshell. However, eggs of *Kinosternon flavescens* .at intermediate stages of calcification have a flexible shell that is indistinguishable from shells of oviposited eggs of chelydrids and emydids; that is, the calcareous layer has numerous small, widely spaced shell units. By the time of oviposition, however, the calcareous layer is a rigid structure characterized by elongate shell units that about tightly. Thus, the differences in morphology of flexible and rigid eggshells may arise simply in consequence of different amounts of time that crystallization proceeds inside oviducts of females.

At the time of oviposition turtle embryos are in the gastrula regardless of how long eggs may have been retained within the maternal oviducts prior to oviposition This consistency contrasts markedly with the variation in developmental stage of oviposition in squamates. The cause for the apparent arrest in development at this stage has not been determined, but hypoxia may be a factor.

Eggs of turtles have a thick layer of albumen at oviposition and this layer holds much of the water that the female parent vests in her eggs for support of embryogenesis. Most of the water passes from the albumen into the vitelline sac during the first 1-2 weeks of incubation, however, so the albumen layer is not prominent in eggs that have undergone much development. Turtle eggs have a large yolk at oviposition, but the yolk becomes even larger during the first 1-2 weeks of development as water flows inward from the albumen. Indeed, by the end of this period, the yolk occupies almost the entire interior of viable eggs. This compartment contains all of the lipid and most (or all) of the protein required to sustain metabolism and growth of embryos. The yolk also supplies part of the calcium used by embryos for osteogenesis, and is the proximate source or water used by embryos.

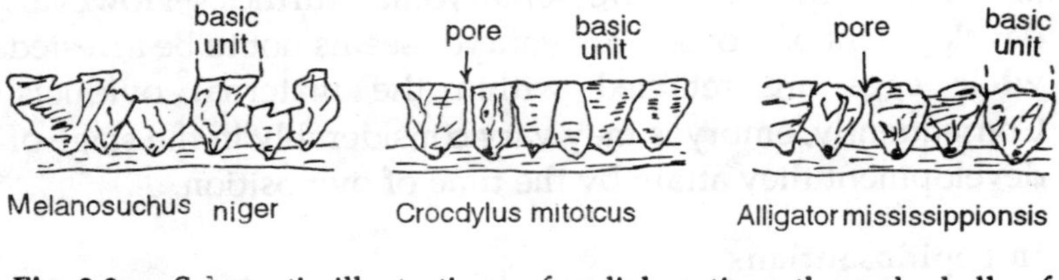

Fig. 3.2. Schematic illustrations of radial sections through shells of crocodilian eggs, showing the relative importance of membranous and calcareous layers and the size and configuration of shell units.

In Crocodiles

In the case of crocodiles the eggs are elliptical having rigid, calcareous shells. The shell is formed of two fibrous membranes overlain by a relatively thick layer of calcium carbonate. The calcareous layer is comprised of crystals of calcite organized into large, wedge-shaped shell units that abut tightly, thereby conferring rigidity to the eggshell. The crystalline layer occasionally is interrupted by pores that provide for an exchange of water and respiratory gases between the enclosed embryo and the nest environment. These pores tend to be more numerous around the equator of the egg than at either of the poles.

Crocodilian eggs contain a distinct layer of albumen at oviposition, and this albumen surrounds a large yolk. The albumen probably contains much of the water required by the developing embryo, because the volume of albumen declines appreciably during the course of incubation. Additional water is contained within the yolk, which seemingly is the source also for all of the lipid and most (or all) of the protein required to sustain the embryo. Although a portion of the calcium required for skeletal ossification apparently is mobilized from the yolk, most of it comes from the calcareous layer of the eggshell.

Embryonic crocodilians in freshly laid eggs commonly have 16-18 somites, and therefore are somewhat more advanced at oviposition than are embryonic turtiles. However, development of crocodilian embryos seems not to be arrested while eggs are retained within the maternal oviducts. Consequently, embryos may vary considerably in the stage of development they attain by the time of oviposition.

In Lepidosaurians

Lepidosaurians usually lay eggs with thin, flexible shells. However, both flexibility and morphology of eggshells very considerably among and within species. The eggshell in some lepidosaurians in comprised of a fibrous membrane that seemingly is devoid of calcareous deposits. In other species, small amounts of calcium carbonate are found as isolated crystals among fibers of the shell membrane or as larger, but widely separated, deposits on the outer surface of the membrane. Finally, in still other species, a layer of calcareous material forms the outer surface of the eggshell. Sometimes this this calcareous layer is formed of numerous thin, flat plaques (or spheres) attached to the surface of the shell membrane but in other instances the calcareous material is formed into columns that penetrate deeply into the membrane. The calcium carbonate usually exists as crystals of calcite but crystals of aragonite are interspersed among the crystals of calcite in eggshells of *Pythom regius* However, variation occurs within some species both in the type of crystal forming the calcareous layer and in morphology of the shell itself, and eggshells of a single species have occasionally been assigned to different categories by different investigators.

Lizards of the family Dibamidae are said to produce eggs with rigid, brittle shells but the original observation has never been repeated. On the other hand, gekkonid lizards in the subfamilies Gekkoninae and Sphaerodactylinae are well-known to produce eggs with hard, calcareous shells. Crystals of calcite in the calcareous layer are arranged into jagged blocks that about tightly and interlock thereby rendering the layer rigid and non-compliant. The proximal portion of these calcareous blocks invests fibers at the outer surface of the shell membrane and anchors the two layers together. Pores have not been observed in the calcareous layer, but all of the studies were preliminary, and pores could easily have been overlooked.

Embryos in freshly laid eggs of most lepidosaurians have attained stage 27-29 (or equivalent) in developmental series of Dufaure and Hubert, and therefore have 20-30 pairs of somites. In *Opheodrys vernalis* and *Sceloporus aeneus,* however, eggs are retained for extended periods within oviducts of females, and embryonic development proceeds well beyond stage 29. Oviposited eggs of these species contain embryos of substantial size, and incubation to hatching may be completed within a matter of days. On the other hand, eggs of *Chamaeleo lateralis* contain embryos in a presomite stage of development and those of *Sphenodon punctatus* apparently contain embryos that are in gastrulation. Considerable variation therefore exists in

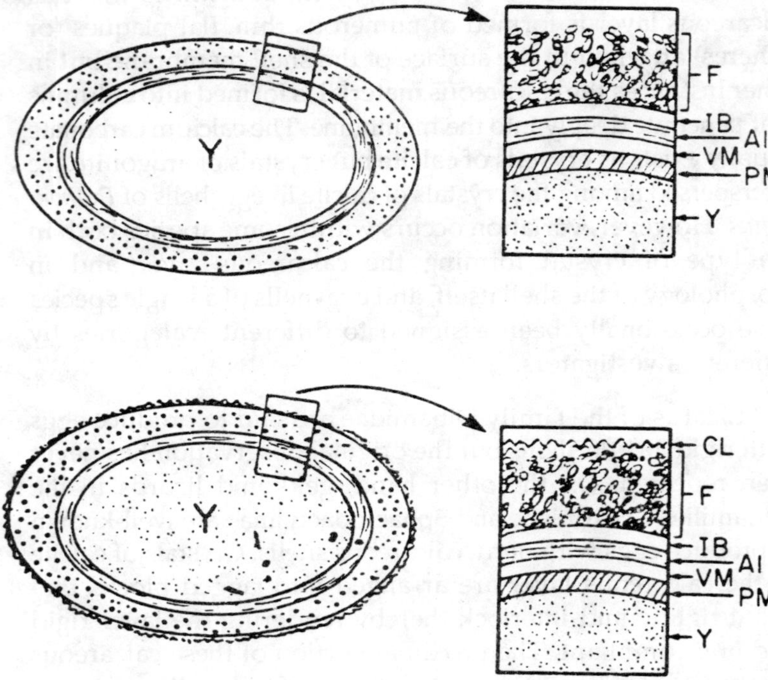

Fig. 3.3. Schematic illustrations of radial sections through shells of lepidosaurian eggs, showing the difference between eggs having no calcareous material in the eggshell and those having a distinct calcareous layer. Insets show enlarged sections passing from the outer surface of eggshell into the yolk. None of the layers is drawn to scale. Cl, Calcareous layer, F, fibers of shell membrane; IB, inner boundary of shell membrane; Al, albumen, VM, vitelline membrane; PM. plasma membrane; Y, yolk.

the stage of development attained by lepidosaurian embryos by the time of oviposition, but these embryos usually are more advanced than are those of turtles and crocodilians in freshly laid eggs.

Lepidosaurian eggs usually lack a distinct layer of albumen at oviposition but eggs of *Conolophus subcristatus* are reported to have an albumen accounting for 28% of their initial mass. However, albumen was separated from yolk of the *Conolophus* eggs by low-speed centrifugation, so the fraction referred to as albumen may simply have been water forced out of the yolk by ultrafiltration.

The albumen of testudinian and crocodilian eggs is an important reservoir of water, so the apparent absence of an albumen layer in lepidosaurian eggs has been taken to indicate that these eggs contain insufficient water at oviposition to support embryos until hatching. Indeed, water represents only 63% of the initial mass in eggs of *Amphibolurus barbatus* and 59% in eggs of *Coluber constrictor*. However, 74% of the initial mass of eggs is water in *Pituophis melanoleucus*. Comparable values for eggs of turtles fall between 68-72%.

The presence of a large layer of albumen in squamate eggs that have undergone substantial incubation has been reported. However, the fraction they identified as albumen may actually have been material from the allantoic sac. The allantoic sac of embryonic *Sphenodon punctatus* becomes greatly enlarged during incubation, owing to the accumulation of a clear, semigelatinous fluid resembling the albumen of an avian egg. Whenever a swollen egg ruptures, this fluid is released onto the surface of the eggshell where it could easily be mistaken for albumen.

The yolk occupies most of the interior of lepidosaurian eggs at oviposition, and presumably provides embryos with almost all of the nutrients including calcium and water required to support development.

TYPES OF NESTS

In Chelonia

Most species of turtiles dig a flask-shaped hole in the ground in which eggs are deposited. In some instances, excavation of the nest chamber is preceded by construction of a *"body pit,"* thereby removing surface debris and loose soil that might otherwise sift into the forming cavity. Among species producing small clutches, all of the eggs rest on the floor of the cavity and therefore have some contact with the substrate. Among species producing large clutches, however, many eggs are supported by points of contact with other eggs and consequently make no contact with the substrate. After all of the eggs have been laid, the female kicks dirt into the neck of the chamber. Some of this dirt may sift downward among the eggs, but air spaces between the eggs at the top of the cavity usually are not obliterated completely. These air spaces persist for the duration of incubation in nests constructed in stable soils, but may be obliterated by infiltration of soil into nests located in less stable substrates. The infiltration of soil or sand into nests sometimes leads to increased mortality among embryos so persistence of air species may be more important to successful incubation than generally is realized.

Some species of the families Chelidae and Kinosternidae depart from the pattern of nest construction characterizing most turtles. For example, *Sternotherus minor* and *Sternotherus odoratus* may simply back up to a fallen log and deposit their eggs beneath the edge; if females fail to scrape leaf litter over these *"nests,"* the eggs remain partially exposed to the elements. Eggs of *Phrynops gibbus* frequently are deposited on the surface of the ground or with only a thin covering of soil. In contrast, *Kinosternon flavescens* from regions of sandy soil in the central U.S. burrow through the ground prior to oviposition, and eggs are simply deposited in the burrow system constructed by each female.

Crocodilia

Crocodilians construct mound nests or hold nests. Mound nests, such as those characterizing *Alligator mississippiensis, Caiman crocodilus,* and *Crocodylus porosus,* are located in piles of mud and/or vegetation scraped together by gravid females. A nest chamber is dug in the top of each mound, and eggs are deposited inside. Eggs at the bottom and periphery of the pile contact walls of the cavity, but those toward the center of the mass do not. The female closes the cavity after all eggs have been laid and may then crawl across the top several times, thereby compressing the loose mud and vegetation. Spaces between eggs occasionally are filled as the nest is closed but pockets of air usually persist within the cavity for the duration of incubation.

Other crocodilians like *Crocodylus johnsoni, Crocodylus miloticus,* and *Gavialis gangeticus,* dig a hole obliquely downward into a sand, gravel, or dirt bank. The narrow tunnel leads to a spacious egg-chamber at the bottom. As eggs are laid, they simply roll to the bottom and form a pile. Eggs in the middle of the pile make no contact with the substrate, whereas those at the periphery have some contact with soil. The female subsequently closes the tunnel with sand or dirt. Some soil and debris occasionally is packed between eggs, but air spaces usually persist inside the egg chamber, presumably for the duration of incubation.

Lepidosauria

Such a diverse array of sites is used for oviposition by different species of lepidosaurians that generalization about nests of these reptiles is virtually impossible. Developing embryos of lepidosaurians encounter diverse ecological and physiological problems. This diversity represents both a challenge and an obstacle: A challenge, in that a great deal of new research is needed to describe nests in the detail required for understanding these problems; an obstacle, in that consequent variation in physiological ecology of embryos may defy simple generalization.

Because snakes lack limbs, most species apparently are unable to construct nests specifically for their eggs, and rely instead on pre-existing sites that are suitable for oviposition. What constitutes a suitable site varies somewhat with the size of the species in question. For example, small snakes, such as the colubrid *Diadophis punctatus,* frequently deposit eggs in small crevices and ant galleries inside rotting logs and stumps. Eggs of larger species, like the colubrids *Coluber constrictor, Masticophis taeniatus,* and *Pituophis melanoleucus,* often are deposited in the subterranean burrows of rodents. The largest of snakes, such as *Python sebae,* usually oviposit in large cavities beneath objects on the ground or in abandoned burrows of aardvarks and anteaters. In the absence of such sites, pythons may lay eggs on the surface of the ground in dense grass or in pockets of dead leaves. Detailed descriptions are not available for nests of any snake, but the lower aspect of some (or all) eggs in a clutch probably contacts the substrate, whereas the upper surface of the eggs probably is exposed to air inside the nest cavity. The proportion of the surface contacting the substrate has important implications for temperature of eggs and for exchanges of water with the environment, so detailed descriptions of egg chambers are badly needed.

One species or another of saurian uses almost every imaginable place to lay its eggs. Limbless lizards, such as *Ophisaurus attentuatus* and *Pygopus lepidopodus,* probably use the same kinds of sites of oviposition as are used by snakes of similar size. Small skinks, anoles, and other lizards lay their eggs in cavities in rotting logs or beneath surface objects whereas others oviposit on the surface of the ground beneath of protective layer of leaf litter. Some geckos lay their eggs beneath fallen limbs and debris but others conceal their eggs beneath loose bark on standing trees or in clumps of dead leaves collecting in crotches of these trees. Many species in tropical regions, including some of the larger varanids, oviposit inside termite mounds. In every case, the site selected for oviposition is a primary determinant of the physical conditions to which eggs will be exposed, but detailed descriptions of such sites are wanting for most of these saurians.

sA number of species of saurians, including many agamids, chamaeleonids, iguanids, lacertids and teiids construct nests specifically for their eggs. A tunnel is excavated at an angle into the soil, and an enlarged chamber usually is located at the end. After the female lays lays her eggs, the nest is closed. Females of some species apparently pack loose dirt into the nest chamber during closure, thereby obliterating air species among the eggs, whereas females of other species do not. The presence or absence of air spaces inside nests has important implications concerning exchanges of heat, water, and gasses between developing embryos and their surroundings. Also, obliterating these air spaces is known to increase embryonic mortality in some species. Thus, it is important to determine whether filling the cavity occurs normally in the species for which it has been reported or whether the published reports are of atypical nests.

Development

Reptilian fertilization is internal. The reproductive cycle is concerned with the relationships of fertilization, ovulation, and the production of fertilized eggs. These relationships are broadly dependent on the interaction of species with their environment. Samples of the remarkable diversity of the squamate reproductive cycles are detailed in a companion chapter.

The number of embryos is species specific; for example, *anolis carolinensis* has two (one in each uterus), *Lacerta vivipara* and *Podarcis muralis* have between eight and 12, *nguis frgilis* bout 20, *Vipernaspis* has between ten and 22, *Natrix natrix* has up to 50, nd *Python reticulatus* has up to 100. In general, young females tend to have fewer eggs than older ones.

The stage of development at oviposition differs interspecifically. Most oviparous species deposit their eggs at an advanced stage, namely when the circulation of the blood is well-established and the limb buds are beginning to develop. This corresponds to stage 30 in the table of *Lacerta vivipara* and to about stage 35 in the table of *Vipera aspis* for lizards and snakes, respectively. Less frequently, the eggs may be laid at the gastrulation stage, as for instance in *Chamaeleo lateralis*.

Oviparous reptiles often deposit their eggs in a warm environment, for instance among rotting vegetation, in sand, or under the bark of trees (tropical geckos); many burrowing snakes and amphisbaenians lay their eggs in anthills or termitaria. Some Boidae incubate their eggs by coiling around them brooding female *Python molurus* can raise their body temperature 7°C by shivering thermogenesis.

In some Squamata, the development of the embryo takes place entirely *in utero.* Such reptiles are called *viviparous,* although this viviparity is not comparable to that of mammals.

In all Squamata, the eggs are ellipsoid *in utero.* They are composed of a large yolk mass, surrounded by a vitelline and shell membrane. In the case of *oviparous* species, this shell membrane is thick, parchment-like, and lightly calcified. In viviparous lizards, snakes, and amphisbaenians, the shell membrane is thin, transparent, and does not become impregnated with calcium. Thickness and complexity of egg shells appear to be adaptive, as the shells retain moisture and protect the egg of oviparous species against predators and microorganisms.

Symmetry and Orientation of the Embryo

In the oviduct, the embryos of lizards and snakes always lie on the dorsal side of the mother, and in 90% of cases the plane of bilateral symmetry of embryos is perpendicular to the long axis of the egg. The orientation of the first furrow of cleavage has rarely been recorded. However, our observations in *Lacerta vivipara* have shown that in most cases the orientation of the first furrow of cleavage is at right angles to the long axis of the egg.

The hypothesis of *Will* (1896) suggests that the orientation of the first furrow may correspond to the orientation of the embryo. On the other hand, Raynaud and *Ancel* (1960) have observed an *in vivo* rotation of eggs in *Anguis fragilis.* The rotation, which stops just before the beginning of the cleavage stage, appears counterclockwise in each uterus when observed on the ventral side of the female moreover, at further stages of development the head of embryos is generally oriented in the direction of rotation. Thus, the orientation of the cephalocaudal axis of the embryo is likely to reflect this rotation.

TABLES OF DEVELOPMENT

General

Normal tables of development are an indispensable basis for experimental work and provide useful bases for comparative studies. They depend upon a detailed description of external morphology and are usually related to measurements of length. Important diagnostic features for staging are provided by the development of the head, optic vesicle, optic vesicle, branchial clefts, heart, amnion, allantois, numbers of somite pairs, limb buds, external sexual organs, pigmentation and scales.

Comparative staging among different species is frequently difficult because the numbering of equivalent stages of development often differs among. Normal Tables. Table 3.1 illustrates this difficulty, as it compares the numbering of stages as defined by *Pasteels* (1953) for *Mabuya megalura* and by *Pasteels* (1956) and *Milaire* (1957) for *Chamaeleo bitaeniatus* with "equivalent" stages for *Lacerta vivipara*.

The duration of embryonic development of a particular species may be markedly modified by incubation at different temperatures. Furthermore, the rates reflect the normal incubation temperatures. This complicates the correlation among the defined stages in different tables, even when they have been incubated at constant temperatures.

The following accounts characterize the tables of development now available in the literature.

Lizards

Lacertidae Lacerta agilis : Complete table comprising 40 stages, each illustrated by a fine set of lithographs. *Podaricis muralis* :Table of 13 stages covering the period after the eggs are laid, very well-illustrated. These 13 stages are characterized in relation to the embryonic development at a specified temperature (table 3.2). *L. vivipara* : Complete Table comprising 40 stages, each illustrated by a photograph of the embryo left in place on the egg. The principal stages of the differentiation of the limbs are presented in a series of close-up photographs. This table is an accepted standard of reference by most authors working on the embryology of lizards. Certain stages are correlated with development in vitro at a constant temperature.

TABLE 3.1

Correspondence Among the Different Numbering Schemes Defined in Three Lizards for Similar Stages of Development

Chamaeleo bitaeniatus	*Lacerta vivipara*	*Mabuia megalura*
18-19	26	
	27	17
20	28	18-19-20
21	29	21
22-23	29-30	
	30-31	24
	31	25
24-25	31-32	26-27
26-27	32	
28-29-30	33	28
31	34	30
32	34-35	
	35	31-32-33
33-34	35-36	
	36	34
35-36	37	35-36
37	38	36
39	39	38
40	40	39-40

2. Iguanidae

Liolaemus gravenhorstii (Lemus 1961): Complete table of 35 stages described and illustrated either by a sketch or by a photograph of the embryo. The timing of these stages has been published separately. *L. tenuis* : Series of 29 intrauterine stages, completed by a series of 14 extrauterine stages.

3. Chamaeleonidae

Chamaeleo bitaeniatus : 10 stages of gastrulation followed by 14 subsequent stages. C. *Lateralis* Series of 44 stages each illustrated by a photograph of the embryo with some indications of timing at constant temperature.

TABLE 3.2

Podarcis muralis* and *Lacrta viridais* Embryonic Development at 26°C

Stages	*P. muralis* (*days*)	*L. viridis* (*days*)
26	0	0
28	1	
30	3	4
31	7	
32	9	10
33	12	18
34	15	
35	17	
36	19	
37	21	28
39	25	
40	42	
Hatching	46	60

*After Dhouailly and Saxod, 1974

4. Agamidae

Golotes versicolor: 26 stages covering the intrauterine period (*Thapliyal et al.,* 1973). 16 stages between oviposition and birth.

5. Gekkonidae

Hemidactylus turcicus, Ptyodactylus hasselquisti, and *Srhaerodactylus argus:* Description of the embryos from time of oviposition to hatching.

6. Scincidaae

Mabuya megalura : Principal stages of the development after gastrulation, illustrated by photographs.

7. Anguidae

Anguis fragilis: Three old, but remarkably well-illustrated, works give a detailed description of the stages of cleavage, gastrulation, and neurulation and the stages following neurulation, up to closure of the amnion.

TABLE 3.3

***Lacerta vivipara,* Embryonic Development at 27°C**

Stage of Development	*Duration of Development (Days)*
16	0
19	1
20	2
24	3
25	4
26	5
28	8
29	9
30	11
31	12
32	13
33	14
34	16
35	18
36	20
37	23
38	25
39	27

*After Hubert, 1976.

Snakes

1. Colubridae

Natrix natrix : Accounts by *Krull* (1906) and *Vielhaus* (1907). *Natrix tessellata* : Brief description of 12 stages illustrated by sketches and photographs. *Thamnophis sirtalis:* A table of 37 stages, unfortunately with the period from gastrulation to neurulation very compressed. Only certain stages are illustrated by drawings. for *Pituophis melanoleucus*.

2. Viperidae

Vipera aspis : The embryonic development is divided into 43 stages, with 18 stages devoted to the period from the beginning of cleavage to the end of neurulation. Each stage is illustrated by a photograph of the embryo.

DEVELOPMENT OF THE EXTERNAL MORPHOLOGY OF THE EMBRYO

A. General

Bearing in mind that some aspects of organogenesis are dealt with in detail in these two volumes and that organogenesis in general has been extensively described, the present account concentrates upon the development of the external features. The main aim of the section is to emphasize morphogenesis, to compare this with the conditions in other vertebrates, and to note its differences among squamate categories. The overall development of the early morphology is remarkably similar in all Lepidosauria.

B. Cleavage

Cleavage is discoidal. A small oval blastodisk divides on a large ellipsoid yolk mass. In most cases, the first cleavage furrow is oriented at right angles to the long axis of the egg. The first blastomere is central, and cleavage furrows radiate from it. Other blastomeres are formed by a progressive segmentation.

In lizards, the cleaving blastodisk passes through a stage

during which the blastomeres cleave more rapidly at one pole than at the other. At the end of cleavage, this difference cannot be detected, and the disk appears to be divided into a homogeneous mass of many small blastomeres.

C. Gastrulation

In all squamates, gastrulation proceeds through a blastopore, which marks the posterior end of the embryonic shield. The blastopore forms in a small, thickened *"blastoporal plaque"* and the reduction of its dimension becomes a measure of the extent of gastrulation.

Initially, the balstopore is straight, then it becomes crescent-shaped, always localized in the embryonic shield. Contrary to the descriptions found in older works, only the crocodilians have a primitive streak, such as is found in birds. The blastopore of other reptiles recalls that of amphibians.

The posterior-anterior movement of the somitic and extraembryonic mesoderm from the blastopore modifies the external morphology of the embryonic shield. The progression of the mesoderm under the ectoderm expresses itself as lateral thickenings, which appear as wings adjacent to the embryonic shield.

D. Neurulation

The blastopore is still visible when the rudiment of the neural groove appears. Subsequently, the embryo elongates and the neural groove encloses the blastopore. However, this occurs only in lizards. In contrast, the blastopore of snakes remains as a posterior enlargement of the neural groove. Likewise, the closure of the neural groove is retarded in the cephalic zone of snakes more than in that of lizards.

In general, four to five somites form during the initial stage of neurulation, both in snakes and lizards. The first blood islands also form in the posterior extraembryonic area during the first stage of neurulation.

POST-NEURULATION STAGES

General

The number of equivalent stages into which Normal Tables are divided may very for different squamates. Therefore, the number of pairs of somites serves as a criterion of development in this section. However, timing of the differentiation of external organs and the formation of somites are not correlated for all squamates. Consequently, the description that follows gives a general rather than a specific, idea of morphological change during the course of embryonic development.

Six to 20 Pairs of Somites

During the period in which the number of somites increases from six to 20, the morphology of the embryo is similar in all squamates. The neural tube is formed by the fusion of the neural folds, first in the cephalic region and then in the posterior part. The brain differentiates and in the embryo with about 20 pairs of somites, the telencephalon, diencephalon, mesencephalon, and rhomobencephalon become distinct. Optic vesicles are formed in embryos with about 15 pairs of somites. Optic cups and otic vesicles are formed when there are 20 pairs of somites, and the mandibular and hyoid arches become distinct to form the bymoandibular branchial cleft. At this stage, the heart begins to form in the midline, and the allantois appears as a small posterior vesicle.

During this period, the embryo is progressively covered by the amnion, and the embryo begins to turn onto its left side. The head turns first (at the stage of six to seven pairs of somites), followed by the trunk; the embryo lies completely on its left side by the time it has 16 to 18 somites.

20 to 35 Pairs of Somites

During the period in which the number of somites increases from 20 to 35, there is still no significant difference between the conditions seen in snakes and lizards. All the branchial clefts are formed, the optic cups appear horseshoe-shaped, and the olfactory vesicle appears. The ventricle and auricle of the heart are distinct, and the heart begins to beat. In lizards, the pineal body is visible when the embryo is transilluminated from below.

The allantois becomes larger and extends to the level of the head. In some lizards, such as *Anguis fragilis*, *Lacerta agilis*, and *Chamaeleo lateralis*, the anterior part of the embryo is entirely covered by the amniochorion. This makes the head and cardiac regions difficult to see in the intact embryo.

More than 35 Pairs of Somites

Once 35 somites have formed, the branchial clefts close, the optic cups are spherical and the eyes, olfactory, and auditory organs start to differentiate. It is during this period that the differences between snakes and lizards become obvious, as do those within each group. The main differences relate to the formation of the limbs, which are missing in snakes, but will develop in most lizards. In all those lizards that normally develop limbs, the limb buds are at first stump-like and have a distinct longitudinal crest. They then become paddle-shaped. Finally, the digits differentiate. In the apodan lizard *Anguis fragilis*, rudiments of the anterior and posterior limbs develop, but regress prior to hatching.

In some boid snakes, rudiments of the posterior limbs develop and persist in the adult. In *Python reticulatus*, a rudiment of the femur forms; in this sense, the limb rudiments of this snake are better developed than those of *Angus fragilis*.

Turning to the differentiation of the genitalia, there are also differences, most obviously among lizards. In some, a penis develops in the females and persists until birth or hatching. For instance, females of *Anguis fragilis* and *Mabuya megalura* show typical hemipenes. In other lizards, such as *Lacerta vivipara* or *Chamaeleo citaeniatus*, a sexual difference becomes apparent during embryonic development. Also, in *Lacerta vivipara* a sexual difference appears at stage 34. A similar developmental pattern is observed in snakes, for example in *Vipera aspis*.

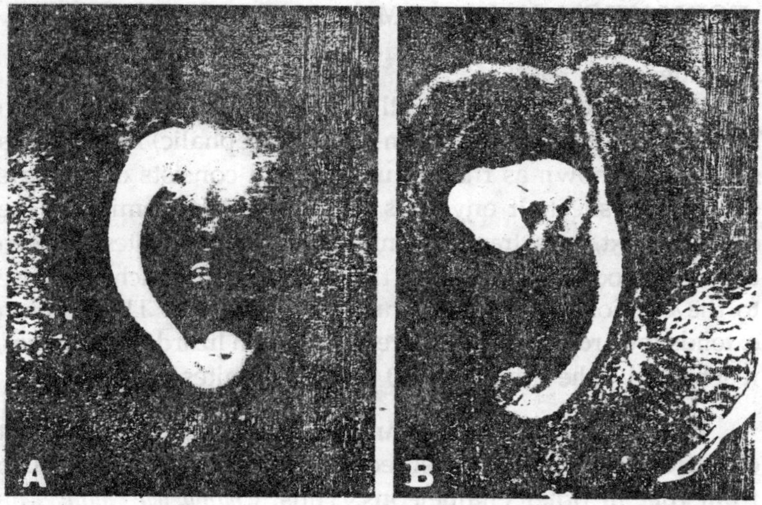

Fig. 3.4. *Anguis fragilis.* The head of the embryo is covered by extraembryonic sheets. (A) Dorsal view; (B) ventral view.

Finally, the coiling of the embryo in the egg is quite different in lizards from that in snakes. The squamates embryo is long, but the egg is relatively short. Consequently, the embryo develops in a coiled position. As the body of snakes is very long, practically all of it is coiled. First, it is corkscrew coiled and later it is coiled in a single plane. The sequence of two types of coiling, has been observed in *Thammophis girtalis, Vipera sppis Hubert* and *Dufaure,* 1968), and *Python molurus* it appears to be characteristic of all snakes. In most lizards, on the other hand, only the caudal part of the embryo is coiled, and the whorls lie in the same plane. Amphisbaenian embryos apparently coil like snakes.

Scales differentiate, and pigment is formed after 35 somites have formed. Just prior to hatching, the embryo shows the morphological features characteristic of its family.

FORMATION OF THE EMBRYONIC MEMBRANES

Amniogenesis

In most lizards and in all snakes, the amnions begin to form at the neurula stage as an anterior (cephalic) fold. At this stage, it is known as the proamnion and consists of a single ectodermal sheet. It envelops the cephalic extremity of the embryo. Next, the extraembryonic mesoderm penetrates between the two ectodermal sheets of the proamnion, which give rise to the chorion and to the amnion *sensu stricto*. The amnion extends toward the caudal extremity and, in lizards and snakes, reaches it by the time 18 to 20 pairs of somites have formed.

In certain chameleons, amniogenesis is very different, not only from the pattern observed in other squamates, but also from that in other chameleons. Thus, *Chamaeleo chamaeleon* and *C. lateralis* from the proamnion prior to gastrulation, and the amnion completely covers the embryonic shield at the end of gastrulation. Moreover, this amnion does not develop from the cephalic to the caudal part of the embryo, as in other Squamata, but forms concentrically around the embryonic shield. In addition, the amnion becomes quickly covered by extraembryonic layers and yolk.

In other lizards, such as *Lacerta agilis* and *Anguis fragilis* the head and the cephralic amnion becomes covered by extraembryonic layers of ectoderm and mesoderm, but never by yolk. This is clearly seen when the embryo has ten pairs of somites, but is particularly obvious in older embryos. In these species of lizards, the trunk region of the embryo remains unaffected by this aspect of the amniogenetic process. According to *Peter* (1904), this process involves precocious downgrowth of the anterior extremity of the embryo toward the yolk at the end of neurulation. However, histological studies of embryos of *Anguis fragilis* with ten pairs of somites have shown that an extraembryonic invagination of endoderm separates the ectodermal and mesodermal layers of the cephalic amnion. This unusual invagination of the endoderm is obviously involved in the covering of the cephalic amnion by the extraembryonic layers.

Formation of the Allantois

The formation of the allantois is very precocious in lizards; it begins to form as an endodermic vesicle at the end of neurulation. At this stage, the embryo still lies flat on the yolk, and the allantoic vesicle cannot be seen in dorsal view. It only becomes visible after the embryo has turned completely onto its left side.

In the snakes studied until now, for example in *Vipera aspis* the allantois is formed later than in lizards, that is, after the embryo has turned completely onto its left side. At a later stage, the allantois extends toward the head of the embryo (Fig. 3.6C). As it develops, the allantois pushes into the extraembryonic coelom. Finally, the superficial layer of the allantois comes into contact with the chorion, and the internal layer of the allantois makes contact with the amnion.

FORMATION OF THE EMBRYONIC AND EXTRAEMBRYONIC LAYERS

The early stages of embryonic development, from cleavage up to the end of gastrulation, have been the subject of various histological studies. The earliest of these, carried out in lizards and snakes over 80 years ago, have given rise to some confusion, especially with regard to the positioning of the endoderm. More recent studies clarify the position of the layers. These studies indicate that the origin of the definitive endoderm is different in lizards from that in snakes. However, it must be noted that only some lizards and a few snakes have been examined in this regard. Further studies of *Sphenodon* and of other Squamata are needed to characterize possible variation.

Lizards

In the course of cleavage, a horizontal furrow separates a layer of superficial blastomeres from a mass of deeper ones.

The superficial blastomeres form an epiblast, a layer of upper cells relatively poor in yolk. The deeper blastomeres form a hypoblast, a mass of round cells, rich in yolk, that forms *in situ* by a process of delamination. Just before the beginning of the gastrulation, the hypoblast is thick and compact in the central area of the embryo; a thin layer extends into the peripheral area, where it runs into the large, round cells of the germ wall.

Near to the posterior margin of the embryonic shield, a small area of epiblast loses its epithelial structure and thickens to form the blastoporal plate, the under surface of which is lined with a layer of unicellular hypoblast. A slight depression, the rudiment of the balstopore, then develops in the blastoporal plate. The unicellular hypoblast is formed from the thin hypoblast, which in the earlier stages occurred in the peripheral zone of the blastoderm. Cells invaginate at the blastoporal groove and accumulate deep to it, causing a thickening of the blastoporal plate. The clearly recognizable layer of hypoblast beneath the plate remains continuous with the central hypoblast.

The blastoporal depression next deepens, and the epiblast invaginates at the dorsal lip of the blastopore. The surface material invaginated by this process is the chordamesoderm, which inserts itself between the epiblast and the hypoblast. The blastoporal plate contracts to form a small appendix at the posterior rim of the shield. This stage marks the start of the differentiation of the endoderm. A thin layer of cells, continuous with the subblastoporal layer, cleaves at the surface of the hypoblast; the cells of this thin layer flatten and lose their yolk, whereas the deeper and contacting elements degenerate. This phenomenon is less pronounced in the axial zone of the shield anterior to the blastopore, because this zone contains a large quantity of tightly compressed hypoblast that has been produced by the intense invagination of the chordamesoderm.

If the volume of hypoblast is smaller, a layer of endoderm becomes differentiated anterior to the blastoporal area. In this case, the only obvious buildup of hypoblast occurs beneath

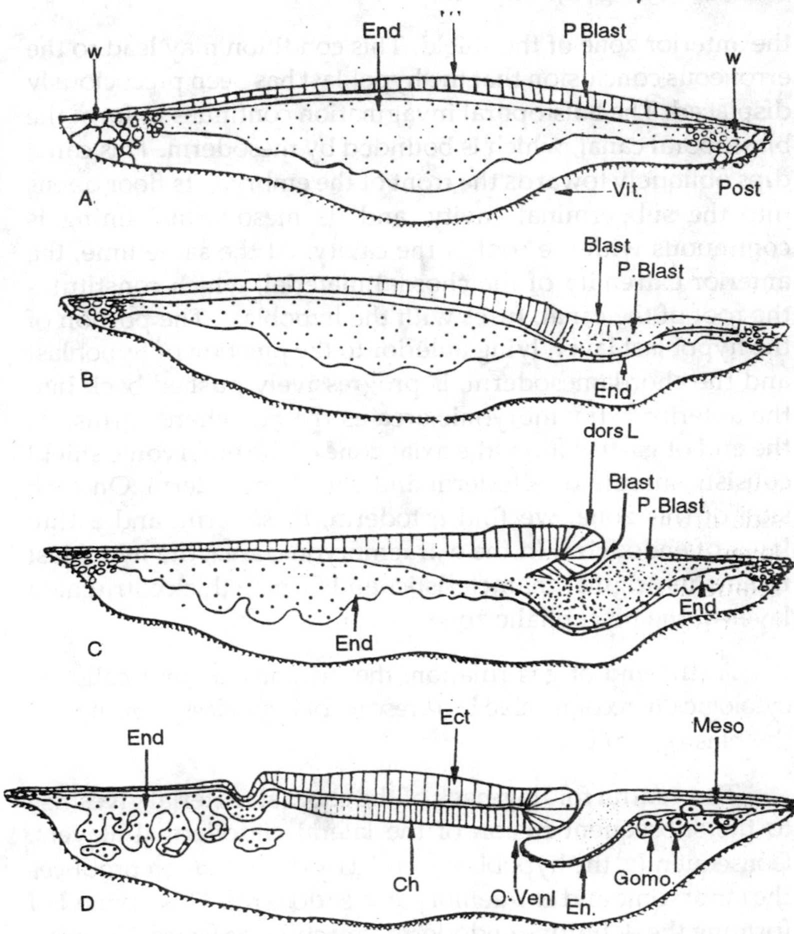

Fig. 3.5. *Lacerta vivipara.* Schematic drawing of sagittal sections of the embryonic shield at different stages of the gastrulation. (A) The blastoporal plate (P. Blast) sketches by thickening of the epiblast (Ect.) in the posterior region of the embryonic shield. A compact mass of hypoblast (End.) lies deep to the epiblast. Post., Posterior end; W, vitelline wall; Vit., yolk, (B) The balstopore (Blast). appears in the shape of a small invagination in the anterior part of the blastoporal plate (P. Blast). (C) The blastopore is well-developed. The epiblast begins to roll in at the dorsal blastoporal lip (Dors. L.) giving rise to the chordamesoderm. (D) The central hypoblast (End.) has been pushed into the anterior extraembryonic area by the chordomesoderm (Ch.). The endoderm (En.) derived from the hypoblast is seen under the blastoporal plate and the gonocytes lie just posterior to the blastoporal plate. The blastoporal canal opens ventrally (O. Vent into the subgerminal cavity. Posterior extra-embryonic mesodermm (Meso.) arises from the blastoporal plate.

the anterior zone of the shield. This condition may lead to the erroneous conclusion that the hypoblast has been precociously displaced. The balstoporal invagination continues to form the blastoporal canal, which is bounded by mesoderm. This canal dips obliquely towards the front of the embryo, its floor opens into the subgerminal cavity, and its mesodermal lining is continuous with the roof of the cavity. At the same time, the anterior extremity of the chordal material, which constitutes the roof of the canal, fuses with the hypoblast. The portion of the hypoblast layer, lying anterior to the junction of hypoblast and the chordamesoderm, is progressively pushed back into the anterior extraembryonic area as the notochord forms. At the end of gastrulation, the axial zone of the embryonic shield consists entirely of ectoderm and chordamesoderm. On each side of this zone, we find ectoderm, mesoderm, and a thin layer of endoderm. In the extraembryonic area, the hypoblast retains its original characteristics and forms a thick corrugated layer in the precephalic zone.

At the end of gastrulation, the primordial germ cells are cytologically recognizable in a crescent of hypoblast surrounding the blastoporal plate.

The ventral folding-back of the chordal material gives rise to the subsequent fusion of the lateral endodermal alayers. Consequently, the hypoblast formed by delamination produces the embryonic and extraembryonic endoderm. This method of forming the definitive endoderm is peculiar to lizards. Turtles, snakes and perhaps *Sphenodon* form the endoderm from epiblastic cells invaginated at the blastopore.

Snakes

The older descriptions of the formation of the germ layers in the embryos of *Vipera berus* and of *Natrix natrix* are confusing, particularly concerning the origin of the endoderm. However, only one snake, *V. aspis,* has been examined since for this purpose. The findings are presented below.

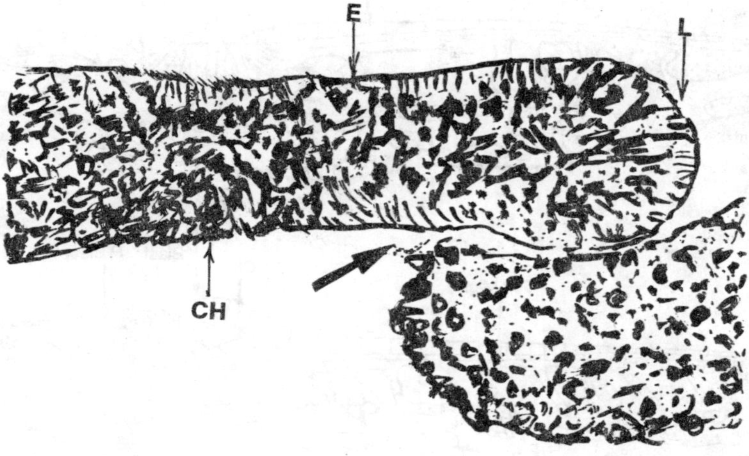

Fig. 3.6. *Lacerta vivipara.* This longitudinal section through the blastoporal region of a late gastrula shows the formation of the chordamesoderm (CH) by a rolling in of the epiblast (E) at the dorsal blastoporal lip (L). The balstoporal canal opens (arrow) into the subgerminal cavity.

As in lizards, the epiblast and hypoblast of *Vipera aspis* are formed by delamination during the course of cleavage. However, the structure of the young embryo is different at the beginning of gastrulation. Only a little hypoblast forms under the embryonic shield; here it is distributed in small scattered groups of cells, most of which lie near the yolky floor of the subgerminal cavity. The remainder of the initial elements of the hypoblast is progressively pushed into the posterior and anterior extraembryonic area by the migration of the epiblastic cells invaginating at the level of the balstoporal furrow. These invaginating cells are distinguished from those of the hypoblast by their smaller size, stellate form, and lower yolk content. They represent the origin of the embryonic endoderm and a small part of the extraembryonic endoderm. The endoblastic cells initially form a cellular sheet that ramifies as cords over the subgerminal sheet. Subsequently, they become organized as an epithelial layer.

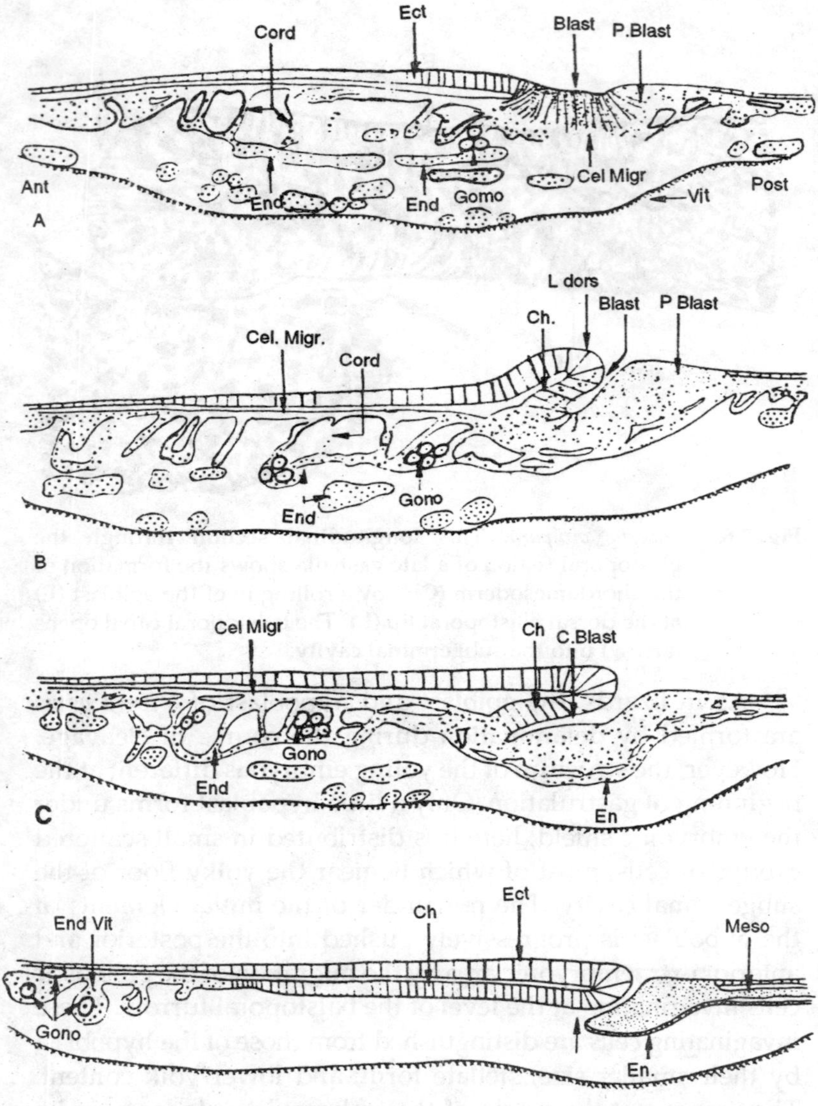

Fig. 3.7. *Vipera aspis*. Schematic drawing of sagittal sections of the embryonic shield at different stages of the gastrulation. (A) At the beginning of the gastrulation, epiblastic cells (Cel. Migr) migrate through the blastoporal furrow toward the

anterior region of the embryonic shield. In the latter, epiblastic migrating cells form cellular cords (Cord.), which hang in the subgerminal cavity. Few of the hypoblastic cells (End.) produced during cleavage lie under the epiblast of embryonic shield. The gonocytes (Gono.) lie close the these hypoblastic cells, just anterior to the blastopore (Blast). Ant., Anterior, Ect., epiblast; Post., posterior; vit., yolk. (B.C.) The epiblast begins to roll in at the dorsal blastoporal lip (L. Dors.) giving rise to the chordamesoderm (Ch.). Migrating cells (Cel Migr) become more numerous and give rise to the endoderm (En.). The migration of endodermic cells from the blastopore towards the anterior region carries away the hypoblast and the gonocytes in the same direction. C. Blast., Blastoporal canal; P. Blast, blastoporal plate. (D) Endoderm (En.) is now clearly seen under the blastoporal plate. In the axial part of the shield, the blastoporal canal opens ventrally (arrow), the hypoblast, the gonocytes (Gono.), and the endoderm (En.), have been pushed into the anterior extraembryonic area by the chordamesoderm (Ch). Ect., Ectoderm; End. Vit. Yolky endoderm; Meso., posterior extraembryonic mesoderm. (After Hubert 1970).

As in the lizards, the chordamesoderm of snakes is formed following the invagination of the epiblast at the level of the dorsal lip of the blastopore. The ventral opening of the balstoporal canal simultaneously unites the chordal material with the axial endoderm. Subsequently, the chordal material pushes the axial endoderm into the anterior extraembryonic area.

At the beginning of gastrulation, the primordial germ cells of *Vipera aspis* lie immediately anterior to the balstopore. Subsequently, they are carried to the anterior extraembryonic area by the successive movements of the endoderm and chordamesoderm.

Thus, the observations made on *Vipera aspis* indicate that snakes resemble birds; both form the endoderm from cells originating in the epiblast. Also, as in birds, the primordial germ cells come to lie in the anterior extraembryonic region, shifted there by the movements associated with the formation of the endoderm. The principal difference from the condition in birds in the occurrence of a blastopore of the amphibian type, instead of an avian primitive streak.

Consequently, snakes form the embryonic layers in a manner that is partly specific to reptiles and partly intermediate between those of amphibians and birds.

DEVELOPMENT UNDER NORMAL AND EXPERIMENTAL CONDITIONS

Natural Environment

In general, the embryonic development of lepidosaurs is slow and takes at least three months for both viviparous and oviparous species. However, in some Squamata it takes much longer. Thus, in *Chamaeleo lateralis,* birth occurs ten months after eggs are laid at the gastrula stage. This exceptional length of time is in part due to the existence of a three-month rest period after the eggs are laid, during which very little development takes place.

Except for a few species of Boidae, in which the eggs are incubated by the coiled female, the temperature of the eggs and embryos is at the mercy of their surroundings and of the thermoregulatory behaviour of the female. The duration of gestation or incubation is markedly affected by temperature; therefore, it can vary with climatic conditions from one year to the next. Thus, *Blanchard* and *Blanchard* (1941) found that in southern Michigan the gestation period for *Thamnophis sirtalis* could vary from 87 days during exceptionally hot summers to 116 days during an unusually cold summer. At high altitudes, embryos of *Lacerta vivipara* develop more slowly than at low altitudes. Embryonic development is also longer for populations living on the southern side of mountains than for populations living on their northern side. In oviparous species, nest to nest variability in temperature also results in differences of duration of embryonic development.

Laboratory

The laboratory can provide favourable conditions for obtaining the embryos of viviparous or ovoviviparous reptiles in constant and clearly defined temperature regimes. Indeed, such conditions and the correlation of the timing of development with definitive stages are an indispensable basis for the progress of experimental research. Two methods are used for sequential analysis. Either gravid females are kept in the laboratory and embryos are regularly observed by laparotomy or the embryonic development is noted *extra-utero.*

Incubation Outside the Uterus

Two operating procedures may be cited—that of *Panigel* (1956) and that of *Raynaud* (1959). (also see review in New, 1966, for the different methods of *extra-utero* development. In the method of Panigel, the pregnant females are killed and the oviducts containing the eggs are removed to a petri dish filled with a sterile solution of 0.6% sodium chloride. The eggs are released from the oviducts with fine forceps and then placed, with the embryo uppermost, into separate petri dishes onto gauze moistened with the saline solution. The addition of chick albumen to the saline solution prevents infection.

Successful culture can be achieved if the saline is replaced entirely by albumen and if the humidity is maintained by sealing the petri dish with a glass cover secured by wax. Using this adaptation, embryos of *Lacerta vivipara,* explanted at the two-pair-of-somites stage, have developed at 27°C until hatching; the young lizards were then kept in the laboratory for two weeks before release to the wild. Older stages (30 pairs of somites) of *Vipera aspis* also developed for two weeks under these conditions. Furthermore. The method of Panigel has been used to maintain the development of the embryos of *Anguis fragilis.* A more successful culture has been achieved by the more elaborate method of *Raynaud* (1959). In this method, the eggs are supported at the surface of a liquid medium in a crystallizing dish, and oxygen is introduced into the dish every 24 hours.

The intact eggs of oviparous species may be incubated by placing them. embryo uppermost, into small dishes containing moistened and sterilized sand. The containers are then closed incompletely with a sheet of glass.

Culture of Explanted Embryos Without Yolk

Only a few attempts have been reported of the culture of squamate embryos without yolk. Such culture has been attempted with snakes (Natrix natrix, Vipera berus) and lizards (*Lacerta vivipara* and *Anguis fragilis)*. However, it appears very difficult to maintain development of squamate embryos explanted without yolk at the cleavage or gastrula stages. Although embryos explanted at older stages sometimes undergo considerable development, further studies are necessary to develop a successful method.

Duration of Development at Constant Temperature

The embryos of tropical squamates need to be incubated at a temperature a close as possible to that characterizing development of this species in the wild. Thus, the embryos of the lizard *Iguana iguana* die when the incubation temperature-varies more than one or two degrees from 30°C and most embryos of *Pythom molurus* die when incubated at 27°C instead of 30°C.

The natural variation of temperature seen in temperate zones may explain why the embryos of squamates from these regions can be incubated in the laboratory over a temperature range of 20 to 30°C; however, it is conventional to incubate these embryos artificially at a constant temperature of 25°C. An increases of temperature decreases developmental time. Development is also more rapid when incubation proceeds at a constant rather than a variable temperature. Under constant (elevated) temperature conditions, the embryos of oviparous species may develop faster in the laboratory than in nature. For example, at a constant temperature of 27° C *Lacerta vivipara* develops twice as fast as in nature, notwithstanding the artificial conditions of incubation morphological and histological studies of different stages of development of the resultant embryos show them to be normal.

At a temperature of 26°C, the embryos of *Podarcis muralis* develop faster than the embryos of *Lacerta viridis*. For *P.muralis*, the duration of embryonic development is 46 days from Stage 26 to hatching; for *L. viridis* it is 60 days. This suggests that the duration of development is genetically determined, Although it can be modified by varying the temperature.

Temporary Arrest of Development by Low Temperatures

Embryonic development is arrested if the eggs are kept at a low temperature and will start again if they are transferred to higher temperatures. This has been reported for embryos of snakes (*Coluber constrictor, Diadophis punctatus, Ellaphe obsoleta)* maintained between 3 and 5°C and for embryos of lizards maintained for 72 hours at 4° C or 6°C. It also applies to embryos of *Lacerta vivipara* kept at 4° of 7°C for several weeks. However, many embryos die if cooled at the beginning of development. In fact, the subsequent development of cold-arrested embryos is usually obtained only in embryos in which the blood circulation was functional.

Cephalic abnormalities have been observed in the embryos of *Lacerta viridis* in which the eggs were kept at 4°C after laying and were then incubated at 25°C. However, the embryos of *Lacerta vivipara* can be kept at Stages 28 or 30 (when the limb buds are beginning to form) for 13 days by maintaining them at 7°C. Subsequently, most of the cold-arrested embryos will develop normally if incubated at 27°C. For these lizards, the method of cooling may be of practical importance, as it allows embryos to be stored until experimental work is convenient. The capacity of *L. vivipara* for prolonged arrest of embryonic development by cooling may reflect an adaptation of this species for living in temperate and cold climates.

Effect of Temperature on the Sex Ratio

Laboratory studies have demonstrated that the temperature of incubation affects the sex ratio of turtles and crocodilians.

However, it seems that in most squamates the sex ratio does not depend upon the temperature of incubation; only in two lizards (*Agama agama, Charnier*, 1966, *Eublepharis macularius, Bull,* 1980) has temperature dependence of the sex ratio been observed. In these cases, a slight increase of the incubation temperature produces males. The response of the lizards thus agrees with that of crocodilians, but differs from that of turtles in which increased temperatures produce more females. However, this aspect has been studied for very few lizards and snakes in the laboratory. Further investigations are obviously desirable.

Production of Twin Embryos

Only a single experiment, on the eggs of *Lacerta vivipara,* has explored the possibility that the reptilian blastoderm may be able to regulate its development. Blastoderms at the end of cleavage and remaining on the yolk were divided, with a fine glass needle, into equal pars perpendicular to the long axis of the egg. The blastoderm was then ligatured with a hair at the level of the cut to prevent contact between the two halves, before being incubated at 27°C by the method of *Dufaure* (1961). Out of the four blastoderms, two failed to develop and two produced twin embryos, which were preserved after 40 pairs of somites had been formed. Except for the cephalic torsion in one embryo, these twin embryos do not show significant abnormalities.

Twin embryos were also obtained by culturing blastoderm on Spratt's medium after dividing them into two halves. However, this method yields poor success; only one of ten trials produced an embryo and this died during neurulation.

Breeding

Difficulties associated with breeding have, more than any other fact, prevented the intensive study of reptilian development. There is the intrinsic problem that most reptiles of temperate climates reproduce, at most, once a year. Generally, the embryonic development is lengthy. However, the principal difficulty reflects the fact that captive reptiles are much more difficult to maintain in a good physiological condition than are other vertebrates. Clearly, willingness to master the technique of laboratory breeding constitutes the main obstacle. The breeding of reptiles requires both care and time, as already shown by *Rollinat* (1934). However, for many species, it seems to be entirely feasible as recently documented elegantly by Langerwerf.

Most species of lizards and snakes may be bred in captivity, as long as they are provided with sufficient space and the critical features of their natural biotope are reproduced as closely as possible. Although UV lamps may be used for indoor terraria to obtain good reproduction, it is preferable to expose reptiles to sunlight. It may also be necessary to enrich the food of young lizards and snakes with vitamin D.

Of the many lizards that have been bred in captivity, the oviparous lizards *Podarcis muralis* and *Lacerta viridis* and the viviparous lizards *L. vivipara* and *Anguis fragilis* are favourite species of embryologists. Many Colubridae and Boidae can also be bred, as can members of the Vireridae, such as *Vipera aspis* and *V. russellii.*

The choice of animal depends on the aim of the embryological studies. An oviparous reptile is a very suitable subject for breeding if the aim of the research is the study of embryos after the eggs are laid. Although it is possible to remove eggs by laparotomy, it is easier to perform experimental operations at a later time. However, the thick parchment-like shell of oviparous species prevents accurate observation, because it is very difficult to remove this covering without damaging the embryo. From this point of view, viviparous species offer a distinct advantage over oviparous ones. Indeed in viviparous lizards, the shell membrane, which is extremely thin and transparent, permits direct observation of the embryo at all stages of development. In species, such as *Lacerta vivipara*, the shell membrane can generally be removed without damage to the embryo. Thus, experimental intervention is fairly easy. The advantages and disadvantages of particular species must obviously be kept in mind when developing research projects.

EFFECT OF PHYSICAL ENVIRONMENT ON DEVELOPMENT

Extremes of temperature, moisture, and gas tensions can lead to abnormal development and mortality of reptilian embryos, as was discussed in preceding sections. There can be little doubt concerning the significance of serious deformity or death to embryos. Within the range of tolerance to each of the aforementioned environmental variables, however, a spectrum of sublethal effects may also provide raw material for the forces of natural selection. Reptilian embryos seemingly incubate longer, consume more of their yolk during development, and hatch at a larger size when they incubate in cool conditions and/or at high water protentials than they do in warm conditions and/or at low water potentials. Large reductions in partial pressure of oxygen from atmospheric levels probably do not occur commonly in reptilian nests, but may be a normal feature of incubation of sea turtles and some crocodilians. Low tensions of oxygen (or high tensions of carbon dioxide) may prolong incubation, inhibit metabolism, and cause hatchlings to be smaller than normal. The possible importance of these sublethal effects is explored briefly in this section.

Size of Hatchlings

One possible product of variation in physical conditions in nests is variation in size of hatchlings, elicited by differences in rates of growth of embryos, by differences in the duration of incubation, or by a combination of these factors. Attainment of large size before hatching may be advantageous to emergent young, by conferring them with a degree of protection against predation. According to this hypothesis, larger hatchlings are more difficult for predators to capture and/or to handle and swallow than are smaller conspecifics, so the former enjoy higher rates of survival than do the latter. Survival in the wild of hatchling lizards and turtles is consistent with this proposition.

Several laboratory studies lend support to this hypothesis. The relatively large snapping turtles *(Chelydra serpentina)* hatching from eggs in wet environments are faster than the small turtles hatching from eggs in dry settings both when running on land and when swimming in water. Superior performance by relatively large hatchlings could be the basis for superior survival by these animals in the wild. Also, larger hatchlings may be more aggressive than smaller hatchlings in pursuing and capturing food, and larger young may therefore sustain higher rates of growth than smaller animals in the cohort. High rates of growth may enable larger hatchlings to pass more rapidly through the range of body sizes in which predation is most intense, thereby contributing also to their superior survival.

On the other hand, larger offspring produced in environments favouring rapid growth and/or longer incubation have less residual yolk to sustain them after hatching than do smaller animals. Because possession of a relatively large nutrient reserve may enhance survival more than large size under some circumstances small size at hatching conceivably is beneficial. For example, young green iguanas *(Iguana iguana)* depend to a large extent on residual yolk to sustain them for the first month of life outside the nest, so survival may be enhanced in those individuals having a large yolk, irrespective of their body size. Furthermore, hatchling of several freshwater turtles may spend their first winter in the nest cavity without feeding during which time they are supported largely by lipids in the retracted yolk. Again, the size of the energy reserve in the retracted yolk may be a more important determinant of survival than body size at hatching. Such possibilities need to be investigated.

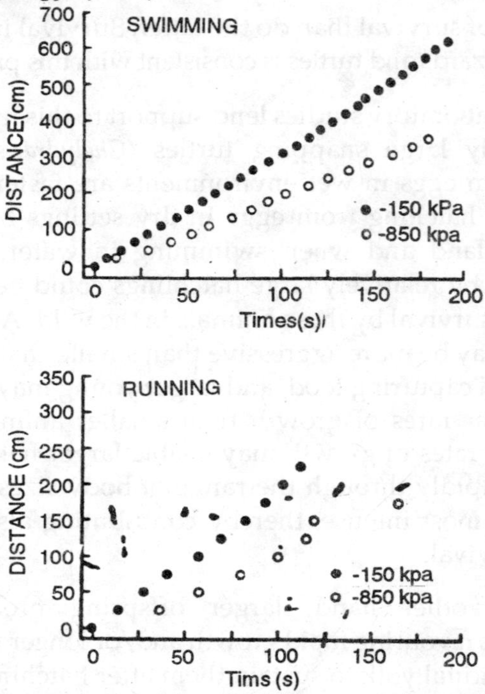

Fig. 3.8. Distance moved by hatchling snapping turtles *(Chelydra serpentina)* as a function of time swimming in water or running on land. Solid symbols represent turtles hatching from eggs on wet substrates (-150 kPa); open symbols represent animals emerging from eggs on dry substrates (-850 kPa). The slopes of the lines represent speeds for animals in the samples. The large turtles emerging from eggs in wet environments were able to swim and to run faster than siblings from eggs in dry environments. (Redrown from Miller et al., 1981).

Length of Incubation

Variation in physical conditions in nests may also influence the duration of incubation. Longer icubation is associated in some istances with conditions that allow embryos to attain large size before hatching (e.g., moist substrate), and in other instances with conditions that inhibit growth and lead to formation of small hatchling (e.g., low partial pressure of oxygen). In all such cases, the question of significance of variation in length of incubation reduces in part to a consideration of the significance of variation in body size, as was discussed previously.

For most species, however, there is another aspect to the question. Long periods of incubation may increase the probability that nests will be destroyed by predators or by inclement weather. Furthermore, if development is not completed before the advent of cool weather in the autumn, advanced embryos or hatchlings may be forced to pass the winter inside the nest cavity. Although embryos/hatchlings may be able to survive a mild winter, mortality is likely to be high under more normal (or severe) conditions of winter weather. Thus, rapid completion of development may be advantageous, even when conditions necessary for this outcome result in hatchling of suboptimal size.

Duration of incubation may establish limits of distribution for oviparous reptiles in temperate or montane regions. Ground temperatures may be too low in parts of the northeastern United States to permit embryos of the terrestrial turtle *Terrapene carolina* to complete incubation during the limited growing season, thereby establishing the northern limits to distribution of this species. Similar arguments have been adduced for distribution of the lizard *Dipsosaurus dorsalis* in deserts of the American Southwest.

Sexual Differentiation

Environmental control of sexual differentiation in crocodilians, many turtles, and some lepidosaurians has important implications for the ecology and evolution of the species in which it occurs. For examples, many species of turtles grow appreciably over the course of their reproductive lives. Shortly after attaining reproductive maturity, females tend to lay relatively small eggs in relatively shallow nests, where generally high temperatures may promote differentiation of females. Later in life, the same females tend to lay relatively large eggs in relatively deep nests, where generally cool conditions may promote differentiation of males. However, the small eggs produced early in the reproductive life of a female have a lower probability of hatching than do larger eggs produced later and the small eggs give rise also to small hatchlings that may suffer disproportionately high mortality. Consequently, first efforts at reproduction by a given female may be devoted to the relatively unsuccessful production of female offspring, whereas later efforts may be devoted to the more successful production of males. Such possibilities, if realized, would have important implications for evolution of life histories. In particular, a high probability of producing female offspring early in reproductive life may explain why maturity is reached at relatively small body size in these species when reproduction is unlikely to be successful.

Turtles and crocodilians with protracted nesting seasons may produce offspring primarily of one sex early in the season and of the other sex later in the year, owing to seasonal changes in temperature at the level of the nest cavity. If catastrophic mortality from heavy rainfall and run-off or from high tides were to occur in the middle of the nesting season, offspring of one sex might suffer disproportionately. Such differential mortality would have consequences for sex ratio of the hatchling cohort and possibly for a subsequent adult population as well.

Environmental control of sexual differentiation has special importance for conservation of sea turtles. For a number of years, a common practice among conservationists has been to remove eggs from nests and to incubate them in styrofoam boxes held inside a shelter near the beach. The rationale for this practice is that nests would otherwise be destroyed by predators, humans or inclement weather. However, the conditions of artificial incubation have been relatively uniform, and relatively cool, and most turtles hatching from eggs handled in this manner probably have been males. Clearly, if use of artificial hatcheries is to continue, more attention must be given to the physical environment to which eggs are exposed so as to assure that young of both sexes are produced.

Finally, sea-turtle rookeries frequently are confined to very restricted stretches of beach. Because thermal conditions vary among beaches, however, offspring of one rookery may be predominantly males whereas those of another may be largely females. Although any single beach may attract large numbers of females for nesting, preservation of that beach alone may not be sufficient to assure survival of the population. Viability may depend instead on provision of enough different beaches as to assure that there is sufficient environmental heterogeneity to elicit differentiation of adequate numbers of both males and females for later recruitment into the breeding population.

4

Placentation and Gestation

Adaptation to Viviparity Among Vertebrates

Viviparity is the live bearing mode of reproduction in which the number of offspring is reduced with high survivorship. Organisms adopting such a strategy may, but need not, display similar anatomical, physiological, and behavioural adaptations, even if they are taxonomically diverse. For instance, many sharks and rays secrete nutrients by the uterine projections (*trophonemata*), and the embryos absorb these via their surface epithelia, especially of the *yolk* sac and *gills*. However, embryos of many poeciliid fishes absorb maternal nutrients by epithelial extensions of the pericardial sac. Although the general function is similar, the gestation chambers are of different origin; in sharks and rays, the embryo develops in the *Mullerian duct*, whereas embryos of the poeciliids develop within the *ovarian follicle*.

Vertebrate viviparity is also widely distributed. Thus, the only extant coelacanth, *Latimeria chalumnae*, is live-bearing and has a well-developed yolk-sac placenta; also a Jurassic coelecanth may have been viviparous. Adaptations for fetal retention are not restricted to structures associated with the genital organs. The teleostean genera *Tilapia* and *Cichlasoma* and the toad *Rhinoderma* exhibit mouth brooding. Parental involvement is not restricted to females. For example, in some cichlids, the male carries the eggs in the mouth and the male seahorse carries the offspring in a ventral pouch. Except for the metatherian and eutherian mammals viviparity occurs sporadically. Sometimes it is characteristic of a family or a group of families. In many viviparous vertebrates, the yolk contained within the egg constitutes the major source of nutrients for the developing embryo. Reduction of yolk, which renders the embryos dependent on nutrients absorbed from maternal circulation, is encountered in some chondrichthyes and teleosts, but it occurs as a general phenomenon only in therian mammals. The exchange of gases, water, nutrients, and metabolic waste products may involve modification of both fetal and paternal tissues into a placenta, which in mammals consists of maternal endometrium and fetal extraembryonic membranes. Placental modifications of various organs occur in viviparous fish, and these may include modification of the yolk sac to form a placenta that resembles that of mammals. However, viviparous reptiles differ from all non-mammalian vertebrates as their placentae also contain chorionic and allantoic elements homologous to those of mammals.

Historical Concepts

The term *viper* derives from *vivus* (alive) and *pario* (to produce), *Prophet Isaiah* (14:29), describe a viper emerging from a snake, can be dated exactly to the year of King Ahaz's death (715 BC). Ancient Hebrew literature contains a legend about a certain philosopher who "wished to know after what period of time a serpent bears. When he saw them copulating he took them and placed them in a barrel and fed them until they bore."

The earliest account on placentation in a reptile is that of *Cesare Studiati* in *Chalcides chalcides*. It includes macroscopic illustrations of an embryo *in situ* showing the amnion, yolk sac, and allantois and also the umbilical and vitelline vessels, and it comments on the functional significance of the finding. The general morphology of the yolk sac and the gravid uterus of the Australian lizard, *Trachydosaurus rugosus*, was described by *Haacke* (1885).

The anatomy of the allantoic and vitelline placentae were first described in *Chalcides chalcides*, which is extremely fortunate as this species exhibits the most elaborate placental structure reported so far in reptiles. Unfortunately, the account figures only the external macroscopic features of the embryo-maternal complex and lacks illustrations of the histological relationship between fetal membranes and maternal tissue. However, a detailed description of oviducal and placental histology in *Chalcides ocellatus*, was published later.

Terminology and Definitions

Oviparity, Ovoviviparity, Viviparity and Euviviparity

Most reptiles are *oviparous* as the female lays large shelled polyleathal and fertilized eggs, and most of the embryonic development occurs after egg disposition. However, many species retain their eggs *in utero* for various period; embryonic development occurs in the shelled egg within the uterus for a considerable time before oviposition. In these "egg retainers," the embryo reaches an advanced stage before the egg is laid, and their incubation period outside the parent may be rather short.

Ovoviviparity, as a mode of reproduction distinctly different from viviparity and oviparity, has been characterized by the following :(1) The development of embryo *in utero* until just before hatching. (2) The presence of non-calcified shell membrane at any time during development. (3) The absence of placenta accompanied by the lack of placental nutrition during gestation.

It would appear that the use of these criteria is not that simple and there is difficulty in accurately defining the dividing line between viviparity and ovovivparity. For example, *Thamnophis sirtalis*, a non-calcified shell membrane surrounds the embryos throughout gestation, so that the snake might be regarded as *"ovoviviparous."* Nevertheless, the snake possesses both choriovitelline and chorioallantoic placentae and transmits nutrients across the placenta. Physiologically, this snake should be regarded as *euviviparous*. *Lacerta vivipara* has been considered as *"ovoviviparous"* as its eggs contain a thin shell membrane that remains throughout gestation and is only ruptured after birth. However, the uterine chamber is extensively vascularized, and the uterine epithelium is flattened during gestation. Although these subtle modifications can hardly be regarded as placentation, electrolytes injected into the mother are found later in the embryo.

The term *euviviparity* as defined by *Smith* (1960) is restricted to cases in which the mother contributes nutritional factors to the embryo during gestation, and the transmission of these factors is enhanced by the development of placental structures. Such a situation is extremely rare among reptiles; in fact, there are only two or three species in which a reduced yolk supply and embryonal dependence on placental transmission has been established.

Placental Structures

The term *placenta* may be defined as "any intimate apposition or fusion of the fetal organs to the maternal (or paternal) tissues for physiological exchange." Lizards and snakes show two kinds of placentae : (1) a chorioallantoic, or allantoplacenta, usually overlaying the embryonal or dorsal hemisphere of the blastocyst, and (2) a yolk sac, choriovitelline or omphaloplacenta, usually laying over the abembryonal or ventral pole. In addition, some species possess a special modification of the allantoplacenta in a restricted elliptical area just under the dorsal artery and vein running along the oviduct.

The term *blastocyst* is used to describe the entire reptilian system, that is, egg, embryo, or fetus, extra-embryonic membranes and their derivatives. This usage differs from that for mammals.

The term *allantoplacenta* (or chorioallantoic placenta) is used for placental structures in which the fetal contribution consists of the chorion (ectoderm and mesoderm), allantois (mesoderm and endoderm), and sometimes the amnion, but does not include any part of the yolk-sac endoderm.

Yolk-sac placenta, choriovitelline placenta, or omphaloplacenta define the placental modification of the chorion and uterine epithelium at the abembryonal pole of the blastocyst that differs from the placental modifications elsewhere over the blastocyst. In addition to the chorion, the omphaloplacenta includes a layer of yolk endoderm; this may be the main bulk of the yolk sac or a layer of yolk isolated by the intravitelline mesoderm. The omphaloplacenta may include allantoic elements. The present definition differs fundamentally, from those of *Weekes* (1930) and *Bauchot* (1965) by encompassing the allantoic elements, but it includes the "stenoplacenta" of sea-snakes, and the omphaloplacenta of *Thamnophis sirtalis.*

PLACENTATION IN REPTILES

Vascularization of the Uterus

In most of viviparous reptiles there is a pair of uterine arteries arising from the dorsal aorta; each vessel passes through the dorsal mesometrium to reach the uterus. However, *Enhydris dussumieri* has two pairs of uterine arteries. Typically, a uterine artery runs along the middorsal line of each uterus and gives off parallel branches that pass transversely around the uterus of ramify into a network of capillaries along its ventral surface. The veins draining this network run parallel to the transverse arteries to open into a longitudinal vein that runs along the middorsal line of the uterus. This main uterine vein empties into the posterior caval vein a short distance anterior to its formation by the two efferent renal veins. The only known exception to this plan occurs in the gekkonid *Hoplodactylus maculatus* in which each uterine vein runs along the midventral line of the uterus, but here it also drains into the postcaval vein.

The extensive ramification of blood vessels on the dorsal part of the uterus, the area which forms the maternal part of the elliptical allantoplacenta in *Leiolopisma entrecasteauxii.* This ramification involves short vessels that are given off by the transverse branches of the longitudinal artery. Another special

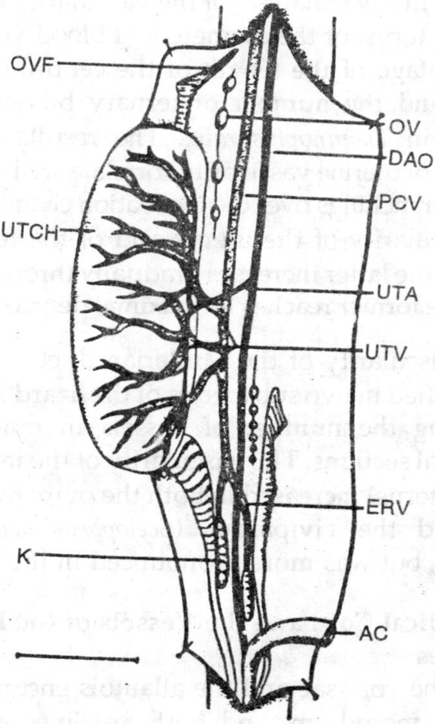

Fig. 4.1. *Enhydrina schistosa*. Uterine blood vessels. AC, Alimentary canal, cloacal region; DAO, dorsal aorta; ERV, efferent renal vein; K, kidney; OV, ovary; OVF, oviducal funnel; PCV, postcaval vein; UTA, uterine artery; UTCH, uterine chamber; UTV, uterine veins. Scale : 40 mm.

ramification of vessels, forming a band-like vascular network, occurs on the ventral aspect of the uterus of *Enhydrina schistosa*; the boundaries of this vascular network correspond exactly to those of the so-called *"stenoplacenta"* (choriovitelline placenta). Tertiary branchings of blood vessels are more apparent over the dorsal part of the uterus in *Thamnophis radix*, than over the poorly vascularized ventral part.

A quantitative analysis of the vascularity over the uterine chamber in terms of the diameters of blood vessels expressed as a percentage of the length of the central one-third of the chamber, and the number of ternary branchings has been attempted in *Thamnophis raiix*. The results of the "linear" estimations of uterine vascularity correlate well with the number of tertiary branchings over the incubation chamber. Comparison of the vascularity of the uterus and of the fetal membranes shows that the latter increases gradually throughout gestation, whereas the former reaches its maximal degree at midgestation.

The vascularity of the Mullerian duct in gravid females was quantified in two subspecies of the lizard *Sceloporus aeneus* by counting the number of vessels in randomly selected longitudinal sections. The vascularity of the incubation region (anterior uterus) increased in both the oviparous (*Sceloporus a. aeneus*) and the viviparous (*Sceloporus aeneus bicanthalis*) subspecies, but was more pronounced in the latter.

The Umbilical Cord and the Vessels of the Fetal Membranes

Both the yolk sac and the allantois encompass a layer of splanchnic mesoderm, and both are lined with endoderm derived from the embryonic hindgut. Upon their emergency form the body of the embryo, they run in parallel along a short "body stalk," in which the vitelline or yolk-sac stalk enters the body just anterior to the umbilical or allantoic stalk. A cross section through the body stalk shows the yolk-sac stalk as a small rounded cavity surrounded-by columnar endodermal cells and is embedded in mesenchyme together with the vitelline artery and vein. The allantoic stalk is accompanied by the right and left umbilical arteries and veins. The vitelline and umbilical arteries arise separately from the dorsal aorta of the embryo, whereas the vitelline and umbilical veins drain into its intrahepatic vein. A short distance from the embryo, the body stalk branches into the ventrally turning yolk-sac stalk and the dorsally turning allantoic stalk. The vitelline blood vessels ramify over the dorsal part of the yolk sac, but do not penetrate it deeply.

The extra-embryonic mesoderm dips into the substance of the yolk at the periphery of the yolk sac. Later, the separation of the somatic and splanchnic layers and the formation of an extra-embryonic coelom cut off an outer layer of the yolk, thereby isolating it almost entirely from the main bulk of the yolk. Apparently, the extension of the extra-embryonic circulation in the embryos of every viviparous reptile examined followed this plan. If the vitelline vessels advance over the yolk sac within the intravitelline mesoderm they would irrigate the main yolk sac, but the complete isolation of the omphalochorionic endoderm (see below) from the main bulk of the yolk sac indicates that it must be vascularized differently. Therefore, it is difficult to understand *Bauchot's* (1965) statement: "Le placenta embryonnaire est donc chorio-vitellin, et la vascularisation est assuree par les vaisseaux vitellins ou omphalomesenteriques." In *Chalcides chalcides,* the capillary bed of the yolk sac is irrigated by vitelline vessels but also receives a vascular strand from the allantoic blood vessels.

TABLE 4.1

Vascular Changes in Embryonic Membranes and Uterine Chamber during Gestation in *Thamnophis radix haydeni*

Approximate day of Gestation :	ca.16	ca.22	24	32	ca.36	41	47	59	75
Zeh Stage:	1-5	6-7	17	20	24	29	32	35	37
Embryonic Vascularity (% linear):	-	-	9	16	21	27	31	35	-
Uterine Vascularity (% linear):	6	7	9	11	12	12	13	11	12
Uterine Vascularity (counts):	16	25	61	125	152	205	188	191	-

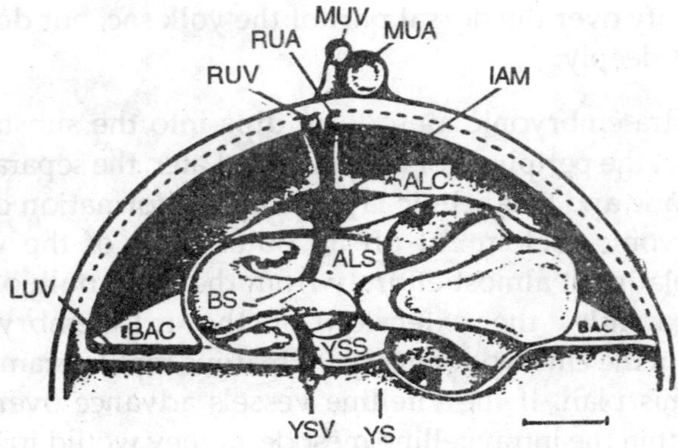

Fig. 4.2. *Egernia cunninghamt* : Section through the embryonic portion of the blastocyst and the surrounding uterus, showing the relations among fetus, umbilical and vitelline stalk, and their blood vessels. ALC, Allantoic cavity; ALS, allantoic stalk; BAC, boundary of allantoic cavity; BS, body stalk (behind the tail); IAM, inner allantoic membrane; LUV, left umbilical vessels; MUA, main uterine artery; MUV, main uterine vein; RUA, right umbilical artery; RUV, right umbilical vein, YS, yolk sac; YSS, yolk-sac stalk; YSV, yolk-sac vessels. Scale : 3 mm.

The umbilical stalk turns dorsally and opens into the allantoic cavity. The right umbilical vessels vascularize the outer allantoic wall. They pass across the cavity of the allantois, where they are enfolded by the allantoic membrane, and in some species, they are suspended by the cellular bridge of allantoic origin. The right umbilical artery and vein branch in the allantoplacenta at the dorsal region of the blastocyst. The left umbilical artery and vein, embedded in the mesenchyme lining the inner wall of the allantois, pass around the posterior end of the embryo.

Hydrophis cyanocinctus also has two allantoic arteries and two allantoic veins. The larger artery and vein, probably the ones on the right side, bridge the allantoic cavity and ramify through the chorioallantoic membrane of the "euryplacenta"

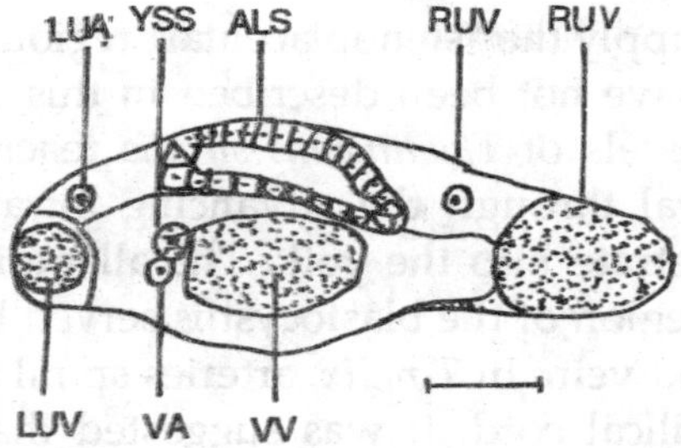

Fig. 4.3. *Sphenomorphus quoyi*. Diagram of section through the body stalk. ALS, allantoic stalk; LUA, left umbilical artery; LUV, left umbilical vein; RUA, right umbilical artery; RUV, right umbilical vein; VA, vitelline artery; VV, vitelline vein; YSS, yolk-sac stalk. Scale : 150 μm.

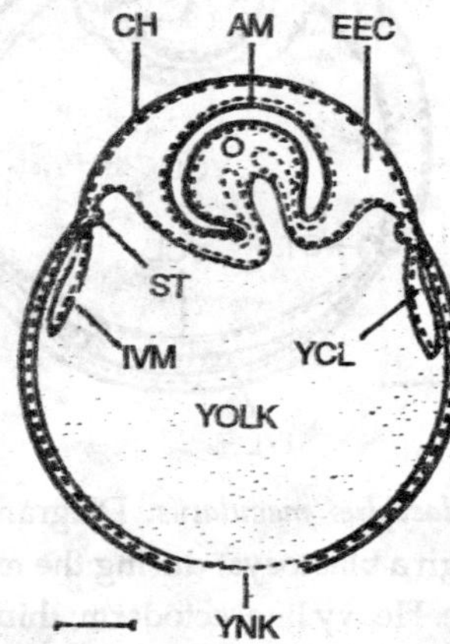

Fig. 4.4. *Hoplodactylus maculatus*. Diagram of transverse section through a blastocyst before the formation of the allantoic vesicle. Heavy line, ectoderm; thin line, endoderm; broken line, mesoderm, AM, Amnion; CH, chorion; EEC, extra-embryonic coelom; IVM, intra-vitelline mesoderm; ST, sinus terminalis; YCL, yolk cleft; YNV, yolk-naval area. Scale: about 1 mm.

(chorioallantoic placenta). The smaller allantoic artery and vein, probably the left umbilical vessels, travel along the inner wall of the allantois to the ventral aspect of the blastocyst; they chiefly supply the "stenoplacental" region. The vitelline blood vessels have not been described in this snake. The vitelline blood vessels of *Thamnophis sirtalis* reach the margin of the yolk naval through the advancing intravitelline mesoderm and penetrate into the yolk. The allantoic circulation in the ventral region of the blastocyst is served by the left umbilical artery and vein. In *T.radix,* arteries spiral around the veins of the umbilical cord. It was suggested that this arrangement may serve as a mechanism for slowing blood flow through the placenta and enhancing the oxygenation.

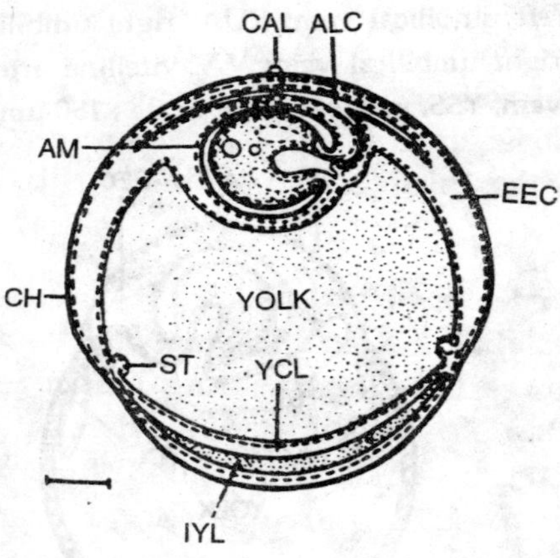

Fig. 4.5. ***Hoplodactylus maculatus.*** **Diagram of transverse section through a blastocyst during the formation of the allantoic vesicle. Heavy line, ectoderm; thin line, endoderm; broken line, mesoderm. ALC, Allantoic cavity; AM, amnion; CAL, chorioallantoic membrane; CH, chorion; EEC, extra-embryonic coelom; IYL, isolated yolk layer; ST, sinus terminalism; YCL, yolk cleft. Scale: about 0.75 mm.**

In nut shell the fetal component of the allantoplacenta at the embryonal pole is supplied by the right umbilical artery and vein. The vascular connection in the omphaloplacental region is more complex : vitelline blood vessels follow the intravitelline mesoderm to irrigate the yolk sac proper, but when the yolk cleft is formed, these vessels are isolated from the chorio-omphalic membrane, which may, at this stage, be supplied indirectly by the allantoic circulation. The placental structures typical to the abembryonal pole in snakes are supplied only with allantoic vessels. In this respect the omphaloplacenta of reptiles differs markedly from that of marsupials or placental mammals, in which it is vascularized by the vitelline blood vessels.

The Omphaloplacenta and its Relationship with the Intravitelline Mesoderm

Lizards

The following description of the structure and development of the omphaloplacenta in lizards is based chiefly on the detailed account on conditions in the gekkonid *Hoplodactylus maculatus.*

At the stage in which the extra-embryonic coelom is limited to the dorsal (embryonal) aspect of the blastocyst, the area vasculosa has already been formed. The sinus terminalis then demarcates the extent reached by the extra-embryonic coelom and the circumference of the area vasculosa. Beyond the sinus terminalis, the mesoderm spreads between the ectoderm and endoderm of the periblast without dividing into splanchnic and somatic layers, which form a trilaminar omphalopleure.

In the lizards studied so far, irrespective of their mode of reproduction the mesoderm at the lower border of the sinus terminalis proliferates and penetrates into the yolk to form the intravitelline mesoderm. This mesoderm extends through the yolk parallel to the omphalopleure. The layer of yolk endoderm outside the intravitelline mesoderm separates from the main bulk of the yolk by a space called the *yolk cleft*. First observed by Virchow, the yolk cleft was assumed to result from the invasion of cells from the superficial endoderm. The yolk cleft of *Lacerta vivipara* and *L. agilis* has been described as a flattened cavity surrounded by cells of uncertain origin. In *Hoplodactylus maculatus*, only the inner face of the yolk cleft is lined by the intravitelline mesoderm; the outer face is bound by the isolated yolk endoderm.

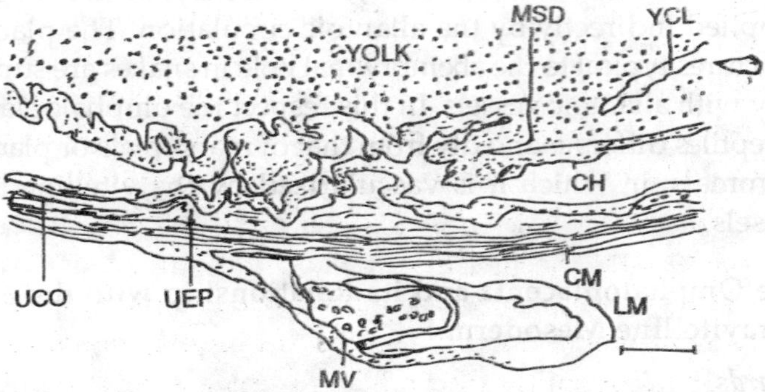

Fig. 4.6. *Hoplodactylus maculatus.* **Transverse section through the pleated part of the omphaloplacenta. CH, Enlarged ectodermal cells of the chorion; CM, circular muscle of the uterus; LM, longitudinal muscle; MSD, mesoderm in chorionic pleats; MV, main uterine vein; UCO, uterine corium, UEP, uterine epithelium, YCL, yolk cleft. Scale: 100 μm.**

At a later stage in *Hoplodactylus maculatus*, the chorionic ectodermal cells lying outside the isolated yolk layer increase in height, and a band of columnar cells develops parallel to the yolk cleft. The latter cells from pleats which, together with mesodermal cells, comprise the fetal part of the omphaloplacenta. Yolk endoderm penetrates the base of the pleats. The adjacent uterine epithelium doubles in thickness, and non-ciliated cells increase in number and size, which compresses the ciliated cells. The largest cells are in close proximity to the main uterine vein, which in this lizard, runs along the midventral line of the uterus. In the uterus, capillaries are quite abundant. However, on the fetal side, the blood vessels nearest to the uterus are those of the intravitelline mesoderm bounding the inner side of the yolk cleft. In later stages, the omphaloplacental modifications shift posteriorly, and the isolated yolk layer is gradually absorbed. Concomitantly, the mesoderm of the omphalopleure splits into splanchnic and somatic layers, thus extending the extra-embryonic coelom over the ventrolateral

side of the blastocyst. The yolk cleft and the extra-embryonic coelom continue to extend toward the abembryonal pole however, these cavities always remain separated from each other by splanchnic mesoderm. The yolk cleft is gradually lost the isolated yolk layer disappears. The allantoic limbs expand ventrally within the extra-embryonic coelom and cover almost the entire yolk sac. Furthermore, the chorionic ectodermal cells flatten, resembling those of the allantoplacenta.

In *Sphenomorphus quoyi* and in *Leiolo spenceri*, the formation and fate of the intravitelline mesoderm may be different. The sinus terminalis has been described as a *node* surrounding the area vasculosa. The mesodermal cells, unable to "force their way" between the chorionic ectoderm and yolk-sac endoderm, "dip into the yolk-sac and continue their growth round the yolk-sac not over its surface as is usually the case, but embedded in its substance." Hence the omphalopleure in these lizards is bilaminar consisting of ectoderm and endoderm only. Weeks (1927) regarded the yolk cleft as the extra-embryonic coelom and described its formation by splitting of the mesoderm within the yolk sac to the somatopleural and splanchnopleural layers. However, Weekes was aware of the abnormal position of the coelom, which lies actually within the yolk sac; thus, the somatopleure is formed from mesoderm, endoderm, and ectoderm.

The first sign of omphaloplacentation in *Sphenomorphus quoyi* is evident by the stage in which the epiblast has only just covered the yolk. The chorionic ectodermal cells around the abembryonic pole enlarge and form a layer of columnar epithelium that is closely attached to the underlying layer of yolk-sac endoderm. In contrast, the apposing uterine epithelium only enlarges at a later stage. By the fourth week of gestation, the omphaloplacenta is fully developed. At this stage the chorionic ectoderm is composed of several layers of cells, sometimes with finger-like projections, and is underlain by a layer of vacuolated yolk-endoderm cells, which, in turn, is bound by a layer of non-vascularized extra-embryonic mesoderm. The uterine wall at this region is thicker than elsewhere, due to hypertrophy of the layer of muscle and the enlargement of the epithelial cells. The remnants of the shell membrane, cell debris, and uterine secretion accumulate as a "coagulum" between the chorion and the uterus. The chorionic cells appear to phagocytize this coagulum. At the periphery of the omphaloplacenta, the height both of uterine components and the chorionic epithelium decreases gradually. At the omphaloplacental region, the uterus is markedly vascularized; no such vascularization is notable in the chorio-omphalic membrane. Weeks was convinced that this omphaloplacenta is actually functional stating : "The maternal secretion absorbed by the chorionic ectoderm must be passed through the narrow layer of yolk-sac endoderm and somatopleural mesoderm and across the cavity of the extra-embryonic coelom before it reaches the blood vessels in the yolk-sac."

Toward the end of gestation in *Sphenomorphus quoyi,* the yolk mass has been reduced and the allantois extends ventrally along the extra-embryonic coelom and attaches to the remnants of the isolated yolk layer; this results in a composite chorio-omphalo-allantoic membrane. The isolated yolk endoderm disappears during the later stages, and the chorion of the omphaloplacenta becomes a part of the chorioallantoic placenta. Also, the uterine epithelium is modified again to form the maternal part of the allantoplacenta at the ebembryonal pole.

The omphaloplacenta in *Leiolopisma spenceri* develops similarly. However, the isolated layer of yolk endoderm is barely distinguished because its yolk spherules have already been absorbed at the stage of omphaloplacental maturity (three weeks of gestation). The chorionic portion of the omphaloplacenta at the base of yolk sac does not modify; here the ectodermal cells are rather flat. As in *Hoplodactylus maculatus,* the uterine epithelium is slightly enlarged mainly at the periphery around the base of the yolk sac. Throughout, this epithelium is underlain by a network of capilaries. The fetal blood vessels nearest to the chorion and penetrating into the yolk sac are those that develop from the splanchnic mesoderm. The extra-embryonic coelom lies between these blood vessels and the omphalochorionic membrane as in *S. quoyi*.

The development of the extra-embryonic mesoderm appears to differ in the three species of *Egernia* studied by *Weekes* (1930). It is possible that their extra-embryonic mesoderm extends from the edge of the area vasculosa, as in *Hoplodactylus maculatus*, beneath the chorion, and also cuts through the yolk. In these lizards, the term *omphalochorion* was used to refer to the united membrane that is composed of chorionic ectoderm and mesoderm and the isolated yolk endoderm (a trilaminar omphalopleure). *Weekes* (1935) clarified this point further in a review published near the end of her studies on various squamates. She states that in lizards and snakes "the mesoderm cells at the circumference of the area vasculosa then divide to form a thick margin of cells, which continue their growth round the yolk sac embedded in the actual substance of the sac, as well as over the surface of the sac beneath the yolk-sac ectoderm."

In *Egernia*, the enlarged chorionic cells are sometimes arranged in patches. The accumulation of debris at the base of the blastocyst may interfere with the growth of ectodermal cells. However, this may not be a general condition as the chorionic ectodermal cells of *Chalcides chalcides* proliferate and form ridges while actually embedded in the "detritus". Also the chorionic epithelium surrounding the yolk-sac of *Chalcide* is linked on its inner aspect with a richly vascularized mesodermal layer of undescribed source, although its blood vessels are connected with the allantoic ones.

Tiliqua nigrolutea and *Sceloporus jarrovi* have a folded chorionic membrane that interdigitates loosely with the uterine epithelium at the omphaloplacenta. The chorionic ectoderm at the omphaloplacental region of *Xantusia vigils* is enlarged and has some fingerlike projections. The uterine epithelium is enlarged as well. A coagulum, consisting of remnants of the

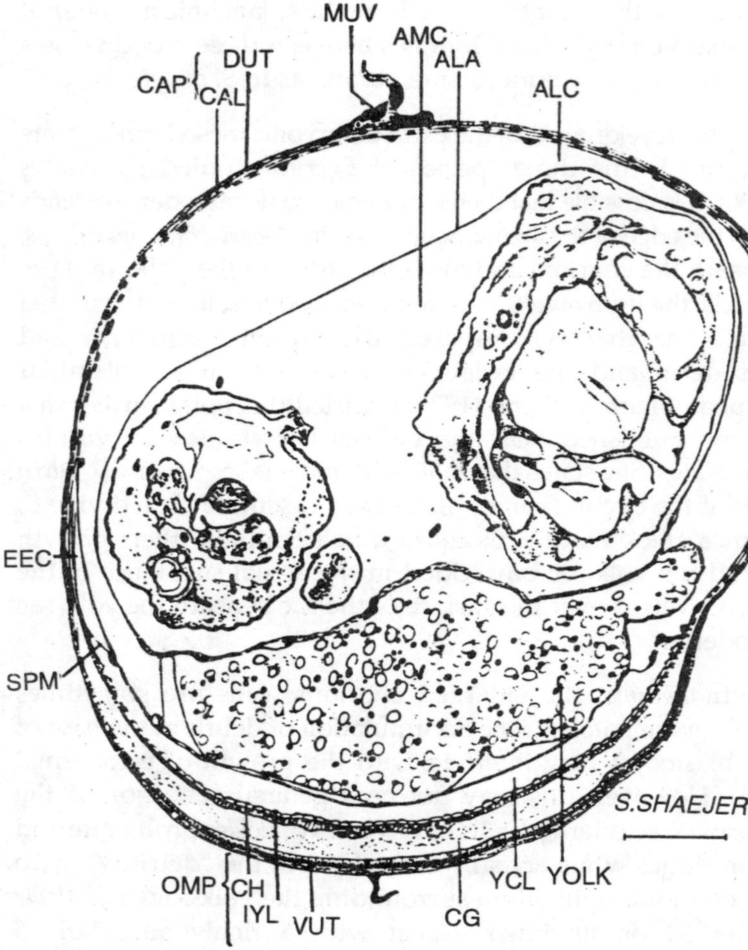

Fig. 4.7. *Xantusia vigilis*. Semidiagrammatic section through a blastocyst *in utero*. The heights of uterine and chorionic epithelia are exaggerated. ALA, Inner allantoic and amniotic combined membrane; ALC, allantoic cavity; AMC, amniotic cavity; CAL, chorioallantoic membrane; CAP, chorioallantoic placenta; CH, chorion; CG, coagulum; DUT, dorsal part of uterus; EC, extra-embryonic coelom; IYL, isolated yolk layer, MUV, main uterine vessels; OMP, omphaloplacenta; SPM, splanchnic mesoderm separating the extra-embryonic coelom from the yolk cleft; VUT, ventral part of uterus; YCL, yolk cleft. Scale: (generally) 1 mm.

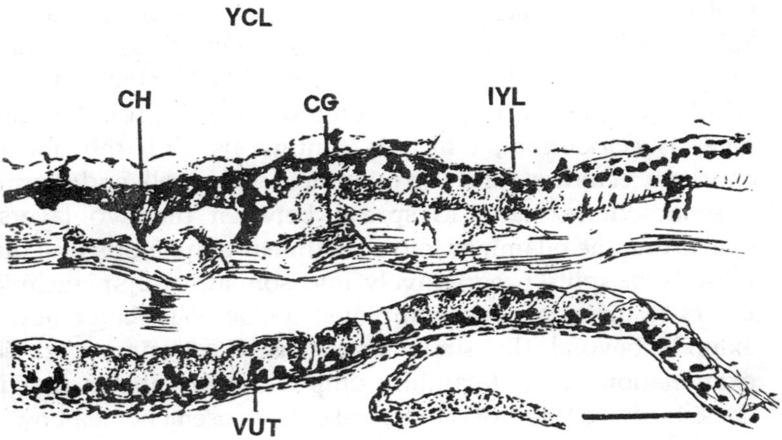

Fig. 4.8. *Xantusia vigilis*. Omphaloplacenta at a stage calculated as five weeks after ovulation. CG, Coagulum; CH, enlarged chorionic cells forming ridges; IYL, isolated yolk layer, VUT, ventral part of the uterus; YCL, yolk cleft.

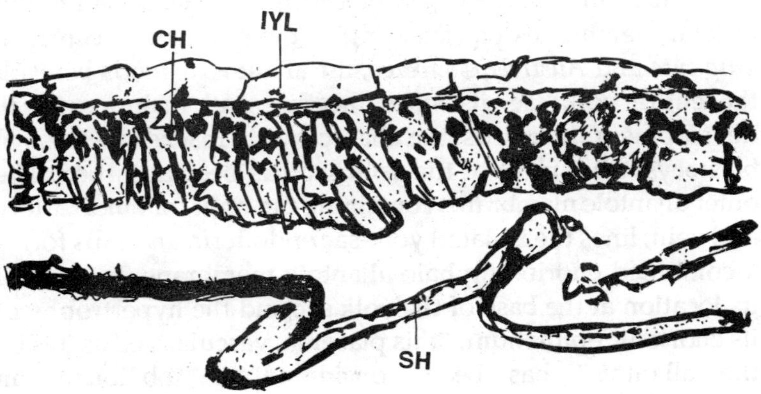

Fig. 4.9. *Xantusia vigilis*. The chorio-vitelline membrane (part of the omphaloplacenta). CH, Enlarged ectodermal cells of the chorion, IYL, vaculated calls of the yolk layer containing remnants of yolk globules; SM, remnants of the shell membrane (coagulum).

shell membrane, separates the chorion ectoderm from the uterine epithelium at the base of the blastocyst. In *Chalcides ocellatus*, the ventral region of the blastocyst is occupied by the yolk-sac, which is enclosed in a bilaminar omphalopleure. The outermost layer is ectodermal, which is continuous with the dorsal chorionic ectoderm and the innermost is the yolk endoderm. The mesoderm starts to spread between the two layers, converting the bilaminar into a trilaminar omphalopleure; the mesoderm splits progressively into somatic and splanchnic layers to form the extraembryonic coelom. The latter never extends beyond the sinus terminalis. Electronmicroscopic examination of the bilaminar omphalopleure shows that it consists of two layers of ectodermal cells, a basement membrane, and yolk endoderm. The innermost ectodermal layer is thin and composed of flattened cells, and the outermost is formed by enlarged cells that send out elongate microvilli or pseudopodia-like processes. The process are embeded in an amorphous substance of low electron density.

Snakes

Snakes differ markedly from lizards in the interrelationship of fetal membranes participating in the omphaloplacenta. In *Suta suta* and *Austrelaps superba*, the allantois extends beneath the entire yolk sac, its sides meeting and fusing early in development. In lizards, the allantois only reaches the base of blastocyst after most of the yolk sac has been absorbed. The outer allantoic membrane, containing the right umbilical artery and vein, lines the isolated yolk-sac endoderm and thus forms a combined chorio-omphalo-allantoic membrane. In spite of its location at the base of the yolk sac and the hypertrophy of its chorionic epithelium, "this placenta, vascularized as it is by the allantois" has been considered as "obviously an allantoplacenta."

In *Storeria*, *Virginia*, and *Thamnophis*, the uterine epithelium adjacent to the base of the yolk sac and the chorionic ectoderm in this region contain very large columnar cells that "cytologically suggest an active exchange of substances." The transition in the region of the yolk sac has been considered to represent a "typical decidual reaction." No details are available on the extent of the allantois in these placentae. However, the chorionic modification at this site differs from the modification at other regions of the blastocyst.

Apparently, a reluctance to use the term *omphaloplacenta* for a structure containing allantoic membrane and umbilical blood vessels led Kasturirangan (1951a, b) to name the long and narrow band of tissue in *Enhydrina schistosa* and *Hydrophis cyanocinctus a stenoplacenta*. In these snakes, the fetal part of the placenta is formed by chorionic ectoderm, an attached thick, spongy layer of yolk endoderm (isolated yolk layer) and a vascularized allantoic mesoderm. The allantoic, endoderm stops short on either side without meeting at the midventral line of the blastocyst. The description of Kasturirangan leave unclear the development of the chorio-omphalo-allantoic membrane, the role of the yolk cleft and the fate of the intravitelline mesoderm, as the study did not include sufficiently early stages of development.

The formation of the yolk cleft and the intravitelline mesoderm has been described in detail in *Enhydris dussumieri*. The mesoderm ventral to the sinus terminalis penetrates into the substance of the yolk sac as well as beneath the ectoderm around the sac. The ingrowth of the intravitelline mesoderm forms the yolk cleft, which is bound by the mesoderm on its inner face and by yolk endoderm on its outer face. The extra-embryonic coelon derives from the spilit of the intravitelline mesoderm. The allantoic llimbs penetrate into the extra-embryonic coelom, and their outer wall attaches to the omphalopleure. Hence, the joint chorio-omphalo-allantoic membrane is composed of chorionic ectoderm, a thin layer of mesoderm, endoderm of the isolated yolk layer, intravitelline mesoderm, outer allantoic mesoderm carrying blood vessels, and allantoic endoderm. The available illustrations leave it unclear whether the allantoic cavity is continuous beneath the yolk sac or closed as it is in *Enhydrina* and *Hydrophis*. The absorption of yolk throughout gestation gradually reduces the isolated yolk endoderm; the vascularized allantoic mesoderm then assumes a position near the chorionic ectoderm. Thus, with the disappearance of the isolated yolk layer, the chorio-omphalo-allantoic membrane becomes chorioallantoic and the allantoplacenta extends through the abembryonal region.

In *Thamnophis sirtalis*, the allantois penetrates beneath the yolk-sac, as it does in the snakes previously described. In the abembryonal region, the allantois grows within a space formed by the intravitelline mesoderm; it is not absolutely clear whether this space is the yolk cleft or the extra-embryonic coelom, thus "the yolk cleft is not continuous with the extra-embryonic coelom at this time as the former space begins as a blind cavity just below the sinus terminalis." No further elaboration of the fate of these cavities is given. The omphaloplacental membrane in *Thamnophis* does contain an isolated yolk endoderm bound externally by ectoderm and internally by a layer derived from the intravitelline mesoderm. A space, probably representing the yolk cleft, separates the omphaloplacental membrane from the outer allantoic wall. The allantoic lumen is constricted by the abembryonal region by the fusion of outer and inner allantoic walls. The omphaloplacenta of *Thamnophis* is attached to the allantois only at the circumference, and allantoic vessels only indirectly supply it by the connection at its margin. The isolated layer of the yolk persists throughout gestation.

In *Thamnophis radix* the fetal part of the omphaloplacenta is formed also by enlarged chorionic ectodermal cells, a layer of endoderm (probably isolated yolk layer) and a layer of mesoderm of unstated origin. The blood supply to the mature omphaloplacenta is rather poor because the separation of the placenta from the yolk sac proper disconnects the vitelline blood supply.

The origin description of placentation in the garter snake suggested that a layer of mesoderm lies between the chorionic ectoderm and the isolated yolk layer of the omphaloplacenta.

This situation would have been similar to that in *Egernia, Hoplodactylus* and *Enhydris* (Parameswaram, 1962). However, electronmi-crographs of this area indicate that the compressed layer is a basal layer of chorionic epithelium, situated on the same thick basal membrane and that it is not mesodermal. A similar situation is reported also in the lizard *Chalcides ocellatus.*

THE CHORIOALLANTOIC PLACENTA

Formation of Fetal Membranes of the Allantoplacenta

The reptilian amnion is formed in a manner typical of that in other sauropsids. The amniotic folds rise on both sides

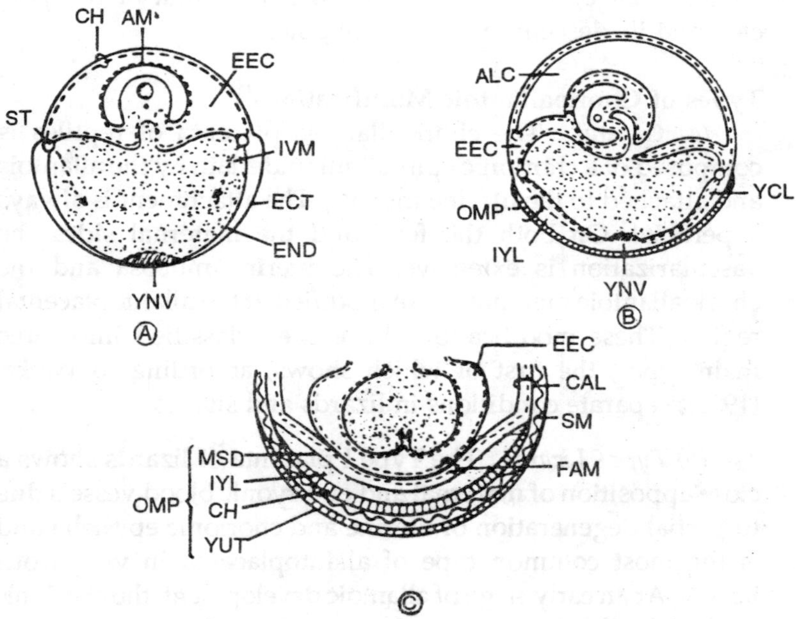

Fig. 4.10. *Thamnophis sirtalis*. Diagram of sections through blastocyst at various stages of development (A, B, C). ALC, Allantoic cavity; AM, amnion; CAL, chorioallantoic membrane; CH, chorion, ECT, chorionic ectoderm: EEC, extra-embryonic coelom; END, yolk-sac endoderm; FAM, fused allantoic membranes, IVM, intravitelline mesoderm; IYL, isolated yolk layer; MSD, mesoderm; OMP, omphaloplacenta; SM, shell membrane; ST, sinus terminalis; YCL, yolk cleft, YNV, yolk naval; VUT, ventral part of uterus.

above the embryo. In this way, an amniotic cavity is formed around the embryo and is separated from the chorion by the extra-embryonic coelom. The allantois is formed as a small vesicular diverticulum protruding from the hindgut. The vesicle expands and continues to enlarge until it fills completely the extra-embryonic coelom at the embryonic region of the blastocyst. Usually, the outer layer of the allantois joins with the chorion to form the chorioallantoic membrane. In lizards, the allantois is initially restricted to the embryonal region and only gradually grows under the yolk sac as the latter is resorbed. In snakes, however, the allantois grows around the yolk sac at an early stage, concomitant with the formation of the yolk cleft and its descent to the ventral pole.

Types of Chorioallantoic Modifications

(a) General : The chorioallantoic placenta of reptiles is composed of a chorionic epithelium underlain by the allantois and attached to the uterine mucosa. The attachment is always superficial. On both the fetal and the maternal sides the vascularization is extensive. The uterine mucosa and the chorioallantoic membrane are modified at the allantoplacental region. These modifications have been classified into three main types, the first of which shows, according to *Weekes* (1955), separate conditions in lizards and snakes.

(b) Type - Lizards : The Type I placenta of lizards shows a close apposition of maternal and embryonic blood vessels due to partial degeneration of uterine and chorionic epithelia and is the most common type of allantoplacenta in viviparous lizards. At an early stage of allantoic development, the chorionic epithelial cells appear to absorb the shell membrane. The uterine capillaries expand and start to invade the epithelium, which is still intact at this stage. The degeneration of the uterine glands the muscle layers, and the epithelial cells reduces the thickness of the uterine wall in the area of the allantoplacenta.

By the second week of gestation of *Sphenomorphus quoyi,* the mesoderm of the allantoic vesicle and the chorion fuses, thus obliterating the extra-embryonic coelom. The capillaries of the chorioallantois are larger than those of the uterus. The blood corpuscles of the fetus (probably erythroblasts) with their round nuclei can be distinguished from the maternal erythrocytes with their void nuclei. By the fourth week of gestation, the uterus thickens again. Numerous capillaries reach the surface of the uterine epithelium and are separated from the underlying fetal tissue only "by their own endothelial walls and perhaps a thin layer of maternal cytoplasm." The capillary walls are thin but definitely recognizable. Some of the uterine epithelial cells degenerate and contain large vacuoles. Clumps of epithelial nuclei are grouped among the invading capillaries. Although the chorionic epithelium is attached to the uterine epithelium, the connection is such that the line of division between maternal and fetal tissues is often visible; chorionic and uterine cells do not mingle. The allantoic cavity may be partly obliterated by fusion of the outer and inner walls of the allantois with the amnion. The area of chorioallantoic placenta extends towards the end of gestation when the allantoic limbs grow beneath the reduced yolk sac.

This type of allantoplacenta occurs in *Egernia cunninghami, E, whitii, E, striolata, Tilqua nigrolutea, T. scincoides, Hemiergis quadrilineatum, Mabuya multifasciata, Liolaemus gravenhorstii*, and *Sceloporus aeneus bicanthalis*. The allantoplacenta of *Hoplodactylus masculatus* is clearly of this type, but it lacks any form of attachment between the chorionic and uterine epithelia.

The study of Giacomini (1960) ,suggests that the allantoplacenta of *Chincides ocellatus* also belongs to this category. Examination of embryos at more advanced stages shows that the smooth close apposition between the uterine wall and the embryonic membranes is interrupted by the formation of an apparently specialized median area at the embryonic hemisphere of the blastocyst. The uterine wall is raised into folds, whereas its epithelium is greatly enlarged and becomes columnar in shape. The chorioallantoic membrane also becomes folded, and the allantoic vessels increase in number and enlarge and resemble the Type III allantoplacenta. However, the figure shows that the chorionic ectoderm is largely eroded, exposing many allantoic capillaries, and the uterine columnar epithelial cells are clustered in pleats without forming an even epithelium throughout. Moreover, an electronmicrograph of the allantoplacenta in *Chalcides ocellatus* shows a close proximity of maternal and fetal capillaries. It seems appropriate, therefore, to include the placenta of this lizards in Weekes' Type I.

In the yucca night lizard, *Xantusia vigilis*, "the center of the allantoplacenta is marked by a large conical plug of villus-like projections covered by columnar ciliated cells." Also "the maternal membrane projects in villus-like ridges covered with columnar epithelium". However, a figure (2D), captioned "the allantoplacenta," shows yolk granules and large vacuoles adjacent to the tall columnar chorionic ectodermal cells, a situation that is generally typical of the omphaloplacenta. Miller does not refer to this description and states that "a chorioallantoic placenta of a non-interdigitating simple type is formed, which lies in intimate contact with the thin glandular wall of the oviduct." The allantoplacenta of *Xantusia* has been re-examined at a stage calculated as five weeks or less after ovulation. The embryonic component of the allantoplacenta then consists of a thin, but highly vascularized chorioallantoic membrane. The uterine epithelium is squamous with capillaries bulging between the epithelial cells. The chorioallantoic membrane lacks columnar cells at the embryonal region of the blastocyst, but such cells do occur in ridges of the omphaloplacenta at the abembryonal region. Hence, this allantoplacenta is, undoubtedly, of Type I.

The allantoplacenta of *Sceloporus jarrovi* shows some hypertrophy of the uterine epithelium in areas immediately surrounding the major blood vessels of the uterus. Nevertheless, the chorionic ectoderm is extremely thin, even in areas facing the hypertrophied uterine epithelium and does not interdigitate with it. It seems appropriate, therefore, to classify this placenta as a Type I placenta.

Even the non-placental *Lacerta vivipara* has well vascularized chorioallantoic membrane and uterine wall. However, fetal and maternal tissues are not in contact as they are separated by the shell membrane (membrane coquilliere) which persists throughout gestation.

(c) Type I - Snakes :Type I allantoplacenta of snakes differs from that of lizards because the chorioallantoic membrane is wrapped around folds of the uterine wall. These folds occur in some restricted areas of the embryonic hemisphere, mainly under the main uterine blood vessels, where the right umbilical artery and vein reach the outer allantoic membrane.

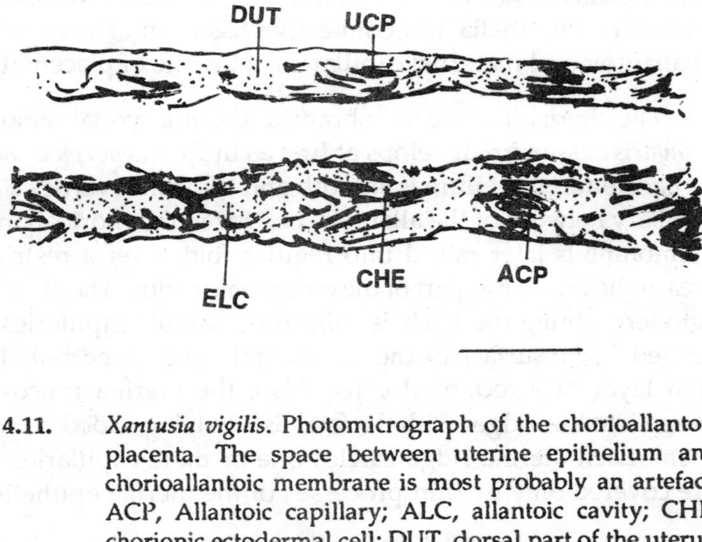

Fig. 4.11. *Xantusia vigilis*. Photomicrograph of the chorioallantoic placenta. The space between uterine epithelium and chorioallantoic membrane is most probably an artefact. ACP, Allantoic capillary; ALC, allantoic cavity; CHE, chorionic ectodermal cell; DUT, dorsal part of the uterus; UCP, uterine capillary. Scale : 20 μm.

Austrelaps superbus and *Suta suta* show two areas of chorioallantoic placentation over the embryonic region of the blastocyst. Over most of this region, the placenta fits the description of the non-ridged placenta Type I of lizards. However, at the posterior end of the embryo, where the right umbilical vessels pass across the cavity of the allantois, the uterine wall is raised into vascular ridges, around which the fetal tissue folds. As in the non-ridged area, the fetal capillaries are in close apposition to the exposed capillaries of the uterus. The histology of the "euryplacenta" of *Enhydrina schistosa* and *Hydrophis cyanocinctus* is similar to that previously described as Type I. However, the "true exposure" of allantoic capillaries" is observed here and there but not uniformly all over the euryplacental region."

The allantoplacenta shown near-term by *Nerodia sipedon* and *N. 'cyclopion* is a simple non-ridged chorioallantoic placenta similar to that described in *Enhydrina* and was regarded as an "endothelo-endothelial" type. According to the classification of mammalian placentae, this would indicate erosion of chorionic and uterine epithelia and connective tissue and the exposure of allantoic and maternal capillaries at the site of placentation.

The chorioallantoic membrane at the embryonal region of *Enhydris dussumieri* develops at first as in *Sphenomorphus quoyi.* Its chorionic epithelium is smooth and thin and overlies a rich capillary network of the allantois. However, the chorioallantoic membrane is later raised into regular folds over a restricted area at the middorsal part of the embryonic region. The chorionic ectoderm lining the folds is thin; the allantoic capillaries are lodged at the surface of the ridges and are covered only by a thin layer of ectodermal cells. Also, the uterine mucosa is raised into low ridges with the fetal tissues lying folded between them. Each uterine ridge carries one or more capillaries that are covered only by thin processes of the uterine epithelium.

In *Thamnophis sirtalis,* the allantois inflates between Zehr Stages 18 and 21. By Stage 21, it has already fused with the chorion above the embryo to form the chorioallantoic membrane, which is highly vascularized, and it is made up mainly of a flattened layer of mesenchyme surrounding the blood vessels. A squamous layer of allantoic endoderm underlies the inner aspect of the membrane, and the outer aspect is covered by the chorionic ectoderm. The chorionic epithelial cytoplasm is thinned where it covers the allantoic capillaries, so that the combined chorionic and endothelial cytoplasmic layers do not exceed 2 μm. The thickness of the uterine wall surrounding the chorionic vesicle is relatively uniform for the first two to three weeks of gestation. The epithelium, which has been secretory before ovulation, is no longer secretory thereafter. On the dorsal and lateral walls of the uterine chamber, the epithelium remains low, and the capillary network becomes denser throughout gestation. Capillaries push up against the epithelium and bulge out slightly, but no "exposed capillaries" or "erosion of uterine epithelium" are observed.

In *Thamnophis radix,* the allinoplacenta is of Type I. It is uncertain whether the uterine and chorioallantoic membranes are raised into ridges *in situ* or whether the ridges that have been illustrated result from unstretching of the tissues during fixation.

In *Vipera berus* the oviduct and the chorioallantoic membrane have interdigitating villi. The uterine epithelium is thinner at the site of placentation than elsewhere. The shell membrane persists around the entire blastocyst at a late stage of pregnancy. it would appear that the placenta is of Type I rather than Type III, as suggested by the authors.

(d) Type II : The Type II allantoplacenta is characterized by the bulging of maternal capillaries, which are covered by a thin layer of glandular epithelium, and by the enlarged cells of the chorionic ectoderm. A Type II allantoplacenta has only been described in the three Tasmanian skinks, *Leiolopisma pretiosa, L. ocellata, and L. metallica*. This type of placenta does not appear to have been re-examined since its original description. The inner face of the uterus is raised into shallow ridges, each containing a blood vessel surrounded by a thin epithelium. The chorionic epithelium is pressed against the uterine ridges; extensions of its cells fit into the grooves between them.

(e) Type III : Type III is the most elaborate type of reptilian placenta. It was initially described in *Chalcides chalcides,* and later in *Leiolopisma entrecasteauxii* and in *Leiolopisma spenceri.*

The Type III placenta is characterized by an elliptical area of apposed, folded, and glandular maternal and embryonic tissues that immediately underlie the main longitudinal uterine blood vessels. The uterine mucosa is thrown into deep and numerous folds or villi that interdigitate in an intricate manner with folds or villi of chorioallantoic membrane. Each uterine villus has a core of capillaries of various sizes and is covered by a single layer of enlarged ciliated epithelial cells. These cells seem to be glandular; occasionally they are shed and may be found in the uterine lumen.

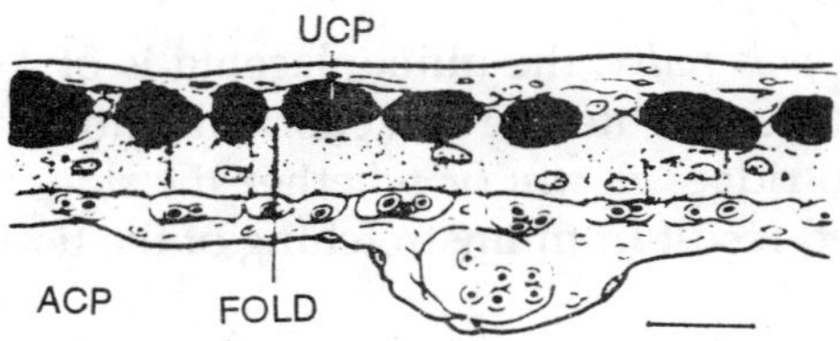

Fig. 4.12. *Leiolopisma ocellata*. Section of mature allantoplacenta (Type II). ACP, Allantoic capillary; UCP, uterine capillary. The contour of the chorionic ectoderm (FOLD) fits into shallow grooves between uterine capillaries. Scale 30 μm.

The fetal part of the placenta is composed of very tall and ciliated chorionic epithelial cells and a relatively thick allantoic membrane. Desmosomes commonly connect the chorionic cells; some of these cells are binucleate, and all are highly vacuolated. At the base of the chorionic epithelium, there are numerous capillaries and a loose allantoic connective tissue. The elliptical placenta of *Leiotopisma entrecasteauxii* is less complex than is that of *Chatcides chalcides;* and the placenta of *Leiolopisma spenceri* is even less developed. However, the epithelia of both uterus and the chorion of all three species contain characteristically high columnar and glandular cells. The difference lies only in the depth of interdigitation between maternal and fetal tissues. At the periphery of the elliptical zone, the allantoplacenta resembles the Type I placenta of lizards.

Summary and General Comments

All placental squamates develop a chorioallantoic placenta over the dorsal hemisphere of the blastocyst. Typically, the maternal and fetal capillaries of the allantoplacenta are in

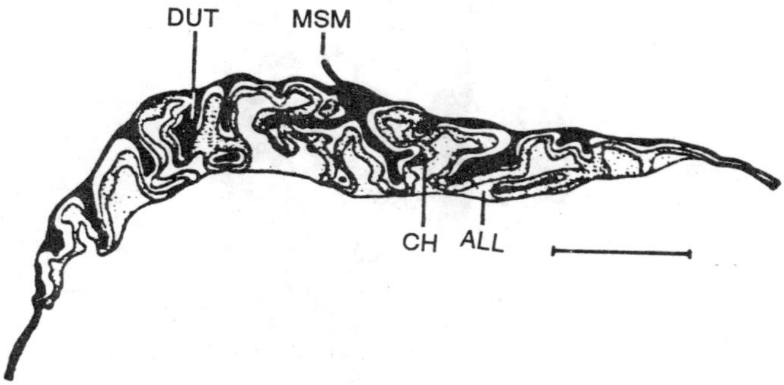

Fig. 4.13. *Chalcides chalcides.* Section through the elliptical chorioallantoic placenta (Type III). ALL. Allantois; CH, chorion; DUT, dorsal part of uterus raised into a fold; MSM, mesometrium. Scale: 1mm.

close apposition and the apposing epithelia may be partly eroded. Some degree of folding and interdigitation between the chorioallantoic membrane and uterine epithelium occurs in an area just below the main uterine blood vessels in several lizards. In these cases, there is some hypertrophy of the uterine epithelium, which generally does not interfere with the close apposition of maternal and fetal capillaries. More material should be examined in order to re-evaluate properly the distinction between allantoplacenta Type I lizards and of snakes. Rarely there is a lower than normal amount of yolk in the egg and an addition "elliptic" placenta situated just under the main uterine vessels. *Weekes* (1930) suggested that the phylogeny of the Type III placenta of *L. entrecasteauxii* passed through a stage now represented by the mature placenta in *Leiolopisma spenceri.* The placenta of *L. ocellata, L. metallica,* and *L. pretiosa* may be regarded as a further specialization of the condition found in *Sphenomorphus quoyi.*

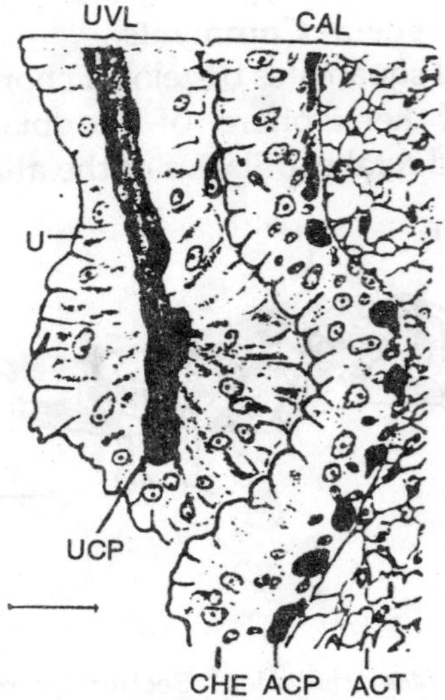

Fig. 4.14. *Chalcides chalcides*. Section through a uterine villus and adjacent chorioallantoic membrane. ACP, Allantoic capillary; ACT, allantoic connective tissue; CAL, chorioallantoic membrane; CHE, chorionic epithelium; U, terine epithelium, UCP, uterine capillary; UVL, uterine villus. Scale: 30 µm.

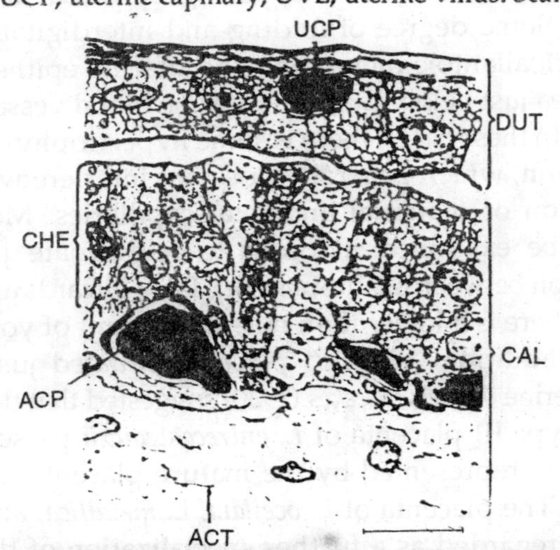

Fig. 4.15. *Chalcides chalcides*. Details of the elliptic allantoplacenta (Type III). ACP, Allantoic capillary; ACT, allantoic mesoderm; CAL, chorioallantoic membrane; CHE, chorionic epithelium, DLT, dorsal part of uterus, UCP, uterine capillary. Scale: 15 µm.

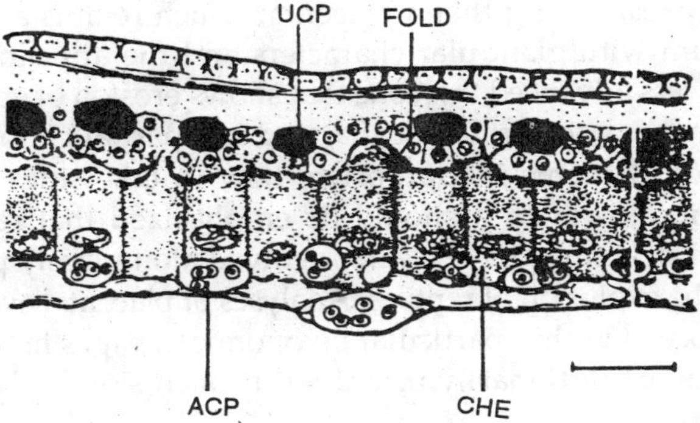

Fig. 4.16. *Leiolopisma spenceri*. Section of the elliptic allantoplacenta. ACP, Allantoic capillary, CHE, chorionic epithelium; FOLD, fold of uterine epithelium; UCP, uterine capillary. Scale : 50 μm.

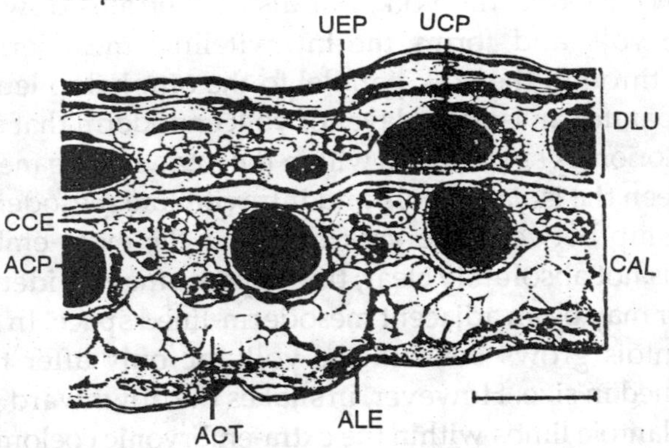

Fig. 4.17. *Chalcides chalcides*. Section of the allantoplacenta beyond the elliptical area (non-ridged area). ACP, Allantoic capillary; ACT, allantoic connective tissue (mesoderm); ALE, allantoic epithelium (endoderm); CAL, chorioallantoic membrane; CCE, syncytial chorion ectoderm; DLU, dorsal-lateral part of uterus; UCP, uterine capillary, UEP, uterine epithelium. Scale : 15 μm.

Bauchot (1965) applies the classification of mammalian placentae to the squamate condition and regard the allantoplacenta of Types II and III as "epithelochorial" and the allantoplacenta of Type I as "syndesmochorial." He disagrees with Weekes, stating that a placenta, which retains a uterine epithelium with glandular characters and the appearance of pleats, is less advanced than one that shows erosion of epithelia, has a tendency to form syncytia, and has partial exposure to capillaries. It seems inappropriate now to take sides in this argument. However, it should be emphasized that reptilian placental structures have clearly evolved along many parallel lines in the evolution of reptiles. Analyses of placental evolution must proceed within particular taxonomic lineages in order to achieve an evolutionarily meaningful conclusion.

The development of the omphaloplacenta at the abembryonal pole is more complex. Unlike the situation in birds, the mesoderm ventral to the sinus terminalis does not grow only around the yolk, but also penetrates downwards into the yolk and forms the intravitelline mesoderm. This extends through the yolk parallel to the omphalop leure and, thus, separates an isolated layer of yolk endoderm that adheres to the chorion. A space forms within the intravitelline mesoderm or between the isolated yolk and intravitelline mesoderm. The relationship between the yolk cleft and the extra-embryonic coelom is not absolutely clear; these spaces may be identical or the latter may be an adjacent mesoderm-lined space. In lizards, the allantois grows beneath the yolk sac only after this has diminished in size. However, in snakes the downward growth of the allantoic limbs within the extra-embryonic coelom occurs at an early stage, and the limbs meet under the yolk-sac. The omphaloplacenta, as defined in the present chapter, is a combination of the hypertrophied chorionic epithelium, underlain or not by chorionic mesoderm, the attached yolk endoderm, the enlarged epithelial cells of the uterine mucosa, and the adjoining blood vessels. The vascular connections of the omphaloplacenta differ markedly between lizards and snakes. Both lack a direct connection between the vitelline blood vessels and the chorion and show only an indirect link with the extra-embryonic splanchnic mesoderm at the periphery of the omphaloplacenta. However, snakes have the parietal layer of the allantois irrigated by the umbilical vessels lying close to the fetal layers of the omphaloplacenta. This unusual arrangement led some authors to argue that the structure seen in snakes is not an omphaloplacenta, although the modifications of its chorionic and uterine epithelia differ markedly from those elsewhere over the blastocyst.

More work is still required to clarify the question of whether the extra-embryonic mesoderm from the sinus terminalis migrates only into the yolk or also under the chorionic ectoderm.

FUNCTIONAL ASPECTS OF REPTILIAN PLACENTAE

Attachment

The placenta attaches the embryo to the uterine wall. In the allantoplacenta Type I of lizards, the chorionic cells insinuate protoplasmic processes into the uterine wall "so securely that, when the attached maternal and embryonic surface are separated, both may be jagged and torn." Although the attachment is secure, it is only superficial and there is no deep penetration of the maternal tissues, such as occurs in many mammals. Folding and interdigitation of the allantoplacenta Type I, which occurs in certain lizards in a restricted area under the main uterine blood vessels, may facilitate attachment of the embryo. In *Hoplodactylus maculatus*, the allantoplacenta is definitely of Type I, but there is no trace of attachment between the uterine wall and the chorionic epithelium. In the snakes *Austrelaps superbus* and *Suta suta*, the chorionic ectoderm is attached to the uterine wall in the way previously described for the Type I allantoplacenta of lizards. In addition, the uterine wall over a restricted area is raised into vascular ridges, which are enfolded by the chorioallantoic membrane.

Neither in *Enhydrina schistosa* nor in *Hydrophis cyanocinctus* do pseudopodial processes of the fetal cells project into the uterine epithelium, although the chorionic ectodermal cells are clearly in strong adhesion to the uterine epithelial cells.

Two ridges area occur in the placenta of *Enhydris dussumieri* : one is middorsal and corresponds to the allantoplacenta, the other is midventral and corresponds to the "stenopalcenta" of *Enhydrina schistosa*. There are no additional attachment mechanisms between the fetal membranes and the uterus.

In the Type II allantoplacenta, for instance, *Leiolopisma pretiosa, L. ocellata,* and *L. metallica,* the attachment of the embryo to the uterus differs only slightly from that described above. The ridges of the uterine mucosa are relatively low, but pseudopodial processes of chorionic cytoplasm do penetrate slightly into the uterine epithelial cytoplasm do penetrate slightly into the uterine epithelial cytoplasm.

The folding and interdigitation is more conspicuous in the Type III allantoplacenta (*Chalcides chalcides* and *Leiolopisma entrecasteauxii*) in which the free surfaces of the chorionic cells are mammilate or ciliated.

Bauchot (1965) suggested that the eggs are spaced and secured during gestation by the successive constrictions in the oviducts, which form incubation chambers; no "nidation" seems to be necessary. The nature of the chorionic-cell attachment to the uterine epithelium requires further attention. Ultrastructural studies of reptilian placental structures did not discuss this issue, which has special implications for the comparison of the placental adhesions in reptiles, metatherian, and eutherian mammals.

Placental Transmission

General : The reptilian placenta facilitates the transmission of oxygen, water, nutrients, and steroid hormones from maternal to fetal compartments and the transmission of carbon dioxide and other waste products in the opposite direction. The information accumulated so far on this subject may be classified into four categories, each derived from a different approach as follows:

(a) Functional deductions based on anatomical observations and blood oxygen affinity.

(b) Changes in mass and composition between ovulated egg and neonate.

(c) Direct measurement of labeled substances transferred across the placenta.

(d) Growth in culture of entire blastocysts removed from the uterus during pregnancy.

Functional Deductions Based on Morphological Observations and Blood Oxygen Affinity

(a) Gas Exchange : The close relationship between fetal and maternal circulation in the simple Type I allantoplacenta and the "exposure" of capillaries because of partial erosion of uterine epithelium have led most authors to assume that this placenta functions only in gas exchange. *Bauchot* (1965) accepted the suggestion of that the simple Type I allantoplacenta functions as a respiratory organ, whereas a nutritive role may be attributed only to the allantoplacenta of Type III. However, he remarked that there is no evidence that water does not traverse the simple allantoplacenta, nor that gases do not diffuse through the omphaloplacenta.

Gas exchange between maternal and fetal compartments has, apparently never been measured directly in reptiles. Consideration of this subject relies, therefore, on data of fetal metabolism in oviparous reptiles. In the Cronodilia and Testudines, embryonic development is arrested while eggs are retained until oviposition in the oviducts. Perhaps, the thick egg shells reduce the rate of gaseous diffusion. The thinning of the egg shell may well be prerequisite for the intra-uterine development that occurs in many oviparous squamates. The oxygen consumption of the embryos in oviparous snakes increases exponentially throughout the incubation period. If this is also the case for the intra-uterine embryos of viviparous forms, it would not only render the fetus dependent on a maternal source of oxygen, but would increase this dependence as development proceeds. Indeed, the embryos of *Elaphe obsoleta* and *E. gratata* either accumulate an oxygen debt during later stages of intra-uterine development, or they show a depressed metabolic rate due to an inadequate supply of oxygen.

Fetal blood of the viviparous snakes *Thamnophis sirtalis* and *T. elegans vagrans* and the viviparous lizard *Sphenomorphus quoyi* has higher oxygen affinity than does that of adults. This higher oxygen affinity facilitates loading of oxygen across the placenta from maternal blood. The exact biochemical or physiological background for this "fetal-maternal difference" in oxygen affinity is not very clear. Bohr effects of fetal and adult garter snake hemoglobins indicate that a different in pH inside fetal and adult cells cannot explain this difference in blood properties. The difference (*Manwell, 1960)* in oxygen equilibria inside the erythrocyte in the garter snake implies "a difference in the intracellular environment of the hemoglobin in adult and in fetal red blood cells." Electrophoretic studies in *Sphenomorphus quoyi* have not revealed different profiles of fetal and maternal hemoglobins, suggesting that the modulation of oxygen affinity may be "effected" by an allosteric co-factor such as one of the red cell organic phosphates. Blood oxygen affinity P5° in fetal blood of *Agkistrodon piscivorus* is about 20 torr, whereas in maternal blood it is about 50 torr. The difference in oxygen affinity is attributed in the lower nucleoside triphosphate level in fetal blood compared with that in the mother. It has been proposed that high oxygen affinity has an adaptive advantage for embryos of oviparous reptiles that lay eggs in soil or sand, in which ambient oxygen is rather low. Hence, the relatively high blood oxygen affinity of embryonic reptiles under conditions of low ambient oxygen is one of the features which "preadapted" reptilian species to viviparity.

Direct measurement of the capacity of gaseous exchange in the oviducts of squamates exhibiting various degrees of egg retention would contribute greatly to the understanding of the mechanisms evolved in egg retention and viviparity.

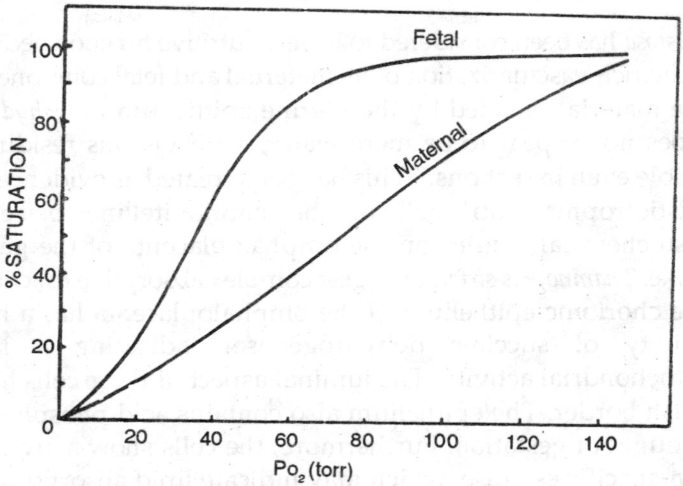

Fig. 4.18. *Sphenomorphus quoyi*. Oxygen equilibrium curves of whole blood from a gravid female and one of her embryos (T + 35° C; Pco_2 = 17 torr). Note the high affinity of the embryo's blood at low Po_2.

(b) Water Passage : The omphaloplacenta may provide the fetus with water "for mixture with the yolk to make it available for transportation to the embryo as food." This is apparently an extrapolation from the structure of the omphaloplacenta, which consists of a sheet of thickened uterine glandular epithelium that is adjacent to the thickened chorionic epithelium covering the yolk sac. The "coagulum" separating the two components is, in this view, considered to indicate the secretory activity of the uterus. However, in *Xantusia* and probably other lizards, both sides of the omphaloplacental components are poorly vascularized. Consequently, an efficient passage of water is unlikely in this region.

(c) Nutrients : The choriovitelline placenta of *Hoplodactylus maculatus* has been claimed not merely to absorb water but also to function "to some extent for food absorption." This claim is based on the almost total absence of debris at the base of the yolk sac of this lizard and enlarged ectodermal cells of the chorion in this region. Also, the *"stenoplacenta"* of *Enhydrina*

schistosa has been considered to have a nutritive function because of the rich vascularization of its maternal and fetal components. The material secreted by the uterine epithelium in *Enhydrina* "does not appear to be mere water, for a viscous residue is visible even in sections." This has been treated as evidence for "histiotrophic nutrition" of the choriovitelline placenta. Histochemical studies on the omphaloplacenta of the garter snake. *Thamnophis sirtalis*, suggest complex absorptive functions. The chorionic epithelium of the omphaloplacenta has a high activity of succinic dehydrogenase, indicating a high mitochondrial activity. The luminal aspect of these cells has a brush border. The epithelium also contains acid phosphatase throughout gestation. Furthermore, the cells show activity of non-specific esterase, which may indicate lipid absorption, as it does in the intestine.

The allantoplacenta of *Xantusia vigilis* shows rich vascularization and a close proximity of maternal and fetal circulations. The maternal and fetal capillaries are separated by a distance not exceeding 3-5μm. However, the choriovitelline placenta has a poor supply of blood vessels, as well as thick epithelia on both sides. A shell-membrane residue (coagulum) lies between the uterus and the chorion at the abembryonal regions. All of these features imply efficient diffusion of gases, and probably other substances, through the allantoplacenta rather than through other regions of the blastocyst.

In the microlecithal viviparous lizard *Mabuya heathi*, the phase of rapid growth during the second half of gestation coincides with the formation of a "placentome" (probably a Type III allantoplacenta). At this phase, the allantois undergrows the yolk sac. It has been suggested that the timing of placental development implicates the chorioallantois in nutrient uptake.

Changes in Mass and Composition Between Ovulated Egg and Neonate

(a) General : It has been generally assumed that live-bearing reptiles are not viviparous in the strict sense of the term as

most of them are lecitotrophs and rely on the ample supply of nutrients stored in the yolk. One way of examining the trophic relations between mother and fetus has been the comparison of mass and composition of the developing egg at various stages. Such studies have been performed in detail on only one snake, *Thamnophis sirtalis* and one lizard *Sphenomophus quoyi*. However, because the results obtained with these reptiles differ in many details they are reviewed here separately.

(b) Thamnophis: In Thamnophis sirtalis, the total weight of fetus and adnexa (yolk and fetal membranes) increases from a mean of 470 mg to a mean of 1311 mg during gestation. The water content increases from 300 to 960 mg or from 233 to 1057 mg. The wet weight of the embryo proper increases markedly during the first 50 days of gestation without any significant decreases in the wet weight of the yolk and yolk sac, indicating a transplacental flow of, at least, water. The fat content of the yolk decreases by about 100 mg at midpregnancy, but only one-third of that fat mass is gained by the fetus. The remaining fat (65.5 mg) appears to be metabolized by the blastocyst with a concomitant yield of 70 mg. water. This gain in metabolic water accounts for only 7% of the total water gain during gestation. It was initially reported that the amount of dry matter in the neonate is 18% lower than that found in the ovulated egg; however, a subsequent paper modified this view. The amount of dry matter in the whole blastocyst increases by 100 mg during the first month of gestation, it then remains stable until day 51, after which it starts to decline, reaching the initial value at day 70. Analysis of the data shows that the trends for increasing dry weight or ash content are not statistically significant. However, the total amount of protein in the whole blastocyst increase gradually during pregnancy; the gain is at least 45 mg per embryo and is statistically significant. Serial measurements of total nitrogen, ammonia, urea, and uric acid indicate that the rate of accumulation of waste products in the early period of development ($K = 0.11$) is less than the rate of wet weight growth ($K = 0.187$) but is

greater than that of dry weight ($K = 0.047$). This points to a relatively greater catabolism of protein early in development. The amount of ammonia in the fetus increases at a constant rate, suggesting that the disposal of ammonia by urea and uric acid synthesis by transamination or even by the presumed transplacental migration may be inadequate for handing the whole amount of ammonia produced. Therefore, urea accumulates in the fetus.

Based on the serial measurements of *Clerk* and *Sisken* (1959), Packard (1966) has pointed out that uric acid comprises only 9.1% of the total nitrogenous waste produced during embryonic life of *Thamnophis* and that such a low fraction can be attributed to purine metabolism. He concluded that embryos of viviparous reptiles are generally uteotelic and only in the last two days (out of the 72-day gestation period) does uric acid exceed 10% of the excretory nitrogen. However, embryos of oviparous reptiles become uricotelic at an earlier phase of their development.

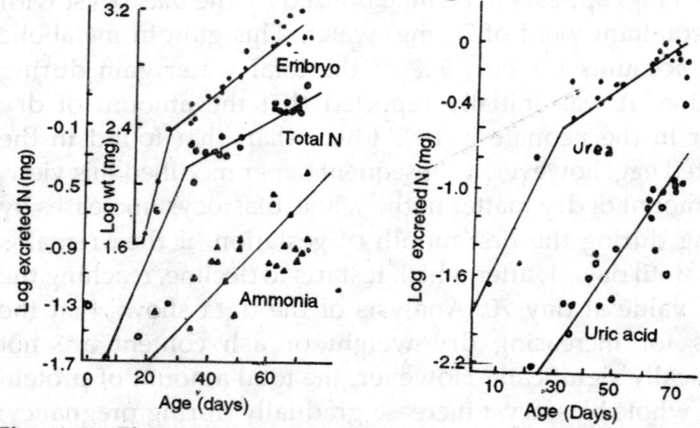

Fig. 4.19. *Thamnophis sirtalis.* Semilog plots of embryonic excretion. (A) Cumulative excreta of embryo in relation to growth. Ordinate : Upper curve, log wet weight; middle curve, long accumulated excreted N of all types: lower curve, log accumulated ammonia N. Abscissa : age in days. (B) Urea N and uric acid N accumulated in embryo. Ordinate and abscissa as in (A). (After Clark and Sisken, 1956).

The amount of nitrogen per gram body weight excreted by the neonate of *Thamnophis* is far less (56%) than that excreted by the neonate of the oviparous *Coluber constrictor*, which is of comparable size and composition. This suggests that some of the excreta of *Thamnophis* (1.11 mg/fetus) may have been lost through the placenta. This assumption is supported by the finding that urea concentrations in maternal blood (5.13 mg% in early gestation and 6.93 mg% in late pregnancy) are higher than in non-pregnant females (5.01 mg%) or in males.

When ^{15}N urea was injected into the yolk sac of the embryonic *Thamnophis*, the tracer was recovered in the uric acid excreted by the female. It has been suggested that the injected urea was hydrolyzed urease and that the resulting ammonia was incorporated into uric acid, probably by maternal tissue. However, the possibility of transplacental passage of urea as a physiological or evolutionary requirement has not been considered, only the inevitable escape of urea from embryonic confines due to its extreme solubility.

(c) Sphenomorphus: The total weight of the conceptus of *Sphenomorphus quoyi* (the embryo, yolk, extra-embryonic fluids and membranes), increases about fourfold during gestation. The most rapid growth occurs between days 35 and 75 after ovulation. The absolute and relative amounts of water increase in the same pattern. The total weight of organic material decreases, but there is a slight, though significant, increase (4 mg) in the amount of ash. The amount of protein in the yolk decreases as that in the fetus increases, but at birth the total amount of protein is less than that initially present in the yolk. There is also a net loss of lipids during gestation. Predictions of the total oxygen consumption of the fetus are 57 ml O_2 at 20° C and 175 ml O_2 at 30° C. The assumption that combustion of 1 mg fat requires 2.03 ml O_2 and that all the oxygen consumed is used in lipid metabolism leads to an estimate that 84 mg of fat would be consumed during gestation at 30° C or 30 mg at 20° C. *S. quoyi* loses 35 to 40 mg of fat per fetus during gestation; this is well within the predicted range. Consequently, it is

unnecessary to postulate sources of energy other than those in the yolk. This is confirmed by the apparently normal growth of embryos in pure saline. The metabolism of 40 mg fat would yield only 45 mg water. This is much less than the measured increase of about 800 mg water/embryo during gestation. Therefore, about 760 mg water must be transferred to each embryo through its placentae.

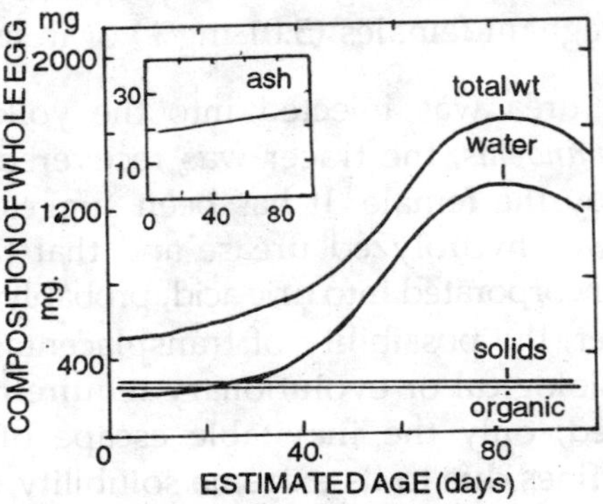

Fig. 4.20. *Sphenomorphus quoyi*. whole blastocyst : change during gestation of total (wet) weight, water content, organic matter, solids and ash (inset).

Comparison of the composition of recently ovulated eggs and of embryos at term indicates that the total ash increases by 23% during pregnancy. The main part of this increase can be attributed to a net gain in calcium, with a small concomitant gain in Na^+ and K^+; the level of Mg^{2+} remains constant. The added ions are incorporated into the fetus proper.

Both urea and uric acid accumulate in the allantoic cavity of *Sphenomorphus quoyi*. Near the end of gestation, the concentration of uric acid increase dramatically; however, it never exceeds more than 40% of the excreted nitrogen. The amniotic fluid also contains considerable amounts of urea. As the concentrations of urea in the emniotic and in the allantoic fluids are highly correlated; it has been suggested that urea diffused readily from the allantoic to the amniotic cavity. However, urea is not transferred into maternal circulation through the placenta; instead, it accumulates in the allantois against a concentration gradient. The total nitrogen content of the entire conceptus does not change significantly during gestation. As these embryos derive most of their energy from the combustion of fat, constancy of total nitrogen during gestation indicates that the transfer of amino acids from maternal to fetal compartments or that of nitrogenous wastes in the opposite direction is not a vital physiological function. However, it should be emphasized that the placenta of *S. quoyi* is permeable to amino acids.

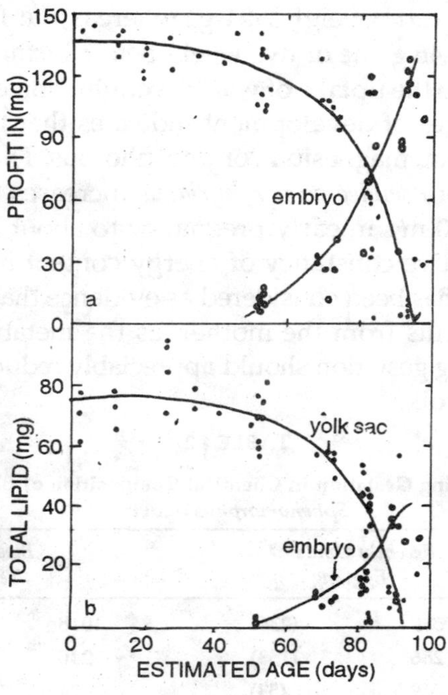

Fig. 4.21. *Sphenomorphus quoyi*. Changes during gestation in the protein (a) and lipid (b) content of the yolk sac (solid circles) and the embryo proper (open circles).

(c) Reptiles : The egg of *Lacerta vivipara* gain about 270 mg of water between Stages 2 and 16; this amounts to about 75% of the total weight at the latter stage. Eggs of *Anguis fragilis*

gain 714 mg of water, which is 72% of their total weight. In an adequately humid terrarium, the weight of eggs of the oviparous *L. agilis* will increase by 2.4 g; this is 90% of the weight at hatching. However, the total dry weight of the eggs does not increase prior to term. Apparently, the dry weight of eggs of *Regina grahami* shows a 30% increase between June and July. These results may be erroneous as they are based on specimens preserved in alcohol and only subsequently dried. In *L. vivipara*, the total dry weight and total nitrogen content of a cluch of eggs remain constant after ovulation. Just after ovulation, the eggs of *Vipera berus* weigh 2.94 g, whereas the fully formed fetus weights 4.46 g one or two weeks before birth; this increase is largely due to the uptake of water. Examination of blastocysts at various stages of development indicates that they had not taken up calcium, magnesium, or phosphorous. The ash content of blastocyst of *Pseudechis porphyriacus* increases significantly from about 530 mg in early pregnancy to about 2.5 g at full-term. The relative constancy of energy content in blastocysts of this species has been considered as evidence that the embryo receives nutrients from the mother, as the metabolism of the embryos during gestation should appreciably reduce the energy stored in the yolk.

TABLE 4.2

Changes During Gestation in Chemical Composition of Blastocyst in *Sphenomorphus quoyi*[a]

Parameter	*Recently Ovulated Egg. mg*		*Blastocyst at Term, mg*	
Wet weight	530	(236)	1018	(149)
Dry weight	286	(153)	240	(92)
Ash weight	19.5	(53)	24.0	(24)
Calcium	2.1	(20)	4.8	(33)
Sodium	0.3	(20)	1.8	(33)
Potassium	1.3	(20)	2.3	(33)
Magnesium	0.23	(20)	0.23	(33)
Total nitrogen	22	(49)	21	(30)
Total lipid	72	(42)	35	(26)

[a]Value were calculated by extrapolation from curves of best fit drawn through the number of data points indicated in parentheses.

The most striking example of embryonic dependence on placental nutrition in reptiles has been reported in the skink *Mabuya heathi.* The ovulated egg is about 1 mm in diameter and measurements of formalin-fixed material show a 538-fold increase in wet weight and 384-fold increase in dry weight during pregnancy.

Direct Measurement of Labeled Substances Transferred Across the Placenta

Radioactive electrolytes were utilized to demonstrate transfer from maternal to fetal tissues of *Lacerta vivipara Nerodia*

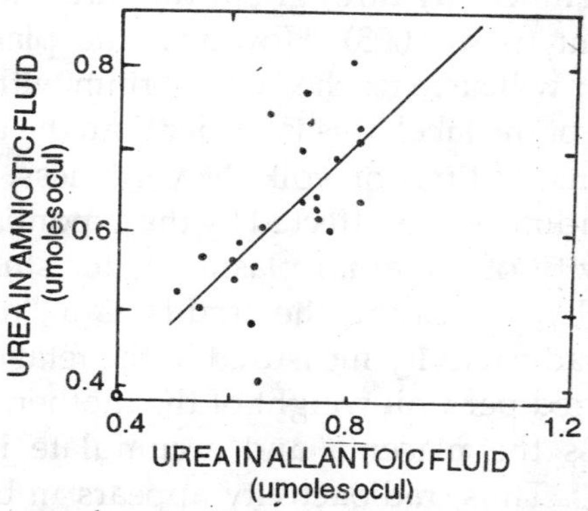

Fig. 4.22. *Sphenomorphus quoyi.* Blastocysts : concentration of urea in the amniotic fluid as a function of its concentration in the allantoic fluid.

sipedon, N. cyclopion, and *thamnophis sirtalis.* These experiments were performed on a limited number of animals that were sampled at long intervals after injection of the label. It is very significant that these electrolytes penetrate into the fetus even in *L. vivipara,* in which the shell membrane persists throughout gestation. The amounts of [22] Na recovered in fetal tissues of *Nerodia* and *Thamnophis* (4-8%) were considerably higher than those recovered in *Laterta vivipara* (~ 2-05%). However, the level of radioactivity in the fetus of the snakes was compared with that in the adult heart or liver, whereas that of the lizard was computed as a fraction of the radioactivity in the blood. It is significant that the fetal thyroid of *N. sipedon* accumulates labeled iodide up to a level of 13 to 18% of that found in the maternal thyroid. Labeled sodium, phosphorus, and iron have been detected in fetal compartments after administration of labeled ions into pregnant lizards or snakes. The low, perhaps negligible, transfer to iron in *T. siralis* may reflect the ample supply of the element in the yolk-naval region.

Tritiated water injected into the maternal femoral vein appears fairly soon in all embryonic compartments of *Mabuya capensis*. If the radioactivity in these compartments is expressed as a fraction of the radioactivity in maternal plasma, it seems to increase significantly throughout the three-hour period of the experiment ($p < 0.005$). However, the plasma and the embryo proper will have reached equilibrium within two hours after injection of the label; this is earlier than the time required for either amniotic fluid or yolk. Nevertheless, this increase may be misleading as it is affected by the expontial decrease in the radioactivity of maternal plasma. Later studies avoided this obstacle by expressing the results as a "transfer ratio" between the radioactivity measured in the fetal compartment and that injected per unit weight of the mother. Amino acids too may cross the placenta and accumulate in embryonic compartments. Thus, radioactivity appears in both amniotic fluid and yolk of *Xantusia vigilis* as early as 15 minutes after injection of [^{3}H]leucine into the mother. Most of the amino acid that disappears from the circulation becomes incorporated into maternal hepatic protein. However, significant radioactivity accumulates in the embryo proper and in the amniotic fluid within four hours, whereas the radioactivity in the yolk at first increases and then decreases. Although the metabolic fate and the physiological significance of the transferred amino acid has not been clarified, there is evidence that both [^{3}H]leucine and [^{14}C]glycine become incorporated into fetal proteins in *X. vigilis*, and *Thamnophis sirtalis*. Glycine and x-amino isobutyric acid, labeled with ^{14}C have been used to trace the transfer of a natural and a nonmetabolizable amino acid analogue across the placenta of *Sphenomorphus quoyi*. The results indicate that the intact molecule is transferred across the placenta to be discribed in various fetal compartments.

Growth in Culture of Entire Blastocysts Removed from the Uterus During Pregnancy

It is possible to assess the extent to which the fetus is dependent on uterine nourishment by modifying the medium in which the blastocysts are cultured. Eggs of *Lacerta vivipara* develop successfully in an aerated NaCl medium (0.6%), both when immersed and when supported on wet cotton wool in a humid atmosphere. Hence, the amount of yolk deposited in the egg is sufficient for complete development of the fetus, and the shell membrane may function as a regulator of ion and water diffusion into the blastocyst. The results were confirmed by other experiments on the same species. Embryos of *Anguis fragilis* have been successfully maintained *in vitro,* but the medium used contained 2/3 (v/v) of chick-egg albumin and Tyrode or Hanks solution. It would appear that the developing embryos of *Anguis* may require albumin, although some other uncontrolled factors, such as oxygenation and pH of the culture medium, cannot be excluded. Embryos of *Chalcides ocellatus* cannot be maintained in culture for more than nine days using only saline. However, the fetuses of *Sphenomorphus quoyi* can be incubated for 44 days, either in the enriched medium of *Raynaud* (1959a) or, more simply, in 0.6% saline. Embryos in both media developed to full term and one hatched.

Summary of Placental Transmission

The data presented above indicate that the reptilian placenta is undoubtedly permeable to water, electrolytes, and amino acids. This establishes only the potential for amino acid uptake from maternal compartments and hence the potential for nitrogen gain by the embryo. However, the protein content of the blastocyst of *Sphenomorphus quoyi* does not change during gestation. Moreover, the evidence indicates that all of the metabolic needs of the fetus are met by fat stored within the yolk. Therefore, *S. quoyi* does not utilize the placental capacity for amino acid uptake. The situation in *Xantusia*

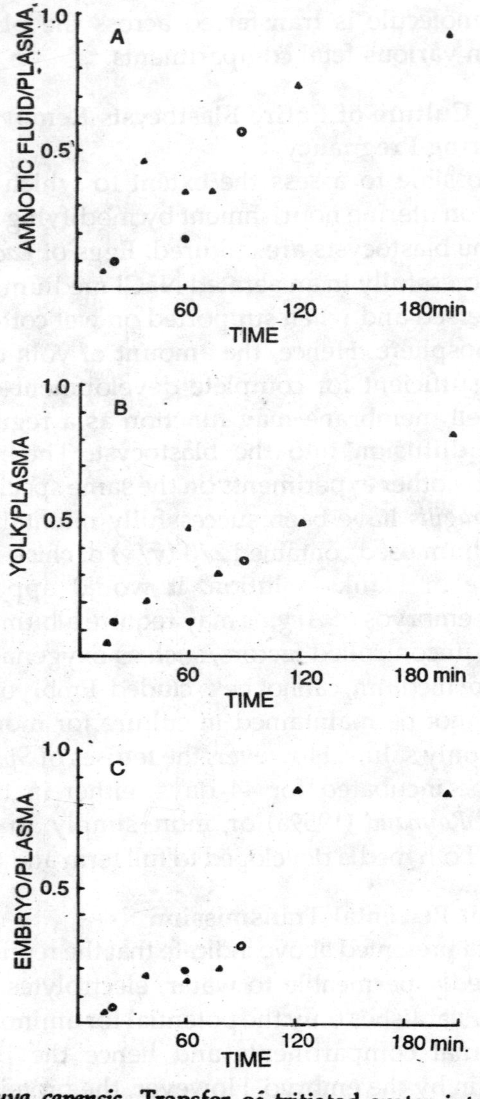

Fig. 4.23. ***Mabuya capensis*. Transfer of tritiated water into amniotic fluid (A) yolk (B), and embryo proper (C) in sham-operated (o) and luteectomized (A) lizards. Luteectomy was performed 4 weeks prior to the experiment. The label was injected into the femoral vein. Results are expressed as cpm in embryonic compartment/cpm in maternal plasma.**

remains unclear for although the potential for amino acid transfer has been established, the chemical composition of the neonate and the ovulated egg have not been compared.

Placental transfer may be different in the garter snake. Amino acids are transferred across the placenta and are incorporated into fetal proteins. Moreover, the total amount of protein in the neonate is larger than that in the ovulated egg. Also, the waste products of embryonic protein metabolism are transferred as urea across the placenta into the maternal circulation. Autoradiographic studies of pregnant *Thamnophis sirtalis* injected with [C]glycine indicate that the amino acid or protein are transferred via the comphaloplacenta. As the label on the fetal side appears in both the amino acid and the protein fractions, it is uncertain whether the amino acid is transferred as such across the placenta or after being incorporated into proteins. Substances of large molecular weight secreted by the uterus may be taken up histiotrophically by the chorionic epithelium of the omphaloplacenta. Indeed, chorionic cells can phagocytoze trypan blue added *in vitro* after the shell membrane surrounding the blastocyst has been removed. Methylene blue injected intraperitoneally into the mother appears in fetal compartments within 12 hours, mostly in the omphaloplacenta. However, it is possible that smaller molecules, such as the amino acid glycin can diffuse across the closely apposed maternal and fetal capillary beds of the allantoplacenta, as the barrier of the shell membrane here is exceedingly thin.

The special arrangement of blood vessels at the ventral region of the blastocyst and the interconnection of umbilical and vitelline vessels cast doubt on the conclusions based on comparison of radioactivity in the yolk and other fetal compartments. Low levels of radioactivity in the yolk and yolk sac do not provide reliable evidence for a poor transport across the omphaloplacenta. Nevertheless in *Xantusia vigilis* the sparse vascularization both of the ventral side of the uterus and of the choriomphalic membrane, as well as the barrier formed by the coagulum between maternal and fetal tissues

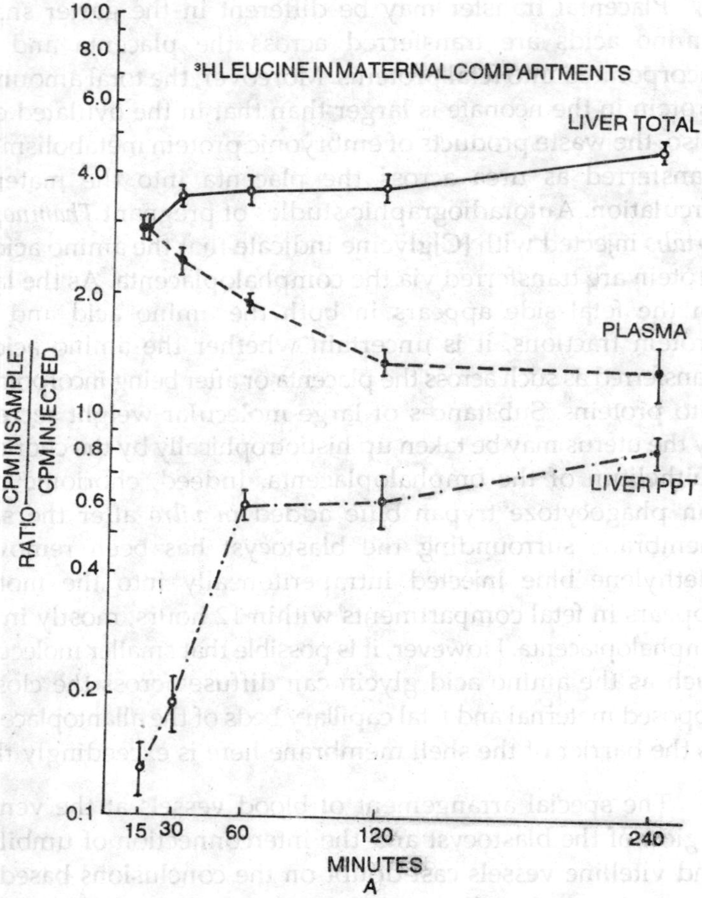

Fig. 4.24. ***Xantusia vigilis*. Distribution of [^{3}H]leucine in maternal (A) and fetal (B) compartments after injection of the label i.m. into the thigh (1 Ci/lizard). Results are expressed (A) as cpm in maternal tissue cpm injected per unit body weight, or (B) as "transfer ratio" (cpm in 100 mg tissue or 100 µl fluid of the embryo/cmp injected per 100 mg of maternal body weight, (n = 6).**

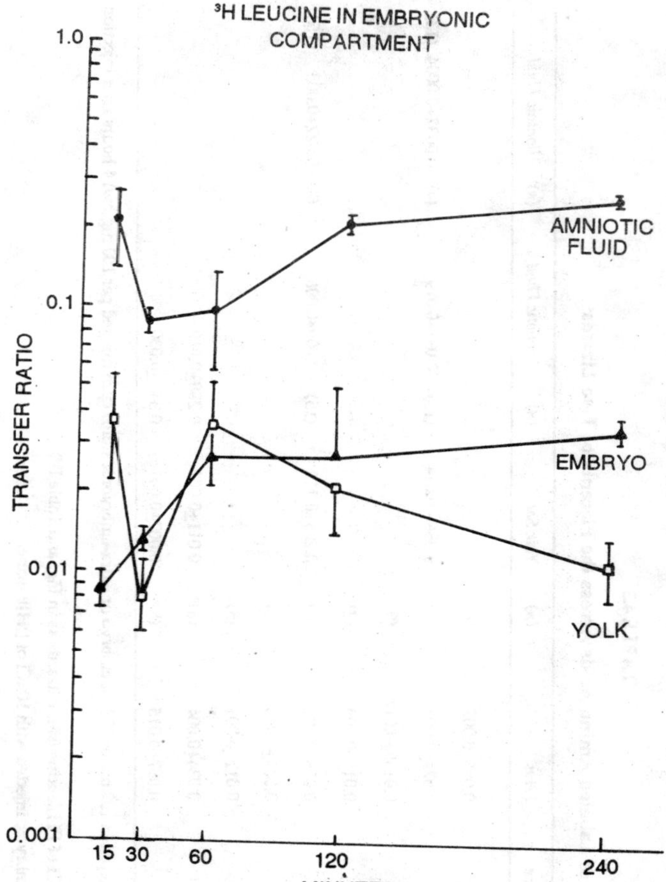
³H LEUCINE IN EMBRYONIC COMPARTMENT
1.0
0.1
0.01
0.001
TRANSFER RATIO
AMNIOTIC FLUID
EMBRYO
YOLK
15
30
60
120
240
MINUTES

B

TABLE 4.3

Transfer of Labeled Amino Acids Across the Placenta of Two Lizards[a]

Species and Source	*Labeled Substance*	*Fetus*	*(n)*	*Yolk Sac*	*(n)*	*Amniotic Fluid*	*(n)*	*Allontoic Fluid*	
Sphenomorphus quoyi	[¹⁴C]glycine	0.046±0.007	(6)						
(Thompson 1977b)		0.074±0.007	(3)	0.056± 0.014	(13)	0.051+0.019	(4)	0.0032+0.0004	(13)
		0.0150±0.017	(5)						
	[¹⁴[C]α-amino	0.018±0.003	(3)						
	isobutyric acid	0.026±0.006	(4)	0.205±0.119	(15)	0.009+0.006	(3)	0.022+0.008	(15)
		0.025±0.003	(3)						
		0.012±0.001	(5)						
Xantusia vigilis[³H]leucine		0.034±0.004	(6)[b]	0.011±0.003		0.268±0.019			
(Yaron, 1977)		0.037±0.015	(5)[c]	0.002±0.001		0.310±0.030			

[a] The transfer was calculated as "transfer ratio" (cpm in 100 mg or 100 of embryo or extra-embryonic fluid/cpm injected per 100 mg bw) 4 hours after injection of label (Mean ± SEM.)

[b] Intact lizards (1.34-2.26 g bw) injected with 1μCi of [³H]leucine (same lizards as in Fig. and Table IV)

[c] Sham-ovariectomized lizards with surviving embryos, injected with 10μCi of [³H]leucine.

indicate that any transfer here will be neither fast nor efficient. The site of gas exchange may also serve haemotrophic nutrition.

In *Thamnophis radix*, the placenta is also permeable to steroid hormones. Labeled testosterone and estradiol injected into the mother reach the embryonic compartments as intact molecules. The fetal tissues convert them to androstenedione and to estrone plus estriol, respectively. It is thus possible that maternal steroids can reach the embryo at stages critical for gonadal differentiation.

ENDOCRINE CONTROL OF GESTATION IN REPTILES

Ovarian Hormones and Gestation

Postovulatory follicles transform into corpora lutea, both in viviparous and oviparous reptiles. Just after ovulation, some cells of the granulosa lie free in the collapsed follicular cavity; later, these cells and the theca re-establish contact. The granulosa cells hypertrophy, their nuclei become large, round, and basophilic and contain a single nucleous; thus, they are transformed into lutein cells. Thecal fibroblasts, with or without accompanying blood vessels, may penetrate into the central mass of the lutein cells. The degree of this penetration does not correlate with the degree of viviparity. However, the mode of reproduction and the longevity of the corpora lutea are correlated in reptiles. In most temperate oviparous species, the corpus luteum regresses shortly after oviposition, which occurs shortly after ovulation. Therefore, corpora lutea persist longer in viviparous than in oviparous forms. A similar condition occurs also in viviparous amphibians (*Amoroso*, 1981). The corpora lutea in both viviparous snakes and lizards resemble those of mammals and persist in the ovary during gestation but may regress previous of parturition. This has led to the assumption that they also produce progesterone and, as those of mammals, serve to maintain pregnancy. This assumption has proved to be only partially correct. Not only lipid droplets or "fat cells" are associated with reptilian lutein cells but steroid converting enzymes such as 3β-hydroxysteroid dehydrogenase (HSD), 17β-HSD, and 11β-HSD have been demonstrated histochemically in the corpora lutea of viviparous and oviparous species. The distribution of 3β-HSD is not restricted to the lutein cells; in *Sceloporus cyanogenys*, it occurs also in the inner thecal elements. Moreover, corpora lutea of various oviparous reptiles convert *in vitro* labeled steroid precursors into progesterone, 17a-hydroxyprogesterone, 17α-hydroxypregnenoion, androstenedione, testosterone, dehydroepiandrosterone, deoxycorticosterone, and estrone.

In spite of the general morphological and biochemical resemblance among the corpora lutea of reptiles and mammals, electronmicroscopy reveals some differences. The mammalian lutein cells show concentric whorls of the tubular endoplasmic reticulum, whereas those of turtles show citernal whorls. Also the latter lack the complex network of microvilli along the border of the lutein cell with the vascular bed. As in mammals, other fine structural features of the reptilian corpora lutea, such as an abundance of smooth endoplasmic reticulum, active mitochondria with lamellar cristae, and associated lipid droplets, are characteristics of steroid synthesizing cells.

Corpora lutea of turtles secrete progesterone *in vitro*. Corpora lutea of the oviparous *Uromastyx hardwichi* contain progesterone (531.0 ± 73.4 ng/g), testosterone (15.8 ± 2.7 ng/g) and estradiol (5.6 + 1.0 ng/g; mean ± SEM,). Corpora lutea of the viviparous *Bradypodion pumilus* contain 56.8 ug/g progesterone, a quantity comparable to that in a variety of eutherian mammals and higher than that of *Uromastyx*. Concentration of progresterone in the corpora lutea of *Thamnophis elegans* increases in the middle third of pregnancy and decreases in the last third. Circulating progesterone reaches highest levels during pregnancy in viviparous lizards and snakes, whereas oviparous reptiles show the highest levels of progesterone just before ovulation. The oviparous *U. hardwickii* differs; the circulating progesterone level of gravid females is eight times the level prior to ovulation. The levels of testosterone and estradiol are also higher in gravid than in prevulatory females. The concentrations of all three steroids are higher in luteal tissues than in follicular ones, both before and after ovulation. In pregnant *T. elegans*, plasma progesterone is at a level of 1 ng/ml in luteectomized snakes; however, it is 4-6 ng/ml in pregnant control snakes, suggesting that the corpora lutea are their main source of circulating progesterone. Conversely, in *Trachydosaurus rugosus*, levels of progestins are unaffected by castration in either sex.

Although reptilian corpora lutea may secrete other steroids, such as testosterone, deoxycorticosterone, and 17β-estradiol, there is a general agreement that they secrete considerable amounts of progesterone during gestation. However, there is less unanimity about the functional significance of the reptilian corpora lutea and their progesterone secretion.

Absolute dependence of pregnancy on the presence of an intact ovary has been demonstrated in *Nerodia* and *Storeria* during early or mid pregnancy. However, the results of the "adequate control" groups mentioned in this work are not referred to thereafter. The oft-cited work of *Fraenkel et al.* (1940), which claims that corpora lutea are essential for the maintenance of pregnancy in several South American snakes, does not provide adequate information on either the number or the stages of pregnancy of the operated animals. Also, none of these animals survived the operation by more than ten days, so that the statement that these "results must remain unconvincing until confirmation is forthcoming" remains appropriate.

Absolute dependence of pregnancy on the presence of intact ovaries or on progesterone replacement has been demonstrated in *Bradypodion pumilus*. However, although there are detailed results on the number of dead, aborted, or surviving embryos for the ovariectomized group, the control results are covered only by the general statement that "sham-operated chameleons did not about and development was normal. Young embryos were absorbed after luteectomy of *B. pumilus*, however, scaled embryos were unaffected by the operation.

The opposite claim that corpora lutea are not essential for maintenance of gestation in *Nerodia* and *Thamnophis* is based on data from three ovariectomized, or 19 hypophysectomized snakes and on the assumption that possible activity of corpora lutea in pregnant garter and water snakes is interrupted by hypophysectomy. Apparently, embryos of operated snakes develop normally but parturition is impaired and embryos are retained *in utero* past term. Also the experiments of Rahn (1939) showed that gestation in most ovariectomized females of *Thamnophis* and *Nerodia* continues normally up to 25 days postoperation.

A carefully designed experiment indicated that luteectomy of pregnant *Thamnophis elegans*, early in the first third of gestation, considerably reduced the number of "birth products" per snake as compared with sham-operated controls. Although some blastocysts about after luteectomy, the surviving embryos develop normally for more than 60 days and thus do not differ from embryos of the sham-luteectomized controls. However, the mean duration of pregnancy is prolonged (97.4 ± 2.6 days in the controls; 106.4 ± 2.4 days in luteectomized snakes; mean ± SEM). Only 14% of the newborn are viable as compared with 84% in the controls.

Lacerta vivipara and *Sceloporus cyanogenys* show delayed and prolonged parturition after ovariectomy or luteectomy, although, embryonic development as such is not impaired.

In *Chalcides ocellatus*, pregnancy is apparently independent of intact corpora lutea. Although eight out of 25 luteectomized lizards in one sample contained at least one dead embryo, this ration (8/25) is not significantly different form that found in females collected in the wild (5/15). Also the ratio of the dead to the total number of embryos in luteectomized lizards (15/93-16.1%) does not differ from that observed in unoperated intact ones (49/316-15.5%).

Luteectomized *Mabuya capensis* show embryonic growth equivalent to that in sham-operated controls for 28 days postoperation. Also placental transfer of tritiated water does not differ between luteectomized and control females. In *M. carinata*, luteectomy has no effect on the number of eggs laid or on the duration of pregnancy.

In *Xantusia vigilis*, the survival of embryos is greatly impaired after any kind of surgery, especially if this is performed during early pregnancy. The proportion of females carrying live embryos, 30 days after ovariectomy or luteectomy, does not differ significantly from that of sham-operated controls. Apparently, neither luteectomy nor ovariectomy produce deleterious effects on gestation in this lizard. Moreover, the rate of amino-acid transfer from maternal to fetal compartments of the surviving blastocysts do not differ among ovariectomized, luteectomized, and sham-operated lizards.

TABLE 4.4

Growth of Embryos in Luteectomized *Mabuya capensis* [a]

Treatment	*Dry Weight of Embryos, mg*			
	Before Operation	*All Autopsy*	*Δmg*	*% Growth*
	8.0	22.0	14.0	275.0
Sham	17.4	109.2	91.8	627.6
	60.0	76.2	16.2	127.0
Mean + SEM				343.2 ± 148.5
	10.2	68.2	58.0	668.6
	13.6	70.2	56.6	516.2
	18.6	31.0	12.4	166.7
Luteectomy	43.0	158.0	115.0	367.4
	56.0	137.0	81.0	244.6
	79.0	138.0	59.0	174.7
Mean + SEM				356.4+ 82.6

a Embryos were removed from the adnexa and dried to constant weight. (Yaron and Tai, 1969, unpublished.)

Replacement of progesterone does not ameliorate the deleterious effects of ovariectomy in most of the species examined. However, when pregnant *Bradypodion pumilus* were ovariectomized and then treated with progesterone, only one female aborted an embryo. Furthermore, treatment with progesterone greatly increases the vascularity of the uterus, and each of the embryos then lies in a separate uterine chamber. Ovariectomized chameleons, which were not injected with steroids, showed a reduced uterine vascularity, and the surviving embryos were sometimes crowded two in a chamber.

HYPOPHYSEAL CONTROL OF REPTILIAN GESTATION AND PARTURITION

General

The effects of hypophysectomy during gestation in snakes and lizards, like those following ovariectomy or luteectomy, are equivocal. In two species of *Nerodia*, also in *Thamnophis butleri and Storeria dekayi*, removal of the pituitary during early pregnancy caused resorption of blastocysts and removal during mid- or late-pregnancy cause abortion of embryos. Another study presents conflicting evidence that embryos continue to develop in most hypophysectomized females. However, as embryos are retained *in utero* past term, the process of parturition appears to be disturbed. Unfortunately, no data on controls are available.

The above results have been criticized on the grounds that hypophysectomy may alter not only luteal function, but the whole endocrine balance of the pregnant reptiles. Consequently, the experiments "may have obscured the true function of corpora lutea by also altering the function of other non-luteal tissues."

In *Lacerta vivipara,* hypophysectomy during early or midgestation apparently has no effect on embryonic growth or development, and the embryos reach normal size. However, some embryos have been found dead in the uterus or in the abdominal cavity past the normal term.

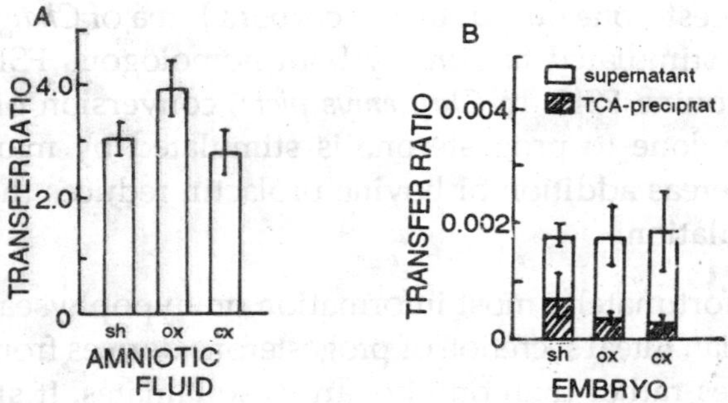

Fig. 4.25. *Xantusia vigilis*. Transfer of [^{3}H]leucine into amniotic fluid (A) and embryo proper (B) of operated lizards. Operations were made on pregnant females about two weeks after the estimated date of ovulation. One month after operation the label (10μ Ci/lizard) was injected i.m. into the thigh. Embryonic compartments were sampled four hours after injection. Results are expressed as mean ± SEM (n=5) of "transfer ratio." SH, Sham-operated, OX, ovariectomized; CX, luteectomized.

Hypophysectomy removes not only the source of gonadotropic hormone(s), thus indirectly affecting the ovary, but also the source of neurohypophyseal hormones that may affect the oviducal musculature and consequently the process of parturition.

Effects of Adenohypophyseal Hormones

Anatomical and histological observations of pregnant *Thamnophis sirtalis* document that hypophysectomy does not interfere with the organization and persistence of the corpora lutea, although the cytoplasmic/nuclear ratio in the lutein cells decreases. Conversely, there is substantial evidence that hypophysectomy in *Sceloporus cyanogenys* diminishes the activity of 3β-hydroxysteroid dehydrogenase (probably in the corpus luteum) and reduces the level of plasma progesterone from 34 to 22 ng/ml. Moreover, treatment of the hypophysectomized lizards with either FSH (follicle stimulating hormone) alone, FSH + GH (growth hormone), or prolactin + GH restores the progesterone level in the circulation. Administration of prolactin alone depresses plasma progesterone.

Progesterone output by the corpora lutea of *Chalgrisenium* may be stimulated *in vitro* by both homologous FSH or LH and by ovine FSH. In *Chrysemys picta*, conversion of labeled pregnenolone to progesterone is stimulated by mammalian LH, whereas addition or bovine prolactin reduces the degree of stimulation.

Unfortunately, most information on hypophyseal control of reptilian luteal secretion of progesterone derives from studies on turtles rather than on viviparous squamates. It should be emphasized that hypophyseal function in the former differs markedly from that of the latter. The pituitaries of Testudines contain both LH and FSH, whereas those of the snakes contain only a single gonadotropin, which bears relatively little resemblance to the FSH or the LH of other tetrapods, including those of turtles and crocodilians. There is rather little information on the gonadotropin of lizards; presumably, it would not differ much form that of snakes.

Neurohypophyseal Hormones and Parturition

Retention of eggs or embryos in the uterus depends, among other factors, on the neurohypophyseal hormones. Perhaps, as in other vertebrates, a surge of oxytocin-like peptide would cause the expulsion of eggs (oviposition) or fully grows embryos (parturition) from the contracting genital tract. Octapeptide principles with typical oxytocic activity, mesotocin, and oxytocin have been found, together with arginine-vasotocin (AVT), in the neurohypophyses of turtles, a crocodilian, snakes, and a lizard. No data are, so far, available on the circulating levels of these principles in reptiles at any stage of their life history. However, administration of neurohypophyseal extracts, oxytocin or vasotocin has been used to induce premature parturition, or parturition at term, in several viviparous snakes and in lizards. Oviposition has been induced in turtles by injection of whole bevine pituitary power, posterior pituitary powder, and oxytocin. Additionally, oxytocin or AVT stimulate oviposition in *S. occidentalis* and *Uta stansburiana*. In the oviparus *S. a. aeneus, Hemidactylus frenatus*, and *Lepidodactylus lugubris*, injection of 0.02 μg AVT induces oviposition and the associated behavioral pattern. Parturition has been similarly induced in *S. aeneus bicanthalis*. However, injection of oxytocin into a pregnant *Barisia coerulea*, in a dose higher than that given to *S. occidentalis*, did not result in premature expulsion of embryos. A similar lack of response to AVT was observed in *Anolis carolinensis*. This may indicate that the sensitivity of the Mullerian duct to the octapaptides is not necessarily similar in all lizards or, alternatively, that the response of the uterus of oxytocin or AVT during pregnancy, or between ovulation and oviposition, is dependent on additional neural or hormonal factors.

There are conflicting statements regarding a possible modulating effect of ovarian steroids on uterine contractility. Low chronic doses of estrone (40 ng lizard $^{-1}$ day $^{-1}$) or progesterone (10 ng lizard $^{-1}$ day $^{-1}$) affect neither the absolute nor the relative sensitivity to oxytocin or AVT of isolated oviducts from the viviparous *Klauberina riversiana*. Conversely, 7-day treatment with estradiol (50 ng g $^{-1}$ day $^{-1}$) or progesterone (16.6 ug g $^{-1}$ day $^{-1}$) has a clear-cut modulatory effect on the contractility *in vitro* of the lower oviducal segment of *Chrysemys picta*. The amplitude of oviducal contractions increases, and the duration between contractions decreases over control level in estradiol-treated turtles. Turtles treated with progesterone show significantly fewer and shorter contractions in response to AVT. Based on these data, it has been suggested that "actions of estradiol and progesterone on turtle oviduct.... are similar to the actions of these hormones on the mammalian uterus." This suggestion has been criticized because of its incompatibility with the steroid profile of this turtle. Although progesterone "block" or myometrial contraction is a reproducible phenomenon in rabbits, it could not be demonstrated in rats or humans. The progesterone "block" may prove to be "a mammalian invention which has not been universally adopted."

In contrast to the results obtained in *Chrysemys picta*, pretreatment with progesterone of ovariectomized *Anolis carolinensis* augments the contractility of their oviducts in response to AVT. Pretreatment with estradiol has the opposit effect. Study of uterine contractility at distinct phases of the reproductive cycle in *Anolis* has revealed that the presence of a corpus luteum inhibits totally the contraction *in vitro* of the ipsilateral uterus in response to AVT. However, the contralateral uterus (on the side where a corpus luteum is degenerating or absent) exhibits rhythmic contractions when exposed to AVT. Unilateral luteectomy removes the ipsilateral inhibition. It has been postulated that the corpus luteum of *Anolis* secretes high amounts of estradiol, which inhibits uterina contraction, and that the unilateral inhibition is achieved by a direct delivery of the steroid from the ovary to the ipsilateral uterus via a special utero-ovarian vascular connection.

In the oviparous *Sceloporus undulatus* and *Cnemidophorus uniparens,* luteectomy elicits early oviposition. Indeed, if progesterone inhibits uterine contractility in viviparous reptiles as in *Chrysemys,* one would expect that this contractility and sensitivity to oxytocin-like peptides would be minimal during pregnancy, when progesterone levels are highest. However, this view is incompatible with the finding that the proportion of oviducts manifesting spontaneous contractions, and the maximal amplitude of the contractions increase progressively during gestation in the live-bearing lizards *Liolaemus tenuis* and *L. gravenhorstii* and declines only after parturition.

The suggestion that progesterone inhibits uterine contractility in reptiles may well explain the observations concerning premature expulsion of embryos or eggs in progesterone-deficient luteectomized reptiles. However, it cannot explain the phenomenon of prolonged gestation reported so often in ovariectomized or luteectomized lizards and snakes. The oppositive suggestion, based on findings in *Anolis carolinensis* that progesterone augments and estradiol inhibits uterine contractility in response to AVT, may explain the observations on delayed parturition in luteectomized or ovariectomized lizards or snakes. However, such explanations should also assume the dominance of progesterone effects over those of

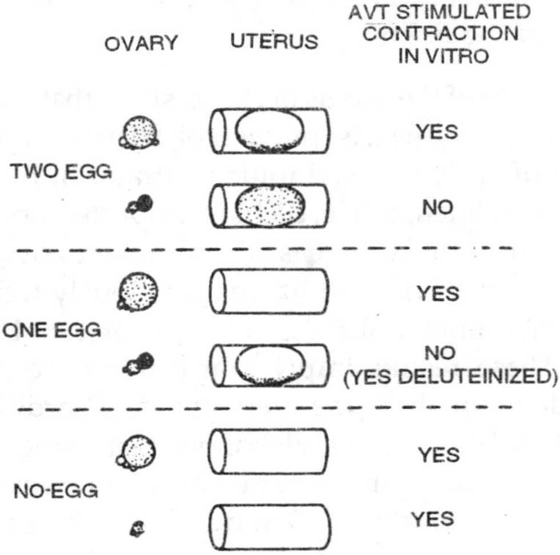

Fig. 4.26. *Anolis Carolinensis*. A diagram of the ovarian and uterine conditions in "two-egg", "one-egg" and "no-egg" females. Fully stippled circle or oval, yolk-laden oocyte in ovary or recently ovulated egg in uterus; shaded oval, shelled uterine egg; dark circle, corpus luteum. The uterus and ovaries of a single animal alternate as a sequence of "one-egg" and "two-egg" conditions that proceed throughout the breeding season. The "no-egg" condition results when captive females cease to cycle. Arginine vasotocin (AVT) fails to induce contractions of uteri associated with an ipsilateral large corpus luteum. When this corpus luteum is removed, the inhibition is lifted.

estradiol, as the extirpation of the ovary or corpora lutea removes both the inhibiting and the facilitating steroids.

Analysis of adrenergic receptors in uterine strips of *Liolaemus gravenhorstii* and *L. tenuis* demonstrates that during gestation both species have more α receptors (inducing contractions) than β ones (inducing relaxation). However, the difference in uterine sensitivity to stimulation of both receptors between pregnant and non-pregnant donors is more pronounced in the viviparous *L. gravenhorstii* than in the oviparous *L. tenuis*. The increase in the spontaneous contractions and the sensitivity to oxytocin during gestation in these lizards could be attributed neither to the number of muscular layers in the uterus nor to the thickness of the fibers.

In vitro studies of the uterus of *Anolis* show that epinephrine or the β adrenergic agonist, isoproterenol, reduce the amplitude and duration of AVT-induced tonic contract. In lizards with regressed corpora lutea, saline, dichloroisoproterenol and AVT alone do not significantly increase the number of ovipositions. However, injection of AVT to lizards previously treated with the β adrenergic antagonist dichloroisoproterenol, result in oviposition. These results imply that β adrenoreceptors are activated endogenously in the uterus of the lizard, inhibiting the action of AVT. Also acetylcholine, histamine, and to a lesser extend 5-hydroxytryptamine cause contraction of the isolated uterus of *L. gravenhorstii* and *L. tenuis*. Taken together, these results indicate that uterine contraction in reptiles is controlled by neurohypophyseal octapeptides and the autonomous nervous system and that this control is modulated by ovarian steroids.

It is also possible that progesterone alone, or in combination with estradiol, controls the hypertrophy of the oviducal musculature. Ovariectomy or luteectomy during pregnancy would lead to partial atrophy of the genital tract. This would keep contractions too weak to expel the neonates, in spite of vigorous compression of the abdominal muscles. This point needs to be clarified by further research.

GENERAL COMMENTS AND CONCLUSIONS

In spite of the disparity of opinions and the shortcomings of some reports, the evidence suggests that ovarian hormones have some effect on gestation, or at least on parturition in most viviparous reptiles. It is not surprising because ovarian hormones (estradiol, progesterone, and probably, also testosterone) have a profound effect on the growth and on the histological and biochemical differentiation of the genital tract prior to ovulation. Estrogen induces the formation of progestin receptor sites in the oviduct of lizards; this may explain the synergism of these steroids in their effect on the differentiation and growth of the genital tract. These steroids also affect the secretion of gonadotropin, the formation and uptake of yolk precursors by growing follicles, and ovulation. It is reasonable to assume that the ovarian control of the genital tract, mediated by these steroids, is not discontinued once gestation commences.

Except in a few species such as *Chalcides chalcides*, *Mabuya heathi*, and possible *Leiolopisma entrecasteauxii*, all other squamates studied so far have embryos with an ample supply of yolk and are mostly independent of placental nutrition during gestation. Still, they are provided *in utero* with water and oxygen. Disruption of uterine integrity due to the removal of ovarian steroids does not necessarily cause cessation of growth. Indeed, the embryos of many species continue to grow following ovariectomy or luteectomy. During the vitellogenic stage, ovarian steroids may be stored within the yolk and may diffuse to the adjacent uterus during gestation. This occurs in the dogfish in which these stored steroids control the maintenance of the uterus until the yolk supply is exhausted. It is also possible that the reptilian placenta may produce the hormones required for the maintenance of the genital tract; these hormones may act locally or via the circulation. However, no firm evidence has been presented so far on the steroidogenic capacity of any reptilian placenta.

When they do embryos die *in utero* after surgical removal of the ovary or part thereof? Studies on the extent of uterine and allantoic vascularity in *Thamnophis radix* show that the size and abundance of the maternal vessels are greatest at Zehr Stage 32. Uterine vascularity does not change appreciably in the last month of gestation. However, oxygen consumption of snake embryos increases exponentially and reaches progressively higher levels toward the end of the incubation period. In *Elaphe guttata*, oviposited eggs of which contain embryos in an early stage, oxygen debt is accumulated at later stages of intrauterine development, or the metabolic rate of the embryos is depressed due to an inadequate supply of oxygen. Also *Caretta caretta* and *Chelonia mydas* show maximal O2 consumption just prior to hatching. The O_2 consumed by the egg increases proportionally to the quantity of tissue produced. Impeded gas exchange prolongs incubation and increases egg mortality. Parturition on term in viviparous reptiles, probably "scheduled" by genetic mechanisms, would then save the embryos from severe hypoxia in the uterus, which has already reached its maximal vascularity. Prolonged gestation or delay of parturition would put the fully grown fetus, with its high and increasing oxygen demand, into a gradually increasing hypoxic environment. As ovariectomy reduces uterine vascularity, it makes this assumed hypoxia even worse.

Fig. 4.27. Oxygen consumption of embryos of an oviparous snake, *Spalersophis diadema* projected on the extent of uterine and allantoic vascularities of a viviparous snake. *Thamnophis radix* in an attempted to show the increasing discrepancy between possible oxygen demand by embryos and the estimated uterine blood supply.

Summary and Concluding Remarks

Live-bearing as a mode of reproduction is an ancient phenomenon in reptiles; the earliest reptiles for which there is evidence of viviparity are the early to mid-Mesozoic ichthyosaurs. Skeletons of young ichthyosaurs have been found inside the body of large individuals "emergent from what would have been the cloacal region in life." However, ichthyosaurs are unrelated to either the synapsid ancestors of mammals or the diapsid ancestors of the squamates. Viviparity in living squamates occurs in genera that have evolved relatively recently and cannot be directly linked with the conditions seen in recent mammals.

In both reptiles and eutherian mammals, the chorioallantoic placenta is the main placental structure in terms of size and duration. However, the yolk-sac placenta is the main structure in most metatherian mammals. The omphaloplacenta also occurs in eutherian mammals, but does not persist throughout gestation. Its nurtritional function is performed histiotrophically as in the omphaloplacenta of lizards and snakes. Some marsupials (peramelids) develop a chorioallantoic placenta, which may become haemotrophic as does that of the eutherian mammals. The chorioallantoic placenta of *Perameles* and that of *Sphenomorphus quoyi* are remarkably similar. In both forms, the uterine epithelia become syncytial and forms nests of nuclei, while the uterine capillaries become superficial and are apposed to fetal capillaries.

The general similarity of the reptilian and mammalian placentation may merely reflect the orthodox arrangement of extra-embryonic membranes and their vascular supply throughout the amniote vertebrates. "The evolution of a vascularized allantois in amniotic vertebrates and its progression into serving a placental function appears to have originated independently in several lineages among the reptiles and the mammals."

The reptilian placenta is undoubtedly permeable, not only

to gases and water, but also to electrolytes, amino acids, and steroids. These substances are transferred haemotrophically across both the omphalo- and allanto-placenta, with more transfer occurring across the latter with its heavy vascularization. There are claims, that the omphaloplacenta are transfers larger molecules, such as proteins secreted by the uterus.

Placental permeability need not have much nutritive significance in viviparous reptiles, because most embryos have ample yolk. Perhaps, the transfer of some electrolytes has some physiological importance. Significant nutritive function can be attributed so far to only *Chalcides chalcides, Leiolopisma entrecasteauxii,* and the "microlecithal" *Mabuya heathi,* in which the yolk supply seems less than that of other forms and probably inadequate to take the embryo to term. Unfortunately, the placental physiology of these species remains unknown. It would be profitable to look for viviparous species with a reduced yolk supply and to study their placental permeability as well as the amount of organic matter they gain during gestation.

In all reptiles, ovarian steroids stimulate the growth and differentiation of the genital tract prior to ovulation. The discharged follicles from corpora lutea in both viviparous and oviparous species. These corpora lutea are actively steroidogenic, secreting progesterone and probably other steroids, such as estradiol, testosterone, and deoxycorticosterone. Their steroidogenic function is stimualted by gonadotropins. In most viviparous species, embryonic development as such appears to be independent of the presence of the corpora lutea, indeed of the entire ovary, but gestation may be prolonged and parturition impaired in the absence of ovarian steroids. Some snakes and one lizard may abort or resorb the embryos following ovariectomy. Mortality of embryos at the end of gestation is attributed to the increased discrepancy between their oxygen requirement and the inadequate blood supply in the uterus. After ovariectomy mortality increases due to reduction in blood supply.

The seemingly conflicting evidence on the relations of gestation and luteal or ovarian hormones in reptiles may reflect some genuine interspecific differences. As viviparity has evolved independently along many parallel lines in squamates, the variability in the extent of endocrine control of gestration is not surprising. Unfortunately, these relations have been adequately examined in only a very few reptilian species. More information is still required on the functional significance of steroids, especially progesterone during gestation. This would allow reconsideration of the view that in reptiles progesterone has only marginal progestational function.

5

Parental Care in Reptiles

Prolonged care of eggs of offspring is a common feature of vertebrate reproduction. Parental care takes diverse forms, and presumably has evolved independently many times. Although parental care is exhibited by only a minority of reptilian species, these include a diverse array of lizards, snakes and crocodilians. The present chapter (a) reviews the occurrence and nature of reptilian parental care and (b) poses and tests hypotheses on the selective forces responsible for the evolution of such behaviour. The only previous comprehensive reviews of reptilian parental care are those of *Noble* and *Mason* (1933) on squamates and of *Cott* (1971) on crocodilians.

I define "parental care" as any form of postovipositional parental behaviour; that is, any action of the parent after oviposition or parturition, that increases the chances of survival of the offspring. This definition differs from the concept of "parental investment" which (a) includes investment prior to oviposition, and (b) is restricted to behaviour that decreases the probable survivorship or future fecundity of the parent. This latter condition is difficult to verify empirically and hard to apply to certain cases. For example, if the "costs" of a female guarding her eggs are independent of the number of eggs guarded, the addition of more eggs would not increase the mother's "parental investment"; however, her behaviour towards those additional eggs clearly would qualify as "parental care."

Parental care may be divided conveniently into two major

categories: care of the eggs and care of the young. Care of the egg is widespread but uncommon in reptiles, and generally consists of the parent remaining with the eggs for sometime after oviposition. This behaviour usually is termed "egg-brooding" or "egg-guarding." However, although protection against predators is probably the major advantage conferred by the parent's presence, actual defense against predators has been witnessed only rarely. Hence, the terms "nest-guarding" and "egg-guarding" should be applied only to species in which active nest defense has been observed. Cases in which the parent remains with the eggs, but without documentation of nest defense, may be termed "nest attendance" or "egg-attendance". Finally, the term "egg-brooding" may be restricted to species in which the female facilitates incubation by raising nest temperatures above ambient. I prefer this terminology to the proposal that the term "egg-brooding" should be reserved for species that actively manipulate their eggs.

THE OCCURRENCE OF PARENTAL CARE

Taxa Reported to Show Parental Care

1. *Testudines* : Parental care is rare or absent in turtles; apparently the only record is that of Hodsdon and Pearson (1943), who described maternal behaviour in the Bahamian emydid *Pseudemys malonei.* According to these authors the female turtles returned to their nest sites immediately before hatching and dug away hard-packed soil above the nest-chambers so that the hatchlings could emerge successfully. Fresh-water turtles have attracted considerable ecological study since the time of Hodsdon and Pearson's observations, and if maternal behaviour of the type described above is a general phenomenon, it is surprising that is has not been noted by other workers.

2. *Crocodilians :* In contrast to the turtles, parental care is common among crocodilians. Descriptions of crocodilian nest defense may be found in the works of Pliny and Aristotle and

more recent scholars have documented parental behaviour in virtually every crocodilian species studied in detail.

Although the frequency of nest attendance and active nest defense varies interspecifically, and even intraspecifically the broad outlines of parental care are similar in all crocodilians for which data are available. The pattern is quite different from that described in any squamate reptile. The female parent remains in the vicinity of the nest after laying, or returns to it at intervals. Potential predators on the eggs may or may not be attacked; species that are not egg predators are ignored. The female opens the nest at the time of hatching, and carries full-term eggs and newly-hatched young to the water. The young remain in a group for days or weeks and are actively defended by the adults. If attacked, even older juveniles will give "distress calls" that will initiate aggressive behaviour from adult crocodilians towards the attacker. Parental behaviours, such as nest-guarding and nest-opening, are probably universal among present-day crocodilians.

3. *Squamates:* Postovipositional parental behaviour by squamate reptiles can take many forms, of which the simplest is concealment of eggs immediately after oviposition. Oviparous squamates generally lay eggs in places with appropriate conditions of temperature and humidity and away from the attention of egg predators. Several species of lizards manipulate their eggs after oviposition, presumably to place them into suitable microhabitats. This behaviour has been reported in lacertids, iguanids sciricids, agamids and gekkonids. Presumably such behaviour is widespread among reptiles.

More complex, or at least longer-term, parental care is shown by many species. Among the squamates, at least 103 species have been reported to have one or both parents remain with the eggs after oviposition. These species include a taxonomically diverse array of lizards and snakes. Unfortunately, some of the cases listed in poorly documented; at least 15 of the 103 examples depend upon single observations of adults found near eggs. Undoubtedly, some of the cases do not indicate parental care, but may be coincidental or reflect incipient oophagy or discovery of a female with a recently oviposited clutch. The force of this object is diminished by the consistency with which the adult is reported to be coiled around the eggs rather than merely close to them, but nonetheless these reports should be regarded with caution. I once found a small snake *(Unechis gouldi)* only a few centimeters from four eggs under a piece of tin; the obvious inference would have been parental care, except that the snake species is viviparous, and the eggs were those of an agamid lizard.

Records of parental care for the Varanidae are particularly difficult to evaluate. No well-documented case is available, but at least three anecdotal reports are suggestive. For example, *Cogger* (1967-60) notes that "on several occasions when adult goannas have been disturbed at termite mounds, newborn young have been found emerging. This suggests that the female may remain near the nest until the young are ready to hatch, at which time she makes a new tunnel to release the young." In each case, the species involved was *Varanus varius*. On two separate occasions, a captive female *Varanus mitchelli* was found with the tail coiled around the eggs. Parental care is suggested by the behaviour of a female *Varanus salvator* that remained at the nest site for several days after oviposition. A less reliable account of possible parental care in varanids, probably a case of misinterpreted eggs predation, has been reported by *Berney* (1936).

Parental care in squamates may be divided into four major categories, the first three of which deal with care of the eggs rather than the offspring: (1) female buries eggs and defends nest site briefly against conspecific females; (2) female coils around eggs, defends them against predtors and warms them by shivering thermogenesis; (3) female remains with eggs after oviposition and may defend them against predation or pathogens; (4) female aids newly born or newly hatched young.

1. Nest-site defense against conspecific females is widespread in iguanine lizards but has not been reported in any other squamate group. In some cases, the nest site is defended even prior to oviposition. The intensity of nest site defense varies considerably among species, and even among geographic areas within a species. Some of this variability may be attributed to the physical condition of the females: only animals in good condition can engage in vigorous nest defense.

TABLE 5.1

Shivering Thermogenesis in Brooding Female Pythons: Disagreement in the Literature

Shivering thermogenesis *Species*	*Occurs*	*Does not occur*
Aspidites melnocephalus	Boos, 1979; Charles et al..,1983	Murphy et al., 1981
Liasis amethystinus	Boos, 1979	Charles et al., 1983
L. mackloti	Charles et al., 1985	Ross and Larman, 1977
Morelia spilotes	Harlow and Grigg, 1983	Charles et al., 1985
Python curtus	Vinegar et al., 1970	Noble, 1935
P. regius	Pitman, 1974; Logan, 1973	Van Mierop and Bessette 1981
P. reticulatus	Hediger, 1968; Muller, 1970	Honegger, 1970; La Panouse and Pellier, 1973; Pitman, 1974; Vinegar et al., 1970
P. sebae	Benedict, 1932; Noble, 1935	Vinegar et al., 1970; Pitman, 1974

2. Egg brooding (shivering thermogenesis) has been recorded only in pythons. The female coils tightly around the clutch, so that the eggs are completely hidden. In at least some species, rhythmic muscular contractions of the mother's body produce sufficient heat to maintain a relatively high and constant temperature in the egg mass. This phenomenon has been described in *Aspidites melanocephalus, Chondropython viridis, Liasis*

amethystinus, L. mackloti, Morelia spilotes, Python curtus, P. reticulatus, and *P. sebae* and may be universal within the pythons.

Certainly the pythons known to show shivering thermogenesis cover a broad taxonomic range, including small as well as large species, and exclusively tropical as well as temperate-zone taxa. Although several authors have stressed that only some python species utilize shivering thermogenesis, the evidence for interspecific differences in such behaviour is weak. Commonly, studies of the same species by different workers are contradictory with respect to whether or not shivering thermogenesis is exhibited. These disagreements are due to several factors, including (a) the type of evidence used (e.g., temperature measurement versus presence of muscular twitching), (b) the temperature-dependence of shivering thermogenesis (if the room is warm, no shivering is observed), (c) the difficulties in interpreting slight temperature differences between the egg mass and ambient, because of the great thermal inertia of these large reptiles, and (d) the low sample sizes (often only one or two animals). Some python species may not utilize metabolic heat production during brooding but convincing data are currently unavailable.

The maintenance of high temperatures during incubation undoubtedly quickens embryonic development and also may increase embryonic survivorship. Python embryos develop normally only at high temperatures and are much more stenothermic in this regard than embryos of most other squamates. One hypothesis is that maternal brooding is an adaptation to this embryonic thermal sensitivity. However, distinguishing cause and effect is difficult in such a correlation. Equally plausibly, the embryonic sensitivity may have evolved because of maternal brooding rather than vice versa. If the eggs are always kept warm by the brooding female, selection to maintain embryonic tolerance of low temperature would be weak or absent. An additional hypothesis is that maternal thermogenesis allows pythons to reproduce successfully under colder climatic conditions than would otherwise be possible.

The apparent restriction of shivering thermogenesis to pythons might be a function of their massively developed lateral musculature. Presumably, most reptiles would lack the physiological ability for sustained muscular contraction and hence, significant metabolic heat production. The substantial increase in metabolic rate of brooding female pythons means that maternal care in these snakes confers a massive energetic cost. Curiously, this group has not followed the boines in evolving viviparity, a strategy allowing egg temperatures to be raised by maternal behavioural (rather than physiological) thermoregulation. One reason may be that an egg-retaining female python could not subdue prey by constriction, because of the risk of damaging the eggs. An egg-brooding female would not face this restriction.

3. The most widespread form of squamate parental care consists of the female remaining with the eggs after oviposition. Her presence presumably serves some function, which has rarely been demonstrated. Functions documented in some species include defense against predators, and regulation of egg temperature or moisture levels. Many of the reported cases of egg-attending are likely to be invalid inferences, because they are based on single observations of adults with eggs. Nonetheless, the consistency with which this behaviour has been reported in particular species clearly indicates that egg-attendance is widespread in squamates.

4. Parental care of offspring rather than of eggs is remarkably rare in reptiles, perhaps because the mother and neonates are rarely in close proximity. One exception involves egg-attending (or guarding or brooding) species in which the female often is present at hatching. In at least one such species a female was observed to help the young hatch, and later "groom" the offspring. Even more remarkable is the observation (unfortunately uncorroborated) of newly-hatched *Python molurus* returning to their empty eggshells at night, with the mother then coiling around them and heating them by shivering thermogenesis. An aborigine from Mudginberri Station

(Alligator Rivers region, tropical Australia) suggested that female water pythons *(Liasis fuscus)* "took the babies down to the water and taught them how to swim."

If the scarcity of posthatching parental care in squamates is due to the lack of physical association between mother and young in most oviparous forms, this problem is overcome in viviparous squamates, which comprise approximately one-fifth of all species of lizards and snakes. Observations of viviparous females at parturition confirm that mothers in some squamate species may stimulate the young to emerge from their membranous sacs by pushing with the snout, or may actually tear open the membranes themselves. This occurs in xantusiid lizards and boid snakes.

However, mothers apparently do not open the fetal membranes at birth in viviparous iguanid, diploglossid, or lacertid lizards. In some colubrid and boid snakes, the mother ingests dead young and birth debris after the living young have emerged from their membranes. A similar phenomenon occurs in oviparous reptiles, the mothers of which ingest spoiled eggs. Such behaviour has been interpreted as parental care, because (a) it may prevent fungal infection spreading to other eggs, and (b) it may prevent odours attracting predators. Alternatively, rotting eggs may have a different odour than healthy eggs, and the females may mistake them for food items.

One further form of squamate parental care, which has received great publicity, is the alleged ability of female snakes to swallow their young for protection when danger threatens. The first published reference to this behaviour comes from the Egyptians at about 2500 B.C. whereas the first reference in the English language is in Spenser's "The Faerie Queene" (1590). The story has been applied to many species, especially to viviparous forms, and often to species that have been observed in detail by other workers who have failed to report the same phenomenon. The extensive literature on the subject has been summarized and evaluated multiple times. The consensus is that no satisfactory evidence of the phenomenon has ever been produced. The stories seem unlikely to be true because (a) the behaviour has never been reported in captivity or from field observations of scientists; (b) undigested young have never been discovered in stomachs of snake dissected for dietary studies; (c) offspring are probably unable to survive for significant periods inside the stomach of the mother. Nevertheless, the occurrence of mouth-brooding and transport of young in cichlid fishes and crocodilians, as well as the recent discovery of gastric brooding in an Australian frog suggest that gastric protection of young in snakes is not impossible. Nonetheless, reliable data are totally lacking.

Other popular tales concerning reptilian parental care include the mother snake nourishing the young inside her stomach, the mother rattlesnake crooning (by rattling mildly) to her newborn young and the mother rattlesnake caring for her young until they are well-grown, and then finding suitable home sites for them. Charming as these stories are, their validity seems doubtful.

Although the present review is concerned only with living reptiles, parental care also may have been shown in a variety of groups that are now extinct.

TAXONOMIC BIASES

General

The incidence of reptilian parental care is not distributed randomly among different taxonomic groups. This taxonomic bias is evident at the levels of order, suborder, family and genus.

Comparisons Among Orders

The three orders of living reptiles are only distantly related to each other, having been evolving as separate lineages for almost 300 million years. This long divergence has resulted in great differences among the living orders in morphology and, as shown by the present review, in the frequency and form of parental care. Parental, behaviour appears universal in crocodilians (and in their relatives, the birds) but seems to be lacking in testudines. Parental care is seen in many squamate taxa, but has been described in only about 2% of all oviparous squamate species (7% of oviparous genera). Of course, many species have yet to be examined in this regard.

Comparisons Between Suborders

Among the squamate suborders, parental care is unknown in amphisbaenians (possibly because of a paucity of field data), rare in lizards (recorded in 41 of approximately 3000 oviparous species, or 1.3%), and more common in snakes. If each species is treated as an independent data point, the proportion of oviparous species recorded to show parental care is significantly lower in lizards than in snakes (2 x 2 table, 1 d.f., $X^2 = 18.5$, $p < 0.001$). However, this may simply reflect a greater tendency for herpetologists to keep and observe snakes rather than lizards.

Comparisons Among Families

Among the squamates, the percentage of oviparous species reported to show clutch attendance ranges from zero in many families (Agamidae, Chamaeleonidae, Cordylidae, Dibamidae, Gekkonidae, Helodermatidae, Lacertidae and Pygopodiae) up to virtually 100% in the oviparous Boidae. The variation in percentages among families is too great to be attributed to chance alone (Combining families to ensure adequate sample sizes; 7 x 2 table, 6 d.f., $X^2 = 103.2$, $p < 0.001$).

Comparisons Among Genera

The 103 squamate species showing parental care are distributed among 13 general of lizards and 33 of snakes. If parental care was randomly distributed with respect to taxonomy, the proportion of generate containing more than one nest-attending species should be low. In reality, many instances of squamate parental care are found in congeneric species; for example, half of all the examples in lizards belong to the single scincid genus *Eumeces*. Parental care is probably ubiquitous in this genus, as well as in others such as *Ophisaurus* and *Phthon*. Such data are clearly sufficient to reject the null hypothesis of taxonomically random occurrence of parental care. Instead, parental care, although rare overall, is very common within certain taxonomic groups.

A strong taxonomic bias is evident not only in the frequency of parental care, but also in its form. Simple clutch attendance is usually the only form of parental care, but different behaviours are exhibited by iguanine lizards (eggs buried, and nests defended against conspecifics), boid snakes (eggs warmed by shivering thermogenesis), and crocodilians (complex biparental behaviour before and after hatching).

NUMBER OF INDEPENDENT EVOLUTIONARY ORIGINS

The tendency for parental care to be widespread within a single phylogenetic lineage suggests by parsimony that in such cases it has evolved only once, early in the group's history, and has been retained during subsequent speciation. The extreme case of the alternative hypothesis is that all species showing parental care have independently evolved this behaviour. Although the strong taxonomic bias in parental care argues against this latter hypothesis, it is difficult to dismiss in specific cases particularly should only a few distantly related species show the trait. This could result from either (a) lack of data on other related species, which show parental care although it has not been recorded, (b) an evolutionary loss of parental behaviour in these related forms, or (c) separate evolutionary origins of parental care in each species showing this trait.

A typical example of such ambiguity occurs in the colubrid snake genus *Elaphe*, two species of which show maternal care for several days after laying *(E. climacophora, E. quadrivirgata)*, whereas a close relative (*E. conspicillata*) does not Nest attendance also occurs in a distantly related North American congener.

In the following analysis, I have adopted a conservative approach has been adopted by considering any cases of parental care in the same subfamily as representing a single evolutionary origin, except in the cases of the speciose snake subfamilies Colubrinae and Elapinae. In these groups, available phylogenetic reconstructions suggest multiple origins of parental care, based on low probability of relatedness among the species involved.

1. Colubrinae

Omitting the doubtful record for *Opheodrys* (Table II), all of the North American colubrids reported to show parental care (*Elaphe, Cemophora, Lampropeltis* and *Pituophis*) probably are derived from a single adaptive radiation from Old World *Elaphe*. Hence, parental care may parsimoniously be regarded as having a single evolutionary origin in these snakes. However, the other colubrine taxa showing parental care (*Farancia, Lycodon, Ptyas*) are only distantly related to each other and to *Elaphe*, suggesting that egg-attending has evolved independently several times in colubrids.

2. Natricinae

The natricine snakes *(family Colubridae)* pose a similar problem. Parental care has been reported in five genera (six species), which appear to be relatively closely related Hence, these cases will be treated as a single evolutionary origin of parental care. This decision may be overly conservative because parental care has been reported from only one or two species within each genus; if it were a primitive character for the entire natricine radiation, one might expect to see more occurrences of the trait. The high probability that parental behaviour has been overlooked in many taxa represents a problem with enumerating the number of independent evolutionary origins by application of a specific method. Hence, it seems most parsimonious to count this as a single example of the evolution of parental care.

This analysis suggests that parental care has evolved independently remarkably few times in extant reptiles; the number may be as low as 21 (Table 5.2). Of these origins, only 5 have occurred among the lizards compared to 15 among the snakes. This ratio of evolutionary origins to the number of present-day oviparous taxa in the suborder thus is much higher (5 in 300 versus 15 in 1700) in the latter group.

TABLE 5.2

The Incidence of Parental Care in Reptiles at the Level of Family Subiamily, Genus and Species With Estimates of the Probable Number of Independent Evolutionary Origins of Parental Care

		Number of taxa in which parental care recorded		*Probable number of evolutionary Origins*
Family	*Subfamily*	*Genera*	*Species*	
Squamata				
Anguidae	Anguinae	1	5	
	Diploglossinae	1	2	1
	Gerrhononnae	1	2	
Iguanidae	Iguaninae	4	10	1
Scincidae	Lygosomirae	1	1	doubtful record
	Scincinae	2	21	1
Teiidae	Teiinae	1	1	1
Varanidae	Varaninae	1	3	1
Boidae	Pythoninae	8	16	1
Colubridae	Boginae	1	3	1
	Colubrinae	8	13	4
	Natricinae	5	6	>1
	Xenodontirae	1	1	doubtful record
Elapidae	Elapinae	6	11	>4
	Laticaudinae	1	1	1
Leptotyph-lopidae	Leptotyphlopinae	1	1	1
Typhlopidae	Typhlopinae	1	2	1
Viperidae	Crotalinae	4	4	1

Testudires				
Embydidae	Emydinae	1	1	doubtful redord
Crocodylia				
Crocodylidae	Alligatorinae	4	6	1
	Crocodylinae	2	10	1
Gavialidae	Gavialinae	1	1	

3. Elapinae

The six elapine genera recorded to show parental care are only distintly related to each other and hence are likely to represent independent evolutionary origins of the trait. The records for Australasian species are questionable and may be omitted, but this still leaves four probable origins within the elapines.

MALE, FEMALE AND BIPARENTAL CARE

Parental care in reptiles is performed almost always by the female sometimes in conjuction with the male. Even in crocodilians, the only group in which male parental care has been reported commonly, the male plays a much less significant role than does the female in nest-guarding or transport of young. However, males may be important in subsequent defense of hatchlings from predators.

Records of male parental care in squamate reptiles are often anecdotal or poorly documented. Egg-attendance by a male *Diploglossus delasagra* was reported by *Barbour* and *Ramsden* (1919), but they did not note how the parent's gender was determined. Egg-attendance has also been reported for a female of this species and for several related forms. *Barbour* and *Ramsden's* (1919) animal may have been incorrectly sexed.

Biparental care has been reported in four species of snakes: the Asian elapids *Bungarus ceylonicus*, *Ophiophagus hannah* and *Naja naja*, and the north American colubrid *Elaphe obsoleta*. The record for *Bungarus* is suspect; two adults of unspecified sex were observed curled up in a hollow with eggs and hatchlings.

Other records of parental care in this genus involve only the female remaining with the eggs. The record for *Ophiophagus* rests upon Wall's statement that "during the incubation period there is no doubt at any rate in some instances, that the male is in close attendance on his mate" (1924:193). This statement is repeated by *Tweedie* (1957), but detailed studies in captivity and collection of adults from 15 nests in the wild provide strong evidence that only the female remains with the eggs.

The record of biparental care in *Naja naja* is based upon, a captive pair; the female usually stayed with the eggs but "the male, which was always near the mound, (took) the place of the female during the three hours that she left the eggs to drink and feed. Both reptiles were particularly vicious during the incubation period" Deraniyagala (1955) reported that both male and female of a captive pair of *Naja naja* attended the nest, but that only the female resumed this behaviour after disturbance. A recent detailed study of reproduction in captive *Naja naja* did not report any assistance by the male to the egg-attending female. A statement by *Jennison* (1931) that the female Naja comes obediently to the whistle of the keeper, also seems inconsistent with observations of captive snakes. Hence, I am skeptical of the reliability of the reported biparental care.

The remaining case of parental care is in *Elaphe obsoleta* at *Jacob's Creek, Pennsylvania.* The eggs were buried in an old sawdust pile, and the two "parent" snakes remained on, in, or near the pile for at least 3 weeks. At one point, "Mr. Medsger secured a fork, and at a depth of twelve inches dug up 44 eggs of the pilot snake. The male snake was coiled around the eggs." Mr. Medsger's talk on this subject was illustrrated by photographs, so there seems no reason to doubt his observations. However, his means of sexing the adult snakes may have been in error. The reported clutch size is very high; usual clutches for this species are 7-12 eggs, and communal oviposition has been recorded. Hence, the clutch was probably a communal one, and the two snakes observed by Medsger may both have been females. Further observations n reproduction of this relatively common snake would be of interest.

HYPOTHESES ON THE EVOLUTION OF PARENTAL CARE

A. Cost-Benefit Models

1. General: The selective forces responsible for the evolution of reptilian parental care may be investigated by an analysis of the possible "costs" and "benefits" of such behaviour. Thus, the fitness (lifetime reproductive success) of a hypothetical reptile showing parental care is compared to that of an otherwise identical animal not showing care, and the ecological conditions or species characteristics conferring a higher fitness to the former individual are considered.

As described in theoretical models for the evolution of viviparity the primary "costs" of parental care are likely to be decrements in the food intake, survivorship, or subsequent fecundity of the parent. Remaining with the eggs may increase vulnerability of the parent to predators, depending on the site chosen and on whether the eggs have an odor detectable to predators. Remaining with the eggs certainly is likely to reduce the feeding opportunities of the parents. This in turn may prevent the accumulation of energy required for production of a second clutch. In males, restriction to the nest site and guarding of young may reduce opportunities for further copulation.

The "benefit" of parental care presumably is to increase offspring fitness, either by increasing survivorship of embryos and hatchlings or by accelerating embryogenesis so that hatching occurs at a favourable time. The increase in offspring survivorship could result from parental protection against many potential sources of mortality (e.g., predation, dessication, flooding, and fungal attack). These factors are considered in more detail below.

2. *Benefits of parental care* : Several of these presumed benefits may be valid, and indeed any single case of parental care may well have evolved for a variety of reasons. The common observations that females are particularly aggressive during the defense of nests suggests that deterring predators is a major function of parental care. This phenomenon is particularly striking in cases in which the individual reptile, or the species to which it belongs, is generally non-aggressive. However, many nest-attending females are not aggressive: females of several pythons and crocodilians, as well as those of *ophisaurus apodus, O. ventralis, Naja melanoleuca,* and *Ophiophagus hannah,* have all been reported to be relatively quiescent (in some cases, almost comatose) while attending eggs Nonetheless, it cannot be claimed unambiguously that any of these changes in maternal behaviour are *adaptations* to parental care. The behaviour of the parents could alternatively be interpreted as a direct response to a changed thermal environment, endocrine modifications associated with oviposition, the physiological stress of oviposition, or some other other factor. This does not mean that the behaviour of the parent fails to protect the eggs, but merely that it may be a direct consequence of the physiology of the parent rather than an adaptation per se.

Although the studies cited that females may actively defend their eggs against potential predators, the only work to compare predation rates on defended versus undefended clutches is that of *Metzen* (1977), who studied 110 nests of *Alligator* in the Okefenokee swamp. Most nests (96) were not defended by females, and almost all of these nests were destroyed by predators, especially bears. However, in 14 actively-defended nests, predation was rare.

An alternative threat to eggs may come from their being dug up by other nesting females. This has been reported to occur in sea turtles and presumably is the selective force favouring nest-site defense in iguanine lizards. Suitable nesting areas are rare in some habitats occupied by these lizards, so that disturbance of are earlier clutch by an ovipositing female is frequent. In these species, nests are defended only until the end of the egg-laying season. A similar situation occurs in *Alligator*. Turtles attempting to nest on their nest mounds may disturb their eggs and nest-attending female *Alligator* attack them under such circumstances.

The only other major function of reptilian parental care is likely to be its effect on the thermal environment of the eggs (*Packard* and *Packard*, this volume). Egg temperatures may be modified by (a) the parent adjusting the depth at which the eggs are laid beneath the soil surface, (b) the parent basking and then returning to the eggs, or (c) the parent producing heat metabolically to warm the eggs. The last of these alternatives is employed commonly by boid snakes; shivering thermogenesis has been reported in at least nine species.

Apart from benefits accruing from defense against predators or disturbance, and from thermoregulation, other functions of parental care have only been hypothesized rather than demonstrated. Most seem intuitively reasonable, but-are likely to be important only in a restricted number of cases. Other hypotheses seem invalid; for example, *Coborn's* (1975) explanations for "shivering" in brooding pythons are that this behaviour (1) promotes circulation of air around the eggs, (2) increases maternal circulation, and (3) is a maternal reaction to the presence of eggs between the coils. Undoubtedly, the posthatching parental care of crocodilians protects offspring from predatory attacks. The parent-offspring bond in crocodilians also may enable the parents to "teach" specific behaviour patterns to their offspring, but no data are available to test this hypothesis.

3. *Costs of parental care* : By staying with the eggs after oviposition, the adult reptile may forego opportunities for feeding. This in turn may reduce energy intake to the point that production of a subsequent clutch of eggs is delayed. Alternatively, a lower growth rate may depress subsequent

fecundity. In either case, the reproductive success of the animal is reduced. A more direct "cost" of parental care may be a decrement in survivorship, due to increased exposure to predators or other hazards. This might be likely if parental care occurs in a habitat different from that normally occupied by adults, or if predators can detect eggs more easily than they detect adults. The high metabolic expenditure of brooding pythons, which raise metabolic rate to warm to eggs, is another clear "cost" of parental care.

A reduced food intake, often a total cessation of feeding has been documented in several egg-attending species both in the field and in captivity. Observations of captive specimens indicate that inanition during this period is due to a specific disinclination to feed rather than a lack of encounters with suitable prey. Apart from the cases cited many other reports of nest attendance refer to the constant presence of the female with the eggs. Presumably, food intake is reduced in most or all of these species. Egg-attending females often are reported as appearing emaciated by the time that parental care terminates. Interspecific differences may occur in fidelity to the nest; females of some species regularly leave the eggs and forage, whereas those of other species never do so. However, specific data on this point are lacking. Females remaining with eggs potentially could obtain considerable quantities of food by consuming potential egg-predators attracted to the nest. Again, data are lacking.

A reduced food intake probably is common, but not universal, in egg-attending females. The same situation occurs with other reproductive modes in reptiles; for example, food intake is reduced during gestation in many but not all, viviparous species.

Without direct experimental manipulation (such as removal of clutches), longer-term costs of parental care are difficult to assess. The hypothesized delay in subsequent reproduction is supported by long intervals (greater than 1 year) among successive clutches in at least four egg-attending species of reptiles. Indeed, a strong correlation has been demonstrated both for reptiles and amphibians, among low frequencies of reproduction and "accessory costs," such as parental care, viviparity, and long breeding migrations. Low reproductive frequencies are common in taxa that to have such "costs." However, this correlation may not reflect the fact that parental care confers a high cost per se, but rather that the cost is relatively independent of fecundity. For example, the reduction in food intake of a brooding female python probably is independent of the number of eggs she is brooding. Under these circumstances, calculation suggest that infrequent reproduction, in each case with the production of large clutch, will be the optimal life-history strategy. A recent study of *Eumeces okadae* confirmed that postovipositional weight loss of egg-attending females was not correlated with their fecundity.

Although the egg-attending species with low frequencies of reproduction provide circumstantial evidence in support of "costs" of parental care, the opposite extreme also occurs; many egg-attending species, especially those from tropical areas, have been reported or suggested to produce more than a single clutch per year. These include *Gerrhonotus liocephalus, Elgaria multicarinatus, Tupinambis tegiuxin, Rhabdophis subminiatus, Xenochrohic piscator, Cemophora coccinea, Lycodon aulicus, Pituophis melunoleucus, Ptyas korros,* and *P. mucosus.* However, these records may be based on females that have lost or abandoned their first clutch.

4. *Conclusion:* Overall, the available data are consistent with hypothesized "benefits" and "costs" of parental care, but are generally circumstantial in nature. The only direct demonstration of the effectiveness of parental care in increasing egg survivorship comes from a study of *Alligator*. Good quantitative estimates of costs of parental care are lacking. These inadequacies in the available data are regrettable, because studies to measure the relevant variables are entirely feasible. Experimental manipulations could provided measurements

both of the benefits of a parental care (what is the survivorship of attended and unattended eggs?) and its associated costs (what are the differences in food intake, growth rate, and in the time of subsequent reproduction, between those females that are attending eggs and females of which the clutches of eggs have been removed?). Sample manipulations might involve (a) removal of part or all of the clutch of an egg-attending female, and (b) removal of the female from the clutch. Studies on parental care in salamanders and insects have progressed much further in this respect, than have those on reptiles.

Intermediate Stages in the Evolution of Parental Care

The vast majority of reptilian species do not show any form of parental care, but a small minority have been reported to remain with the eggs after oviposition. In most cases, the parent is believed to remain with the clutch for the entire duration of incubation, and only to leave after the eggs have hatched. This situation raises the question of intermediate stages in the evolution of parental care. Presumably, a species without egg-attending behaviour does not give raise by a single mutation to individuals that remain with eggs throughout development. Instead, the two more likely scenarios are as follows :

1. The female remains with the eggs for only a short time after oviposition and then leaves. This could occur if the female was "exhausted" by oviposition. Given the chance association between parent and eggs, selection could then favour "parental" behaviour and ultimately prolonged nest-attendance. The feasibility of short-term egg attendance as an intermediate stage towards prolonged parental care is supported by records of short-term egg attendance in several squamate species. Unfortunately, many cases of this phenomenon have been reported in captive animals, even in species known to attend eggs throughout incubation in the wild state. This suggests that nest desertion in captivity may be an artifact arising from human disturbance. Variation in the duration of nest defense in iguanine lizards may reflect material condition; emaciated

females defend nests only briefly, then leave to feed (*Wiewandt*, 1982).

2. The female leaves the eggs after oviposition, but returns to them regularly because they are located under a favoured refuge. This intermittent proximity of parent and eggs provides the opportunity for selection to favour "parental" behaviour, and perhaps leads eventually to more persistent nest attendance. We do not know how commonly this phenomenon occurs in nature, although *Mell* (1929) suggests that there is a continuum among species from intermittent to constant egg attenders.

Both pathways may well have been involved in the evolution of parental care, as is suggested by the apparent occurrence of both putative "intermediate stages" in modern reptiles. However, such strategies in modern reptiles may be secondary modifications of prolonged nest attendance, and not intermediates in the evolution of this trait.

Factors Increasing the Benefits of Parental Care

1. *Prediction* : If the primary benefit of parental care is an increase in egg survivorship, then parental care should evolve most readily in situations in which it has a large effect on survival of the eggs. This may occur in species or environments in which the attending parents are unusually successful at reducing mortality rates of eggs, either because of the abilities of the parents, or the high susceptibility of the unprotected eggs.

2. *Exposed versus hidden nest sites* : Eggs that are laid under superficial cover on the surface of the ground, rather than being deeply buried, may be particularly vulnerable to predation. Hence, parental care might evolved more readily in species that do not bury their eggs. This hypothesis has been used to explain the presence of parental care in *Tupinambis teguixm*, and its absence in the synonymous *T. nigropuctatus*, which oviposits inside termite mounds. However, the argument can be generalized to predict that parental care should be more

common in snakes, which typically do not bury their eggs (probably because of the lack of limbs with which to excavate a nest-hole) than in lizards, which typically dig nest-holes. For the same reason, parental care should be more common in limbless (non-burrowing) lizards than in species with fully-developed limbs.

Data presented earlier support the prediction that parental care is more common in snakes than in lizards (3% of oviparous species versus 1%) and has evolved more often in the former group. Limblessness has evolved many times among lizards, but is shown by only a small proportion of taxa. Hence, the parental care of the limbless *Ophisaurus* offers further, although weak, support for the hypothesis that limblessness favours the evolution parental care.

3. *Ability of parents to defend eggs* : If a major benefit of parental care is the repulsion of potential egg-predators, then parental care should evolve most often in species in which the parent is capable of deterring predators. This is most likely to be true of large and of venomous species. No such trend is evident among the lizard, except for *Tupinambis* and the tentative records for *Varanus*.

However, in snakes there is a strong bias for such parental care in large and venomous species. The only major taxon in which parental care is common is the Pythoninae, and this group contains the largest oviparous snakes of the world. Parental care also occur in 4 of the 8 oviparous genera of the Viperidae (50%) and 7 of the 17 oviparous genera of the Elapidae (41%). In contrast, the much larger (>200 genera) Colubridae, which consists primarily of non-venomous (or less venomous) species, shows parental careless commonly. Several of the colubrids reported to show parental care are unusually large or belong to the minority of venomous forms within the family. The prevalence of parental care in crocodilians—most of which are large and formidable—also is consistent with the prediction that this behaviour should be shown most frequently by groups in which parents are well able to defend their eggs. Interestingly, many taxa of marine invertebrates show the opposite trend; parental care is most common in the smaller species of marine invertebrates. Nonetheless, extended parental care may also be common in highly venomous forms.

4. *Limited availability of nest sites* : If nest sites are scarce relative to the number of nesting females, older nests may be excavated and destroyed by new arrivals. This situation has been described in sea turtles and island populations of iguanine lizards. Active defense of nest sites by postovipositional females is common in iguanines. Intraspecific variations in intensity of iguanine nest defense are correlated with the degree to which suitable nest sites are available, and the ease of digging burrows. The selective advantage of parental care in this situation is increased by the relatively synchronous nesting of the population; thus even brief nest defense is effective.

Although these arguments are consistent with the behaviour seen in iguanines, they are unlikely to be of general importance because scarcity of suitable nesting sites may be a rare phenomenon. The only parallels to iguanine nest defense may be in alligators, in which the nest-guarding female prevent ovipositing turtles from digging into the nest mound, and tuataras, in which resting females defend nesting sites against other females.

Factors Reducing the Costs of Parental Care

1. *Prediction* : If the major "costs" of parental care are a reduction in food intake probable survivorship, or subsequent fecundity of the reproducing female, then parental care, would be expected to evolve most often in species and environments in which such costs were minor or insignificant.

2. *Selection for "Risky" life-history strategies* : Life-history theory predicts that small, short-lived species are more likely to pursue "risky" reproductive strategies than are large, long-lived ones. Although parental care was suggested to be "risky,"

a test on fishes revealed no clear trend for more parental care in smaller species, possibly because the smaller species were not capable of deterring egg-predators. Similarly, a compilation of data on 49 species of lizards showed a trend reverse to that predicted; parental care was more common in late-maturing species than in those that matured early (2 of 35, or 6%). This contradiction has been explained by suggesting that parental care is not as "risky" as it appears; "lizards that practice it usually remain with their eggs in a well-hidden site in which they may be less exposed to dangers than a female that does not tend her eggs". Indeed, following a general discussion of the evolution of lizard life histories, Tinkle (1969) predicted that parental care will generally be found in long-lived iteroparous species, especially those with a short, annual breeding season.

To the degree that large body size is correlated with late attainment of maturity the trend for parental care in large reptiles is consistent with *Tinkle's* (1969) prediction. However, it is consistent also with the hypothesis that parental care by small species is ineffective against predators, or the alternative hypothesis that in small species, parents themselves are too vulnerable to predation to permit the evolution of parental care. Certainly, the data on reptiles are inconsistent with the prediction that parental care will be most common in small, short-lived species.

3. *Frequency of reproduction* : Egg-attending reduces food intake of the reproducing female, and thus may delay the time at which a subsequent clutch can be produced. Hence, parental care is likely to evolve in species that produce only a single clutch of eggs per year. The costs of remaining with the eggs may be independent of the number of eggs guarded, whereas the benefits increase with the number of eggs. Hence, parental can might be more likely to evolve in species that produce large but infrequent clutches rather than in those that produce small and frequent clutches. Both hypotheses predict that parental care should be most common in species that produce

clutches only once a year or less often. Data on lizards support the prediction, although there are some puzzling cases of multiple-clutching, egg-attending species. The correlation between parental care and extremely low frequencies of reproduction also is consistent with this prediction, but is open to the other interpretations discussed earlier.

Because the annual production of multiple clutches is common only in the tropics parental care should typify temperate rather than tropical species. The available data are biased by the concentration of scientific study in the temperate zone, but even so, the prediction is refuted; parental care is widespread in tropical reptiles. The reptilian subfamilies in which parental care is most common are all primarily tropical groups. Overall, the predicted bias toward temperate-zone species is conspicuously lacking.

4. *Suitability of habitat for clutch attendance* : Parental care might be unlikely to evolve whenever eggs are laid in a habitat different from that usually occupied by the adult. In such a situation, the parent may be unusually susceptible to predation or physiological stress. This hypothesis suggests that parental care should be rare in aquatic or arboreal species. It is an obvious explanation for the lack of testudinian parental care, but is unconvincing because (a) many turtles are terrestrial; and (b) many other aquatic reptiles show parental care (e.g., Crocodilians, laticaudid sea snakes)

5. *Harsh and unpredictable environment* : Prolonged attendance on the clutch might be most likely to be favoured in environments in which resources for the adult, at about the time of egg deposition, become increasingly scarce or unpredictable so that searching for such resources becomes a high-risk endeavour. Under these circumstances, parental care might strongly benefit egg survivorship while conferring only a minor cost on the opportunities of the adult for future reproduction. This hypothesis predicts that parental care will most often be found in environments in which resources for

adults are limited at the time of egg deposition. The prediction is difficult to test, but seems inconsistent with the strong taxonomic bias in the distribution of reptilian parental care. If specific environmental variables are important, one might expect to see parental care restricted to particular habitat types rather than to all species within a given taxon. However, available data are insufficient to convincingly refute the prediction.

6. Brief incubation periods : Parental care might be associated with brief incubation periods for three reasons:

(1) the "costs" of remaining with the eggs depend upon the incubation period: if the eggs hatch soon after oviposition, the female is burdened only briefly by egg attendance.

(2) Natural selection may favour prolonged oviducal retention of eggs in species with parental care, because the costs of this retention may be no higher than the costs of parental care. In contrast, species without parental care are less likely to evolve egg retention; in this case, the physical budening of the gravid female imposes too high a cost on survivorship and subsequent fecundity.

(3) Parental care and prolonged uterine retention of eggs are both examples of increased maternal investment, and may be favoured by the same selective forces. Hence, they are likely to occur in the same species and the same environments.

A recent review of incubation periods showed "at least an indication that egg-guarding species may have slightly shorter development times than those that do not guard eggs." More detailed analysis of the data from *Tinkle* and *Gibbons* confirms that the mean incubation period of egg-attending species is lower than that of non-attending species, but the variances are so high that the difference fails to reach statistical significance. One confounding variable in this test, however, is the trend for parental care in larger species large species tend to have larger eggs, which in turn have longer incubation periods.

An alternative test involves the examination of the embryonic stage of development at oviposition in egg-attending and non-egg-attending species. "Visible embryonic development at oviposition" has been reported more commonly in species with parental care. However, a more recent study, which used objective criteria to stage embryonic development, found that relatively prolonged retention of eggs is the rule rather than the exception in oviparous squamates. No evidence exists to show that uterine retention is more prolonged among egg-attending species.

A recent analysis of parental care in salamanders (*Nussbaum*, 1985) argues that parental care should evolve in species with long incubation periods; that is, it predicts the reverse of the above-discussed prediction. Nussbaum notes that salamanders with parental care tend to have large eggs, and the such eggs take a long time to develop. From these data, he argues that parental care has evolved to reduce the otherwise high rate of mortality of these slowly-developing embryos. An alternative interpretation of the same data is that natural selection has favoured an increase in egg size in species with parental care, because the offspring may thereby be kept for longer in a low-risk situation (protected eggs) rather than as unprotected free-living juveniles. The association of parental care with large offspring is less clear in reptiles than in many other animal taxa, possibly because of the concentration of most reptilian embryonic mortality to a short postovipositional period.

Which Sex should Show Parental Care?

One consistent feature of reptilian parental care is that involvement by the male is non-existent (squamates) or relatively minor (crocodilians). In contrast, male parental care is common in amphibians, fishes and birds. The evolutionary basis for sex differences in the tendency to show parental care has been the subject of several recent discussions. Two hypotheses for the selection of parental care have received wide attention:

1. Selection against male parental care, because internal fertilization results in a delay between insemination and oviposition, making the paternity of any given clutch uncertain. This low reliability of paternity may select against male parental care in species with internal fertilization including reptiles.

2. The unlikelihood of selection for male parental care in species with internal fertilization, because the time delay between insemination and oviposition means that a male and his offspring may never be in close proximity.

Both of these hypotheses correctly predict that reptilian parental care should be performed by females rather than males. Although the reptilian data therefore do not permit a test between the two hypotheses, the latter (parent-offspring proximity) model seems more accurately to predict the distribution of parental care in teleosts and amphibians. The paternity hypothesis also has been criticized on the grounds of faulty logic.

An alternative approach is to consider the effects of reproductive activities on the parents. If the female is "exhausted" by oviposition, she may be likely to stay with the eggs until she recovers, protoadapting the species for female parental care. However, exactly the opposite prediction is made by *Maynard Smith* (1977); because the female is exhausted, she has a greater need than the male to recommence feeding as soon as possible. Thus, Maynard Smith predicts that this situation should favour the evolution of male parental care. The reptilian pattern is consistent with the prediction of Noble rather than with that of Maynard Smith, but does not provide strong support for the former hypothesis, because the predominance of maternal care is consistent with several alternative theories.

A related question is why biparental care is virtually unknown in squamates, and probably rare even in Crocodilians. This result is consistent with a general trend for biparental care to be less common in ectotherms than in endothermic vertebrates. Biparental care may be most likely to evolve when the form of care is such that two parents are much more effective than one; for example, feeding or guarding mobile young, rather than merely guarding eggs. This hypothesis predicts that biparental care in reptiles should be restricted to crocodilians (the only group to guard offspring after hatching), and that the contribution of the male should commence only after the eggs hatch.Why Is parental Care Rare in Reptiles?

An analysis of parental care at the familial level reveals major differences among reptiles, teleost fishes, and amphibians in the frequency and form of parental care they exhibit (Table 5.3). The proportion of families containing care giving species is highest among amphibians, in terms of parental care by either sex, or overall. Reptiles differ strongly from the other two groups in their low frequency of male parental care. This probably reflects the lack of externally-fertilizing reptilian species. Although the overall proportion of families showing parental care is similar in fishes and reptiles, this direct comparison is misleading. Most fish species produce pelagic eggs, so that parental care is impossible. The proportion of species showing parental care would be much higher in demersal-spawning teleosts than in reptiles. This difference may be attributable to (a) the high incidence of external fertilization in teleosts, protoadapting for male parental care, and (b) the brief incubation period of teleost eggs, requiring only a short duration of parental care.

Although parental care is proportionally less common among reptiles than among other ectothermic vertebrates, an alternative form of increased parental investment—viviparity—is strikingly more common in reptiles. This may reflect two reptilian features: internal fertilization and behavioural thermoregulation. The former protoadapts the species to viviparity whereas the latter enables rates of embryonic development to be accelerated greatly by uterine retention of eggs. Because parental care of eggs may serve as a protoadaptation to viviparity, the incidence of reptilian parent care may be lowered by a trend for viviparity to evolve in care-giving species.

TABLE 5.3

The Distribution of Parental Care and Viviparity Among Families of Teleost Fishes, Amphibians, and Reptiles

	Teleosts	*Amphibians*	*Reptiles*
Number of families for which data available	182	35	43
Number of with male parental care (and proportion with male parental care)	63 (.35)	16 (.46)	1 (.02)
Number with female parental care (and proportion)	28 (.15)	17 (.49)	13 (.30)
Number with parental care by either parent (and proportion)	68 (.37	23 (.66)	13 (.43)
Number with viviparity (and proportion	11 (.06)	6 (.17)	18 (.42)
Number with parental care and/or viviparity (and proportion	77 (.42)	25 (.71)	24 ((.51)

6

Parthenogenesis in Reptiles

The study of parthenogenesis in reptiles parallels in many ways the study of clonal reproduction in fish and amphibians. Parthenogenesis in vertebrates was too starling to be more than tentatively proposed in many of the earliest works, but our knowledge of this remarkable phenomenon has grown ever more rapidly since the early 1930s. Clonal reproduction is now firmly established, new forms have been discovered, new questions raised, and new techniques applied. We provide here an overview of clonal reptiles, especially the parthenogenetic taxa of the *Lacerta saxicola* group. We use the basic data to discuss some of the current controversies in the study of clonal vertebrates, and finally we point toward areas that are expected to yield significant new data and concepts in the near future.

Three forms of clonal reproduction can be distinguished within the vertebrates: *parthenogenesis,* in which neither genetic contribution nor activation of the ovum depends on sperm; *gynogenesis,* in which sperm activate the ovum but normally make no nuclear genetic contribution and *hybridogenesis,* in which sperm both activate the ovum and contribute to the zygote. Gynogenetic and parthenogenetic taxa are predominantly or exclusively females; the reproductive capacities of the few males reported are uncertain. Hybridogenetic taxa can be of either or both sexes. One genome is inherited clonally. The other is introduced at fertilization, but is lost before the completion of the next meiotic division.

Within the vertebrates, clonal anamniotes may be either hybridogenetic or gynogenetic; no clear cases of parthenogenesis in *Wilson's* (1925) sense are known, although it may exist in some populations of *Ambystoma* studied by *Downs* (1978). Clonal taxa of reptiles, however, and presumably of other amniotes in which clonal reproduction occurs, are exclusively parthenogenetic in their usual reproduction. If this dichotomy between amniotes and anamniotes is real, it may reflect the need of anamniote ova for a particulate non-nuclear contribution from the sperm. This has been demonstrated, at least for *Rana pipiens*, by Shaver (1953); the contribution is possibly the centriole.

Ploidy is usually one of the first features of clonal vertebrate taxa to be determined, and it has been determined for most; sometimes, in fact, ploidy provides the only real evidence that a taxon reproduces clonally. Among the 50-odd clonal vertebrate taxa, some 26 are known to be diploid, whereas 19 are triploid. Within reptiles, some 15 forms are diploid, whereas 13 are probably or certainly triploid, for four, the ploidy is presently unknown.

Many of the clonal taxa of vertebrates can be shown to have genomes contributed by more than a single species, although this has not been determined for many forms. The role of hybridization in the origin of clonal reproduction has been a matter of considerable argument. It seems rather clear that hybridogenetic taxa are a result of hybridization, because they can be reproduced in the laboratory. Those diploid clonal taxa that are known to be hybrid must also have arisen in this way, because a diploid clonal individual within an otherwise sexual species would produce diploid gametes; these might become hybrid by addition of a foreign genome to form triploids. Triploid clonal forms that contain complete genomes from three different species also must surely have arisen by hybridization. Many clonal taxa, however, are known to be diploid but are not known to be hybrid; of the triploid taxa, only one is known to have three different genomes. In any of these, clonal reproduction could conceivably have arisen within a species without recourse to hybridity. It is probably no coincidence that all the best studied clonal taxa of vertebrates are known to contain genomes from two or more parental species. The 23 for which hybridity (as opposed to hybrid origin) has not been convincingly demonstrated include eight of the nine for which chromosome number has not been established. Nor is it a coincidence that some of the strongest arguments made for non-hybrid origin of clonal taxa are based on the more poorly known taxa.

Not until a laboratory synthesis of several clonal taxa, with a study of whatever genetic variation is necessary for that synthesis, has been carried out will it be possible to prove, if it is indeed true, that hybridization is the cause of clonal reproduction. Most of the data that we present, however, are drawn from taxa in which hybridity if not hybrid origin has been demonstrated. Much of the discussion will therefore seem to assume that all clonal taxa of vertebrates are hybrid. This may be true, but we are aware that it has not been demonstrated and may in fact be false.

For clonal taxa that are diploid and hybrid one can ask which features of their biology depend on their clonal reproduction and which depend on their hybrid make-up. For triploid taxa, one can ask which features depend on their ploidy, which depend on their clonal reproduction, and (if they are hybrid), which depend on their hybridity. Often the data available to us have been presented as observations on a clonal taxon or a series of clonal taxa, with little or no comparison to biparentally reproducing species especially to the parental species for those of known hybrid composition. Such observations on known hybrids are difficult to use for synthetic purposes, because it is often necessary to compare data collected by different workers for different purposes, with resultant variation in quantification and presentation.

A BRIEF SURVEY OF THE KNOWN CASES OF PARTHENOGENESIS IN REPTILES

Of the 32 parthenogenetic taxa of reptiles that have been at least tentatively identified. Some have been extensively studied, but almost nothing is known about many. The descriptions below serve as an introduction to the various taxa.

Hemidactylus (Gekkonidae)

The genus *Hemidactylus* includes about 75 species distributed in Africa, southern and southeastern Asia, Polynesia, Central and South America, and southern Europe.

1. *Hemidactylus garnotii,* unlike most members of this genus has a very wide, mostly insular, range that to a considerable extent is the result of its fortuitous or deliberate introduction by man. It has been reported from eastern India southern Burma, Thailand, China, the Philippines from a number of Indonesian islands and the Indo-Australian Archipelago, New Caledonia, Melanesia, northern Australia (?" Loyalty Islands, New Zealand, Polynesia, the Hawaiian Islands and Florida (United States). The wide distribution of this species is apparently one of the major reasons for its having at least eight different names.

2. Specimens of *Hemidactylus* from Vietnam earlier considered to belong to *H. garnotii* represent a new all-female triploid species *H. viernamensis.*

Heteronotia (Gekkonidae)

Heteronotia currently has three recognized species, all confined to Australia, Karyotype examination of the most widespread, *H. binoei,* has revealed that it is a complex with two sexual diploid and three probably parthenogenetic triploid forms. The triploid clones may be lineages formed by backcrossing a diploid hybrid between the diploid chromosome types, once to one diploid type, twice to two karyological variants of the other diploid type. This suggests (1) that the diploids called *H. binoei* are at least two distinct species, and (2) that at least two species of triploid parthenogenetic *Heteronotia* exist. The holotype of *Heteronota binoei* Gray is a male and this name is thus probably not available for any triploid form.

Lapidodactylus Lugubris (Gekkonidae)

The genus *Lepidodactylus* contains about 12 species, *Lepidodactylus lugubris* is as widely distributed as the genus, occurring on Sri Lanka, on the Nicobar and Andaman islands, in Burma, on the Malay Peninsula, on the Indo-Australian Archipelago, and on many islands in the Indian and Pacific Oceans. Introduced populations have been reported from the Panama Canal Zone and New Zealand. Recently, it has been discovered in Ecuador and in Colombia. *Kluge* (1968) suggested that some poorly known species of this genus should be united with *L. lugubris*. The parthenogenetic nature of *L. lugubris* was deduced when only four males were found among numerous specimens from Guam, the Solomon and Hawaiian Islands, and some other parts of the species range.

Leiolepis Triploide (Agamidae)

The brightly coloured "butterfly lizards", *Leiolepis*, were long referred to one polytypic species, *L. belliana*, but the genus actually includes at least four bisexual diploid, one triploid all-female species described recently as *Leiolepis triploida* by *Peters* (1971) and possibly a diploid parthenogenetic form. The distribution of the triploid parthenogenetic species is not clear because all known specimens came from some ill-defined locality near the Malaysia-Thailand border. Peters also reported one possible specimen in the Copenhagen Museum that came from Singora in southern Thailand on the shore of the Bay of Siam, the locality of Bohme's diploid form.

Lepidophyma Obscurum (Xantusiidae)

The night lizards of the genus *Lepidophyma* include nine species. The genus ranges from central Mexico in the north to Costa Rica and Panama in the south. Most widely distributed is *L. flavimaculatum;* its subspecies occupy most of the range of the genus. The all-female species (apparently the taxon described as *L. f. obscurum* by Barbour) inhabits the extreme southern border of the range in central Panama and Costa Rica *L. obscurum* is so far the only viviparous form known among parthenogenetic species of lizards.

Lacerta (Lacertidae)

All-female species of the genus *lacerta* belong to *Lacerta* part II. Fourteen distinct species are found in the Caucasus proper, most of which have been, until recently, considered to be subspecies of *L. saxicola* Eversmann. Darevsky (1967) divided the species *L. saxicola* into four biparentally reproducing species *(L. saxicola, L. rudis, L. caucasica, L. mixta)* and four parthenogenetic ones *(L. armeniaca, L. dahli, L. rostombekovi, L. unisexualis)*. More recent studies indicate that many of the forms that remained as subspecies of *L. saxicola* are themselves distinct biological species. These include *L. valentini, L. portschinskii, L. parvula,* and *L. raddei* with two races, *L. r. raddei and L. r. nairensis, L. uzzelli* is a fifth parthenogenetic species.

The common range of the above unisexual species occupies a vast territory in the mountains of the Lesser Caucasus within the borders of northern Armenia, the mountainous part of Western Azerbaijan, southern Georgia, and contiguous areas of northeastern Turkey. Individual species distributions follow.

Lacerta armeniaca

Lacerta armeniaca is distributed in the mountainous regions of central Transcaucasia, the southern border of its range running along the northern and northeastern slopes of Aragatz Mountain in Armenia; the northern border runs along the southern spurs of the Trialet ridge in Georgia and the northern foothills of the Lesser Caucasus in western Azerbaijan (Fig. 6.2). An isolated population occurs on the Yalnizcam ridge in northeastern Turkey, another in the Zigana pass on the Trabzon ridge in northern Turkey. The vertical distribution is between 1400 and 2000 in above sea level.

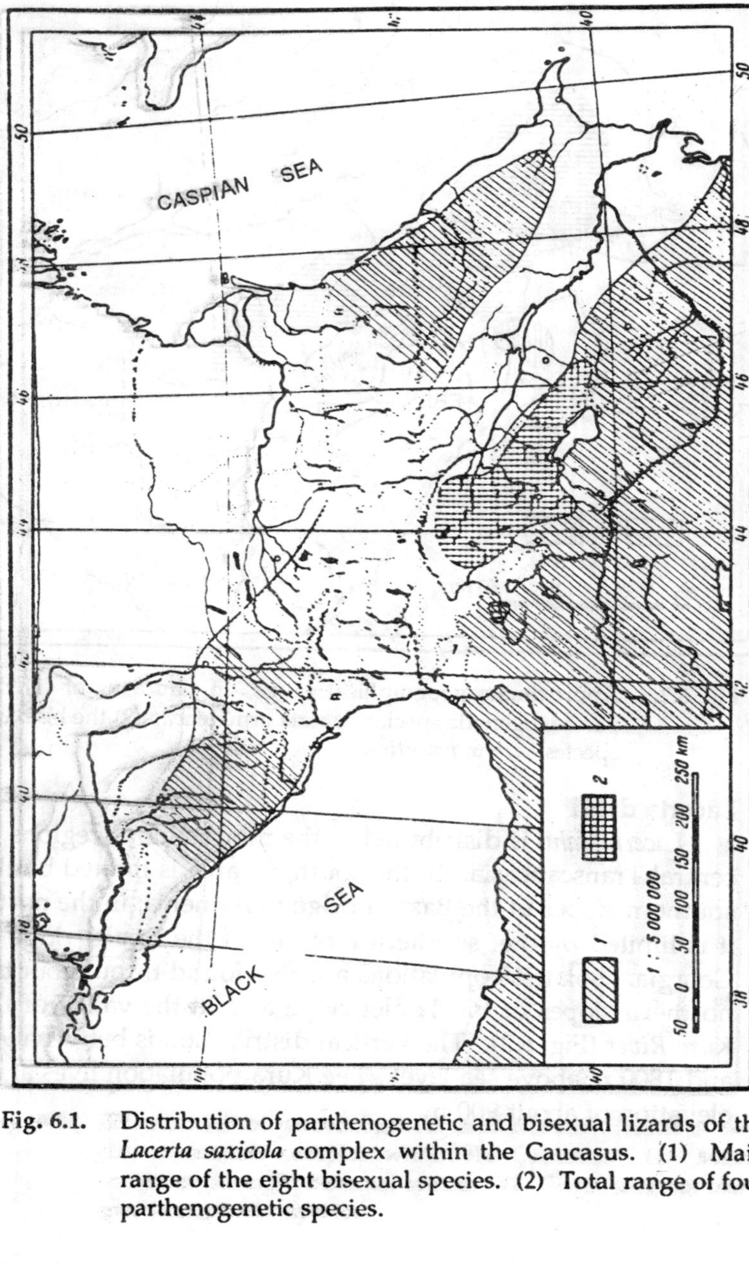

Fig. 6.1. Distribution of parthenogenetic and bisexual lizards of the *Lacerta saxicola* complex within the Caucasus. (1) Main range of the eight bisexual species. (2) Total range of four parthenogenetic species.

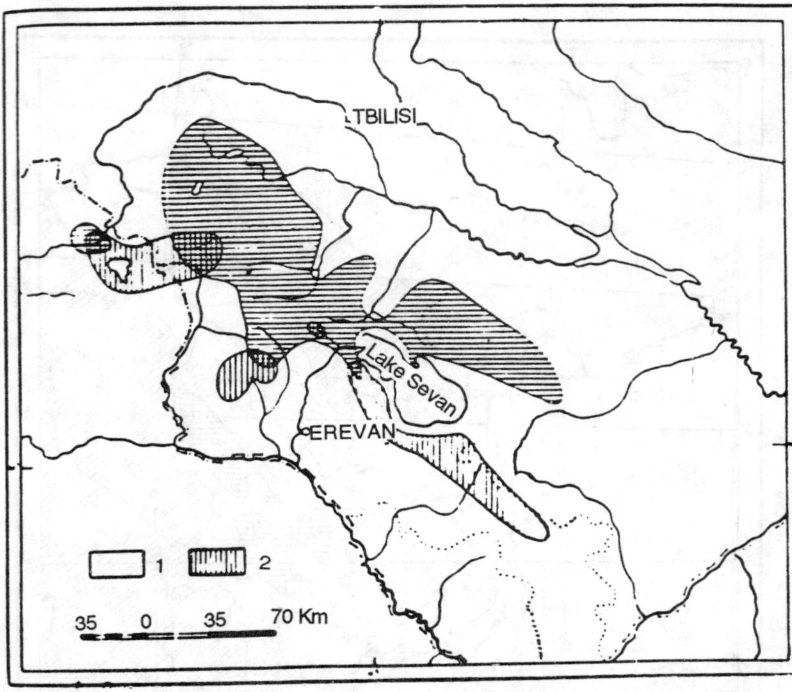

Fig. 6.2. Approximate geographical ranges in Caucasus of (1) the parthenogenetic species Lacerta armeniaca. (2) the bisexual species *Lacerta valentini.*

Lacerta dahli

Lacerta dahli is distributed in the mountainous regions of central Transcaucasia. In the south, its area is limited by the southern slopes of the Bazum ridge in Armenia; in the north, it is limited by the southern foothills of the Trialet ridge in Georgia. Isolated populations are also found throughout the northern slopes of the Trialet ridge and in the valley of the *Kura River* (Fig. 6.3). The vertical distribution is between 900 and 1800 m above sea level. The Kura population lives at an elevation of about 900 m.

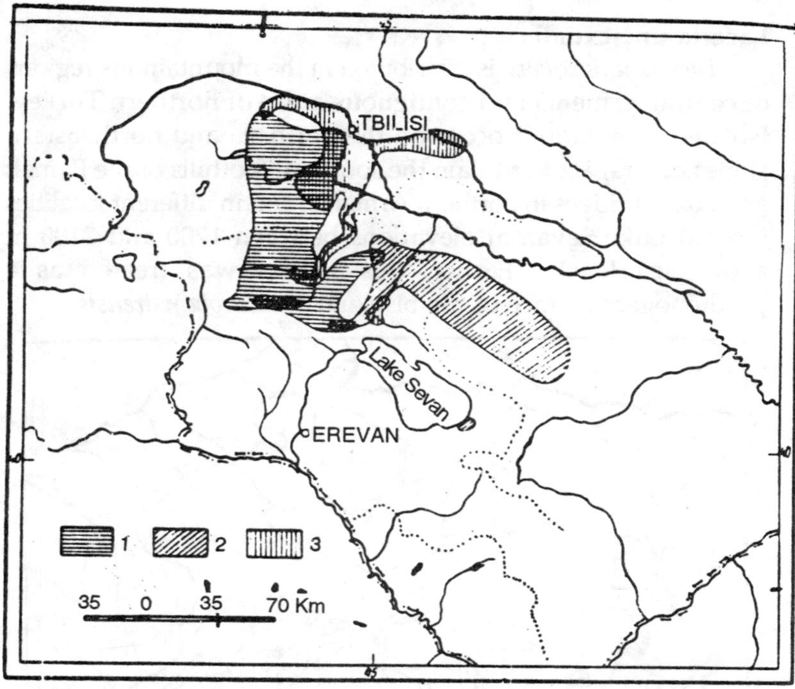

Fig. 6.3. Approximate geographical ranges in the Caucasus of (1) the parthenogenetic species *Lacerta dahli*, (2) the parthenogenetic species *Lacerta rostombekovi;* and (3) the bisexual species *Lacerta portschinskii.*

Lacerta rostombekvi

Lacerta rostombekovi inhabits the mountainous regions of central Transcaucasia. The southern border of its range passes along the southern foothills of the Bazum ridge in Armenia, the northern border running along the eastern foothills of the Somkhet ridge and northern foothills of the Lesser Caucasus in western Azerbaijan. A small isolated population occurs on the eastern shore of Lake Seven in Armenia. The vertical distribution is largely between 600 and 1500 m above sea level. The Seven population occurs at nearly 2000 m above sea level.

Lacerta unisexualis

Lacerta unisexualis is distributed in the mountainous regions of central Armenia and contiguous areas of northern Turkey. Isolated populations occur on the northern and northeastern slopes of Aragatz Mountain the southern foothills of the Pamak and Bazum ridges in northern Armenia, and in different localities around Lake Sevan at elevations between 1700 and 2100 m above sea level. Earlier, this species was treated as a parthenogenetic race of the bisexual *L. saxicola nairensis*.

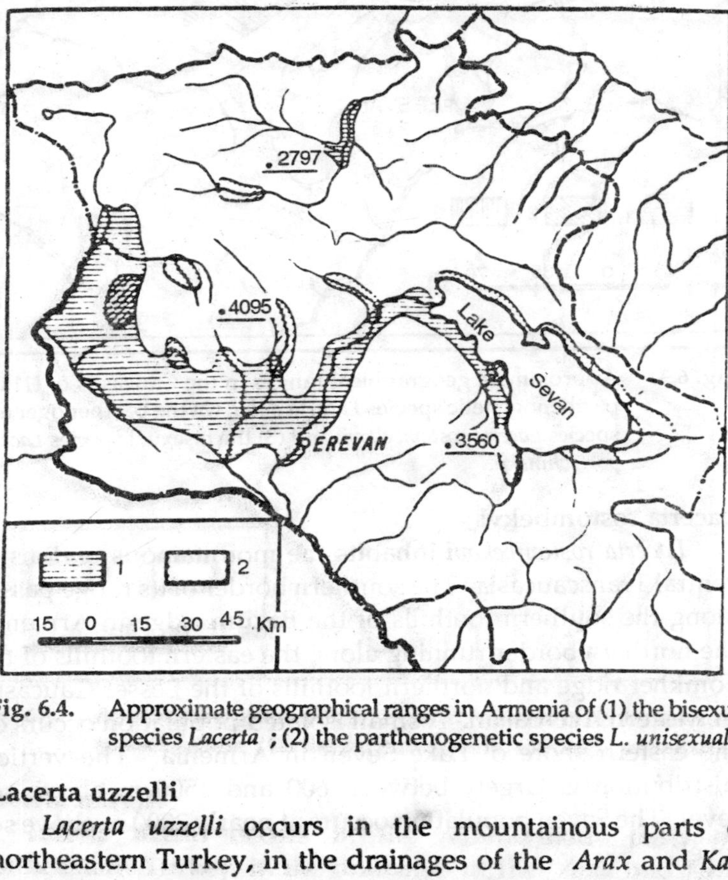

Fig. 6.4. Approximate geographical ranges in Armenia of (1) the bisexual species *Lacerta* ; (2) the parthenogenetic species *L. unisexualis*.

Lacerta uzzelli

Lacerta uzzelli occurs in the mountainous parts of northeastern Turkey, in the drainages of the *Arax* and *Kars* rivers. Somewhat atypical specimens are also known from west of Ercic, some 100 km to the south.

Cnemidophorus (Teiidae)

Because of their pronounced variability and wide geographical variation American whiptail lizards of the genus *Cnemidophorus,* like lacertid lizards of the *Lacerta saxicola* group, attracted the attention of herpetologisis long before the discovery of parthenogenetic reproduction and polyplo.

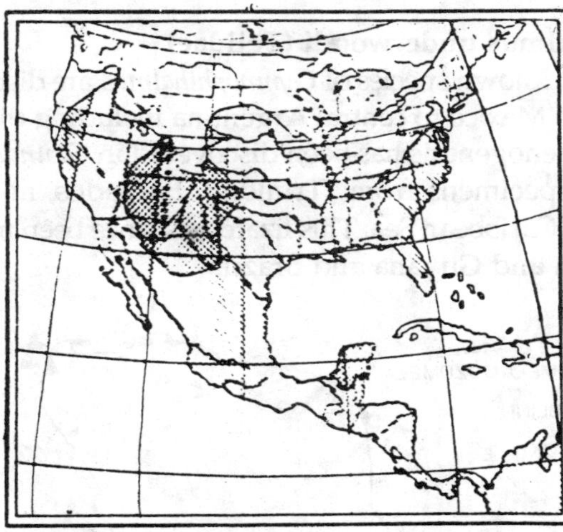

Fig. 6.5. Approximate geographical distribution of the genus *Cnemidophorus in North America.* Cross lineation indicates parthenogenetic species.

At present, about 40 species and numerous subspecies are known. On the bases of morphological and karyological characters, the biparentally reproducing forms are arranged into four groups. The geographic range of the genus extends from Oregon, Idaho, the Dakotas, Minnesota, Wisconsin, and Virginia (United States) in the north to southern Brazil, Bolivia, and northern Argentina in the south. Most whiptail lizards occur in Mexico and neighbouring regions of Central America and the United States; only a few species and subspecies occur in South America.

As seen from the summary map 11 of the 15 all-female species of the genus *Cnemidophorus* are distributed in the elevated regions of western United States, where they are more or less sympatric at elevations between 550 and 2500 m above sea level. Three other parthenogenetic taxa inhabit southern Mexico and bordering regions of Guatemala on the Yucatan Peninsula and adjacent islands. One apparently is widely distributed in northern South America. The distribution of each parthenogenetic species of *Cnemidophorus*.

Gymnophthalmus underwoodi (Teiidae)

The sevan known species of *Gymnophthalmus* are distributed from southern Mexico to central Argentina including the *Lesser Antilles*. Parthenogenesis has been discovered in *G. underwoodi*; based upon specimens from Trinidad, Barbados, and Saint Vincent in the Caribbean Sea. This lizard has since been reported from Surinam and Guyana and brazil.

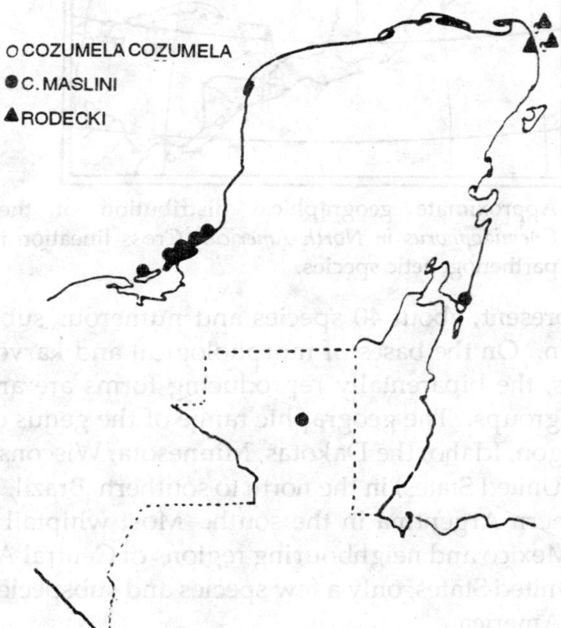

Fig. 6.6. Distribution of the *Cnemidophorus cozumela* complex on the Yucatan Peninsula.

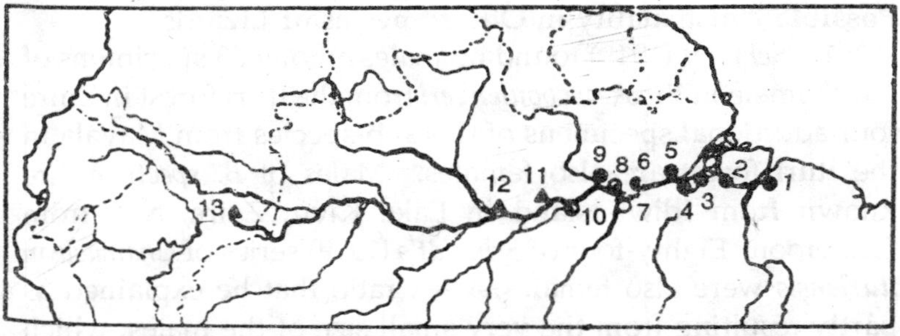

Fig. 6.7. Geographical distribution in the Amazon Valley of *Cnemidophorus lemniscatus* and the unisexual taxon associated with it. (1) Belem; (2) Breves and Corcovado; (3) Gurupa; (4) Porto de Moz; (5) Almeirim; (6) Monte Alegre; (7) Santarem and Alter do Chao; (8) Obidos; (9) Oriximina; (10) Parintins; (11) Itapiranga; (12) Manaus; (13) Mount of Amplyacu (1-6.9) Unisexual populations; (7) bisexual population; (8) shift from bisexual in 1965 to unisexual in 1968; (10) bisexual population in 1863, not explored recently; (11-13) no. *C. lemniscatus.*

Leposoma percarinatum (Teiidae)

The genus *Leposoma* includes seven species in South America and Panama. Scarcity of males of *L. percarinatum* was reported by *Ruibal* (1952), Additional specimens raise the total known to 30; all are female. This additional material provided such a strong contrast to the closely related sympatric *L. guianense* (7 males, 10 females) that *Uzzell* and *Barry* (1971) suggested that *L. percarinatum* is a unisexual species, a view supported independently by *Hoogmoed* (1973). *L. percarinatum* occurs in the Guianan uplands in Venezuela, the Guianas and adjacent Brazil.

Kentropyx borckianus (Teiidae)

The genus *Kentropyx* contains perhaps nine species; it occurs widely in South America east of the Andes, but is absent from the far south of Argentina. Absence of males in *Kentropyx borckianus* was reported by Hoogmoed (1973), who also noted that this is the proper name for the taxon usually called *K. intermedius*. The taxon is known from the three Guianas, Trinidad and Barbados.

Possible Unisexuality in Other Species of Lizards

1. *Schmidt* (1919) found no males among 63 specimens of the chameleon *Brookesia boulengeri* from the Ituri forest in Zaire four additional specimens of this sub-species from Mayala in the Ituri forest are also females. Males of *B. spectrum* are known from Idjwi Island in Lake Kivu, Zaire, and from Cameroon. Eighty-four of *Schmidt's* (1919) series of *Chamaeleon ituriensis* were also females, a sex ratio that he explained as partly resulting from the very small size of the males, which might have enabled them to avoid discovery and capture.

2. *Makino* and *Momma* (1949) reported 2n = 63 chromosomes for a female gecko identified as *Gehyra variegata ogasawarasimae; Hall* (1970) suggested that this might represent a triploid parthenogenetic taxon. *Shigei* (1971) synonymized *G. v. ogasawarasimae* with *Lepidodactylus lugubris*. The type of *G. v. ogasawarasimae* was reportedly a male. The proper interpretation of Makino and Momma's data is not clear.

3. *Bohme* (1975) reported that three captive females of *Basiliscus basiliscus* (Iguanidae) from Colombia, which were kept in isolation from males laid eggs over a period of years. All the hatchlings were females. These data suggest that these three females may have been derived from an all-female population.

4. *Schwaner* (1980) found no males in a sample of 51 specimens of the gecko *Cytodactylus pelagicus* collected in January and July on Ta'u Island, American Samoa.

Ramphotyphlops braminus (Typhlopidae)

The worm snake family Typhlopidae is worldwide in tropical and subtropical regions; it probably contains over 200 species. Boulenger's genus *Typhlops* has been subdivided by Robb (1966), on the basis of male genitalia, into two genera, *Typhlops* and *Ramphotyphlops*. *Typhlops* occurs throughout most of the range of the family, although it is absent from Australia and from many of the Pacific islands; *Ramphotyphlops* occurs from the Malay Peninsula and the Philippines to Australia. Assignment of *T. braminus* to *Ramphotyphlops* is based on uncertain features of the external morphology, the features of the male genitalia that distinguish *Typhlops* and *Ramphotyphlops* could not be determined, because all the specimens that he examined were female.

Ramphotyphlops braminus has an enormous distribution, based largely on introduction. It is widely distributed in the East Indies and occurs on many other Pacific islands as well. It is known in Asia from Malaysia west to Saudi Arabia, India, and Pakistan and also occurs in southern Africa, western Mexico, and at many other localities.

EVIDENCE FOR PARTHENOGENETIC REPRODUCTION IN REPTILES

Evidence of parthenogenetic reproduction is direct (virtual absence of males, absence of mating, and reproduction without males and mating) or indirect (genetic homogeneity shown by acceptance of skin grafts, or fixed heterozygosity revealed by electrophoretic or karyological data).

Absence of Males in Populations of Parthenogenetic Species

In most cases of unisexuality in reptiles the suggestion that they reproduce parthenogenetically derives from the absence of males in populations. *Darevsky* (1966, 1967) examined more than 5000 specimens of four unisexual species of rock lizards. *Tinkle* (1959), *Maslin* (1962), and *Zweifel* (1965) studied 65, 223, and 200 specimens, respectively, of *Cnemidophorus tesselatus*. *McCoy* and *Maslin* (1962) and *Fritts* (1969) dissected more than 533 specimens of *C. cozumela*. According to *Cole* (1975), different authors have examined over 600 specimens of *C. neomexicanus*, about 750 of *C. uniparens*, and over 400 of *C. velox*. *Kluge* and *Eckardt* (1969) examined 218 specimens of *Hemidactylus garnotii*, *Thomas* (1965) and *Hoogmoed* (1973) studied over 100 specimens of *Gymnophthalmus underwoodi*, *Telford* and *Campbell* (1970) examined 55 *Lepidophyma obscurum, Peters* (1971) examined 30 *Leiolepis triploida, Sabath* (1981) examined 102 *Lepidodactylus lugubris*, whereas *McDowell* (1974) and *Nussbaum* (1980) studied a total of 146 individuals of *Ramphotyphlops braminus*. With a few exceptions discussed below, all of these specimens were female.

Information in the literature on the presence of males in some unisexual species tends to be based either on erroneous determination of sex or on misidentification. For example, *Darevsky* and *Kulikova* (1961) noted that males of *Lacerta valentini* were mistakenly identified as *L. armeniaca*, and *Kluge* and *Eckhardt* (1969) noted that males of some bisexual species of house-gecko (probably *Hemidactylus frenatus)* were misidentified as *H. garnotii*.

Because the males of normal bisexual lizards are collected more commonly than females, the absence of males from large samples provides strong evidence of unisexuality. The argument is often convincingly supported by the absence of males among young reared from eggs laid in the laboratory. *Darevsky* (1958, 1966) reared 588 hatchlings from eggs of *Lacerta armeniaca, L. dahli, L. rostombekovi*, and *L. unisexualis;* all were female. Only females hatched from 171 eggs of *Cnemidophorus exsanguis, C. neomexicanus, C. tesselatus*, and *C. velox*. No males were found among a large number of young wild rock lizards and of whiptails.

Maslin (1966) discussed the problem posed by potential similarity between males of bisexual species and putative males of sympatric unisexual species. Although individuals of different sympatric species may be rather similar morphologically (for example, unisexual *Cnemidophorus tesselatus and bisexual C. tigris)*, the bisexuals form small parts of mixed populations, and there are too few males in mixed populations to explain the absence of males of the unisexual taxa as a result of their misidentification as males of sympatric bisexual species.

In parthenogenetic *Lacerta*, especially in *L. dahli*, male embryos, identifiable by presence of hemipenes, develop in some eggs, but perish at different stages of embryogenesis; the males usually are lethal monsters. There is thus good reason to assume that males are eliminated early in ontogeny through lethal genetic factors.

Absence of Mating

Regular reproduction by females in the apparent absence of males may have either an ecological or a genetic basis. The ecological hypotheses suggested are not supported by observations of rock lizards or of whiptails. The hypothesis of an extremely cryptic mode of life of males and their rare occurrence must be rejected. Specimens were caught throughout the activity period of the lizards, at any time of the day or night and in various habitats. Although among the extensive material examined there were found single males and hermaphrodites, their extreme scarcity precludes their playing a significant role in the maintenance of the population of unisexual species.

In many lizards, spermatozoa can be retained in the female reproductive tracts long after mating and can easily be detected under the microscope, thus providing reliable evidence of copulation.

The first searches for retained sperm in both bisexual and all-female rock lizards were undertaken by *Darevsky* (1958). It was shown that spermatozoa are always absent from the oviducts of females of *Lacerta armeniaca, L. dahli,* and *L. rostombekovi,* whereas spermatozoa are abundant in oviducts of related bisexual forms caught at the same time. Similarly, no spermatozoa were found in the vaginal and infundibular regions of 36 specimens belonging to the parthenogenetic species *Cnemidophorus tesselatus, C. exsanguis, C. neomexicanus, C. velox, and C. uniparens,* although sperm were detected in six out of nine females of the bisexual *C. tigris.* No spermatozoa were found in the oviducts of 19 mature females of the gecko

Lepidodactylus lugubris, which indicates that this species does not mate. The absence of spermatozoa in unisexual females reasonably excludes gynogenesis, cryptic male behavior, and differential male and female activity periods as interpretations for the all-female condition, and it suggests parthenogenesis.

The absence of mating in parthenogenetic *Lacerta* is also indicated by the absence of the marks that male's jaws usually leave on the bellies or thighs of females when holding the females during copulation. Such marks are always observed in females of bisexual species during the breeding period.

Reproduction in the Absence of Males and Mating

Reproduction in the absence of mating was demonstrated by *Darevsky* (1958), who obtained progeny from females known to be unfertilized. Thirty female *Lacerta armeniaca* of different ages were caught in September, 1956, in northern Armenia, half were immature. The lizards were put into a cage and lowered into a deep, damp crevice in a rock. Next April, after hibernating, the females were placed in a terrarium and later in an open air cage, where they remained until the beginning of the breeding period. Thus, the possibility of copulation was precluded. At the beginning of July, the 16 surviving females began laying eggs. By the end of the August, 56 of these eggs hatched; all hatchlings were female. Nine of these progeny were obtained from six females in their first reproductive period. In a similar experiment, progeny were obtained from a female of *L. armeniaca* brought from Armenia a year earlier. Females of bisexual species caught early in the spring and maintained for a long time without males lay inviable eggs that usually perish early in the incubation period. In wall lizards, particularly in *Podarcis muralis*, each egg laying period is preceded by mating.

Specimens of *Lacerta armeniaca* were released in 1964 on rocks along the upper Teterev River in the Zhytomyr region of Ukraine, about 1200 km north of their range. The introduced population has existed there for about 16 years; males are completely absent (unpublished data).

There is other evidence that all-female *Lacerta* reproduce parthenogenetically. Mating between parthenogenetic females and males of closely related bisexual species, both in nature and under laboratory conditions, results in sterile triploid hybrids. This phenomenon is apparently irreversible; return to bisexual reproduction by such sterile triploids seems impossible.

Conclusive evidence of parthenogenesis has been provided for four whiptail species. Laboratory hatchlings of *Cnemidophorus neomexicanus, C. tesselatus,* and *C. uniparens* have been raised to maturity. Their eggs contained embryonic tissue or embryos at various stages of development. Three eggs of *C. uniparens* hatched naturally, and thus the cycle, from hatchling, was completed in the absence of males. The cycle has been completed for several successive generations of *C. exsanguis.*

Acceptance of Skin Grafts

Except for members of highly inbred strains, the members of a species in which genetic recombination occurs reject tissue transplants from other conspecifics. That individuals of parthenogenetic species accept tissue grafts from other individuals is good evidence for parthenogenesis. Individuals of bisexual species usually reject tissues transplanted among themselves, because to accept them, donors must possess the same antigenic or immunologic properties as the recipients; this is rarely the case. Parthenogenetic species, in contrast, produce progeny genetically similar or identical to those of the mother. Thus, homografts transplanted among parthenoforms are not regarded by them as immunologically foreign. Similarity for histocompatibility alleles in some parthenogenetic species is characteristic only of members of single clones, although the similarity (and thus, the clone) may be observed in many different local populations.

Maslin (1967) corroborated parthenogenesis in *Cnemidophorus tesselatus* by the acceptance, within classes, of skin transplants; he detected one clone at two localities 350 km apart. Similar results were obtained for the gecko *Hemidactylus garnotii* and for *Lepidodactylus lugubris* it is possible that two histocompatibility clones exist in *L. lugubris* on Oahu.

Recent studies have examined histocompatibility in three parthenogenetic species of *Cnemidophorus*. In *C. uniparens*, 171 of 175 grafts were accepted, indicating that at the two localities sampled, some 120 km apart in central New Mexico, most, if not all, individuals belong to a single histocompatibility clone; the four grafts that were rejected perhaps represent technical errors. *C. neomexicanus* also has one widespread histocompatibility clone stretching through three New Mexican localities along the Rio Grande Valley. One individual from San Antonio, however, rejected a transplant from one individual at Tijeras, although the reciprocal was accepted, and reciprocal transplants between six other pairs at these localities were also accepted. *C. velox*, in contrast, has three well-marked clones, one each in the upper drainages of the Colorado, Rio Grande, and Pecos rivers. In addition, one individual from Espanola, New Mexico, twice rejected grafts from another individual from Espanola and one from Ojo Caliente, New Mexico, although the reciprocal grafts were retained, and seven other pairs from these two localities also accepted reciprocal grafts.

Of a total of 135 transplants among 35 individuals of *Lacerta unisexualis*, representing five central Armenian populations, 96.0% were permanently accepted. This indicates that these populations represent mostly the same clone.

Karyology

Although karyological heterozygosity has been used primarily to demonstrate hybridity in parthenogenetic reptiles and to restrict the possible candidates for parental species, fixed heterozygosity is in itself indicative of clonal reproduction. Similarly, fixed triploidy indicates clonal reproduction.

The biparentally reproducing lizards of the genus *Cnemidophorus* form four species groups on the basis of karyology and morphology. Several of the parthenogenetic taxa arose as a result of intergroup hybridization. *C. neomexicanus* contains one chromosome set from *C. inornatus*, a member of the *C. sexlineatus* group, and one from *C. tigris of the C. tigris* group. *C. tesselatus* contains one chromosome set from *C. tigris* and one from *C. septemvittatus* of the *C. sexlineatus* group; the triploid derivatives have a *C. sexlineatus* set in addition. That all individuals of these taxa examined show nearly identical karyotypic heterozygosity is indicative of clonal reproduction, although in the case of *C. neomexicanus*, which is sympatric throughout its range with both parental species, F_1 hybrids with the same karyotype, which may or may not be parthenogenetic, would not be distinguished in such an examination.

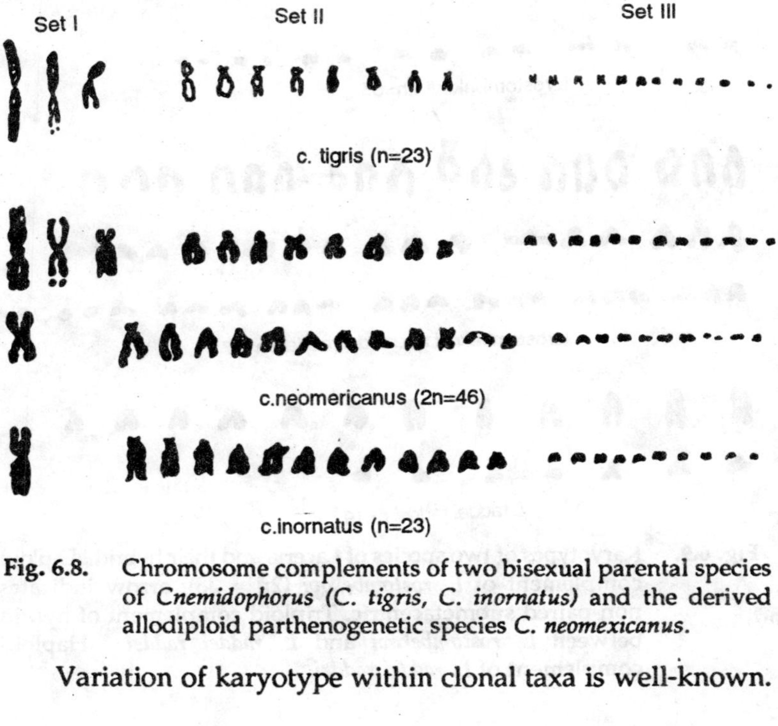

Fig. 6.8. Chromosome complements of two bisexual parental species of *Cnemidophorus (C. tigris, C. inornatus)* and the derived allodiploid parthenogenetic species *C. neomexicanus.*

Variation of karyotype within clonal taxa is well-known.

Three different karyotypes have been reported by Peccinini-Seale and *Frota-Pessoa* (1974) in the parthenogenetic populations called *Cnemidophorus lemniscatus*. Individuals have different karyotypes at different localities. Karyotypic heterogeneity has also been reported in *C. cozumela maslini* by *Fritts* (1969), in *C. sonorae* by *Lowe et al.* (1970a), and in *C. exsanguis* by Cole (1979). All individuals of four populations of *Lacerta rostombekovi* have a karyotype distinct among members of the *L. saxicola* group in that there is a heteromorphic pair, but all have the same karyotype. At least one distinctive karyotype within a parthenogenetic taxon has been shown to have clonal inheritance.

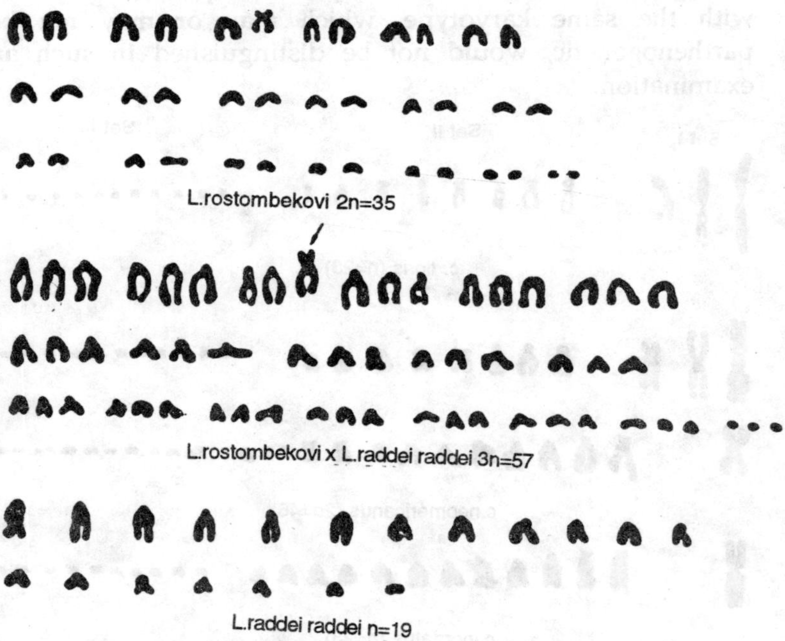

Fig. 6.9. Karyotypes of two species of Lacerta and their hybrid. Diploid complement of *L. rostombekovi* (2n = 38); arrow indicates non-paired submetacentric. Triploid complement of hybrid between *L. rostombekovi* and *L. raddei raddei*. Haploid complement of *L. raddei raddei*.

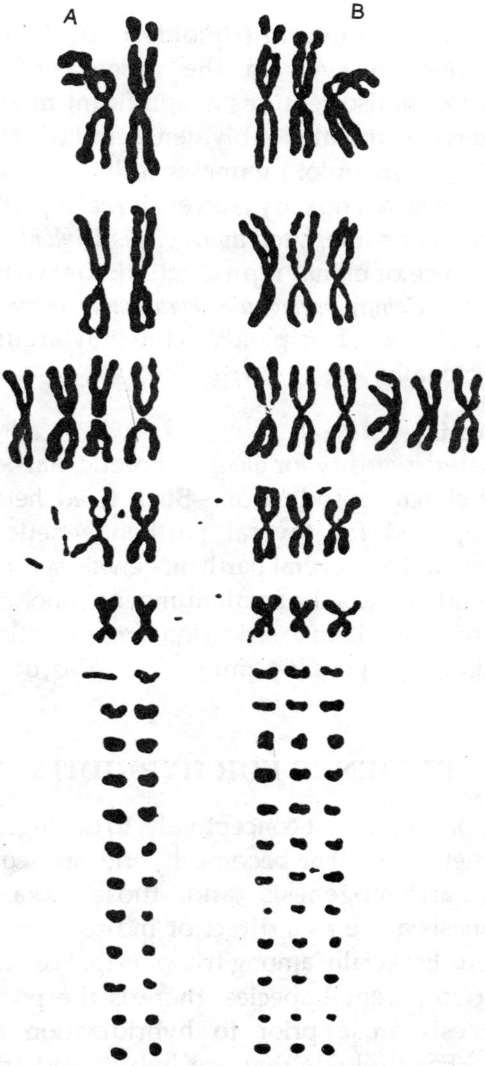

Fig. 6.10. (A) Somatic karyotype of a female *Leiolepis belliana* from Bangkok, 2n = 36; (B) somatic karyotype of female from Singapore, 3n = 54. The probable homologous are arrayed opposite one another for purposes of comparison, *n* indicates the conspicuous secondary constriction, usually well-developed in only a single chromosome of each cell.

The presence of uniform triploidy is strong evidence of clonal reproduction even in the absence of karyotypic heterozygosity. It also requires a significant modification of meiosis; otherwise univalents, bivalents, and trivalents would occur, leading to aneuploid gametes following meiosis. For all of the triploid taxa. Triploidy is an evidence of parthenogenetic reproduction. For some, such as *Leiolepis triploida* triploidy is the best evidence of clonal reproduction; for whatever taxon was reported as *Gehyra variegau ogasawarasimae* by *Makino* and *Momma* (1949), probable triploidy is the only argument that it reproduces clonally.

Electrophoresis

Fixed heterozygosity for electrophoretic markers is strong evidence of clonal reproduction. Such fixed heterozygosity has been reported for several parthenogenetic species of *Cnemidophorus* and for several parthenogenetic species of *Lacerta*. A detailed study revealed a small number of anomalous alleles suggesting that in addition to strict clonal reproduction mutation, recombination, and possibly multiple origins occurred in *C. Tesselatus*.

EVIDENCE FOR HYBRIDITY

It is important, at least conceptually, to distinguish between parthenogenetic taxa that become hybrid subsequent to the origin of parthenogenesis and those taxa in which parthenogenesis arose as a direct or indirect consequence of hybridization. Especially among triploids that contain genomes from only two parental species, there is the possibility that parthenogenesis arose prior to hybridization rather than afterwards. For diploids that are hybrid and triploids that contain three different genomes, however, it seems almost a contradiction to say that parthenogenesis arose before hybridization occurred.

It is not only possible that some of the parthenogenetic

triploids that are hybrids between two species were parthenogenetic before they were hybrids, but it is also possible that some parthenogenetic taxa are not hybrids at all. This seems unlikely, however, because hybridity is clear in each well-studied case; doubt persists only in the less well-known cases. We review below the evidence bearing on the hybridity of the various forms. The important evidence is karyological and electrophoretic, although skin grafts, mitochondrial DNA, and morphology all supply additional information.

Karyology

In *Cnemidophorus*, four groups of biparentally reproducing species can be recognized by morphology and karyology. Three of the parthenogenetic taxa have genomes from two of these groups and thus clearly are hybrids if not of hybrid origin. These include the diploid forms *C. neomexicanus* and *C. tesselatus*, which both have the very distinctive X-chromosome of *C. tigris*. *C. dixoni*, which is externally very similar to *C. tesselatus*, will probably also be shown to be an intergroup hybrid. Two different karyotypes were observed in *C. maslini*; one of them indicated that the population is also an intergroup hybrid. Extensive data on the karyotype of *C. lemniscatus* and its diploid parthenogenetic relative suggest to us the possibility that the parthenogenetic taxon is a hybrid between *C. lemniscatus* and some taxon with a diploid chromosome number of 46, such as a member of the *C. sexhneaus* group. Among the triploid species, only the form called *Cnemidophorus tesselatus* is certainly an intergroup hybrid based on Karyology.

In contrast to *Cnemidophorus*, in which several species groups and thus intergroup hybrids can be distinguished by karyology, biparentally reproducing members of the *Lacerta saxicola* complex all have rather similar karyotypes, so that evidence of hybridity from karyology is lacking. Even less has been reported on the chromosomes of other parthenogenetic taxa and of their close relatives that reproduce biparentally.

Electrophoretic Data

As evidence of hybridity, electrophoretic comparisons of proteins from biparentally and parthenogenetically reproducing are presently our most powerful tool. Using the enzymes adenosine deaminase, lactate dehydrogenase, 6-phosphogluconate dehydrogenase, and NADP-dependent malate dehydrogenase, it has been possible to show that two diploid species of *Cnemidophorus* and four triploid species (the triploid called *C. tesselatus, C. exsanguis, C. uniparens and C. velox)* are fixed heterozygotes for one or more electrophoretic markers. Using the electrophoretic data plus karyological data, it was able to assign probable parents to each of the diploid forms and to identify the three parents of the triploid derivative of *C. tesselatus.* For the other three triploids, the source of at least two genomes could be clearly identified, but in each case the source of at least two genomes could be clearly identified, but in each case the source of the third genome is uncertain. Neaves suggested that these genomes may be derived from Mexican species not included in this study, although the problem may result from undiscovered alleles in species that were examined. The origins of *C. tesselatus* and its triploid derivative have been abundantly confirmed by a careful study of 21 loci.

Examination of 15 loci demonstrated that *Cnemidophorus laredoensis* represents a diploid hybrid between *C. gularis* and *C. sexlineatus.* Examination of five loci of four parthenogenetic and six biparentally reproducing *Lacerta* of the *L. saxicola* group suggested that in each case, a unique pair of biparentally reproducing taxa could account for the heterozygosities observed in the parthenogenetic forms.

On the basis of electrophoretic and karyological data, we can now be certain that at least 15 of the 32 parthenogenetic species of reptiles are hybrid if not of hybrid origin. The remaining parthenogenetic taxa have not been examined in this way to search for evidence of fixed heterozygosities or to compare them with possible parental species.

Skin Crafts

Reciprocal skin grafts within and between pattern classes A, B, C and E or *Cnemidophorus tesselatus* and its triploid derivative have been made. Transplants within pattern classes were accepted; transplants between classes, except of A or B to C and E, were also accepted. Classes A and B are triploids containing a *C. sexlineatus* genome in addition to the two present in *C. tesselatus*. The presence of this third genome accounts for the rejection of grafts from these classes by Classes C and E. Similar logic has been applied to a study of hybrids between parthenogenetically and sexually reproducing *Cnemidophorus*.

Mitochondrial DNA

Although data on mitochondrial DNA by themselves do not provide evidence of hybridity, they confirm the identification of one parental species for taxa otherwise shown to be hybrid, and they identify the maternal species.

Restriction enzyme analysis of mitochondrial DNA from three parthenogenetic species of *Cnemidophorus* has been carried out. Both *C. neomexicanus* and *C. tesselatus* contain mitochondrial DNA from *C. tigris marmoratus*, indicating that this subspecies was the maternal parent for both in the originally hybridizations. A similar analysis of *C. laredoensis* indicate that *C. sexlineatus* was the original maternal parent.

Morphological Intermediacy

Morphological intermediacy appears to be a weak indicator of hybridity in clonal reptiles, yet careful comparative studies can yield significant information. Fifty-one samples of *Cnemidophorus tesselatus* and four samples of its triploid derivative have been studied. For comparison with the parental species, five samples of *C. septemvittatus*, ten of *C. tigris, and* four of *C. sexlineatus* were used. Five of the 14 characters are size related; nine scale counts are not.

Cnemidophorus tesselatus resembles *C. septemvittatus* in body size, relative head length and relative hind leg length; in none of these characters, however, does the mean value for *C. tesselatus* fall between the means for the two parental species. The number of teeth in the upper and lower jaws also falls outside the range of means for the two parental species. Four of the nine scale counts fall between the means for the two parental species, but five do not.

A discriminant function analysis using all nine scale characters indicated that *Cnemidophorus tesselatus*, while intermediate, is much closer to *C. tigris than to C. septemvittatus*. For the four size-related characters, the two parental species are poorly separated by a discriminant function. For all 13 characters together, the two parental species are well separated, and *C. tesselatus* is intermediate but more like *C. septemvittatus*.

When the triploid derivative of *Cnemidophorus tesselatus* is compared with *C. tesselatus* and *C. sexlineatus*, the triploid taxon is usually intermediate, but in general closer to *C. tesselatus*.

The surprising feature of Parker's results is the absence of intermediacy of phenotype for *Cnemidophorus tesselatus*, quite in contrast to other results on clonal vertebrates. The greater intermediacy usually seen probably reflects other workers choosing those characters that best distinguish the parental species.

Although no detailed comparisons have been made in other cases of parthenogenetic reptiles, it is apparent that many of the mean values for characters of parthenogenetic *Lacerta* fall between the means of their two parental species. *Kentropyx borckianus* is intermediate in many morphological features between *K. striatus* and *K. calcaratus*.

PATTERNS OF GEOGRAPHICAL AND ECOLOGICAL DISTRIBUTION OF PARTHENOGENETIC SPECIES OF LIZARDS

Studies of the ranges and habitats of parthenogenetic species reveal certain peculiarities of their geographical and ecological distribution that distinguish them from related biparentally reproducing forms. The most extensive data are for *Lacerta* and *Cnemidophorus*.

The present distributions of parthenogenetic species of *Lacerta* are markedly sympatric. Much of the distribution of *L. dahli* and of *L. rostombekovi falls* within the range of *L. armeniaca*. The distribution of *L. unisexualis* also coincides with the range of *L. armeniaca* in a number of places.

In Armenia and Georgia, there are also areas of sympatry between some parthenogenetic and biparentally reproducing species. Such contact zones, occupying very small parts of the periphery of the ranges of all-female species, are much smaller than sympatric zones between parthenogenetic species. The most extensive sympatric zones occur between *Lacerta armeniaca* and the bisexual species *L. valentini* and *L. portschinskii* and between the latter and *L. dahli*.

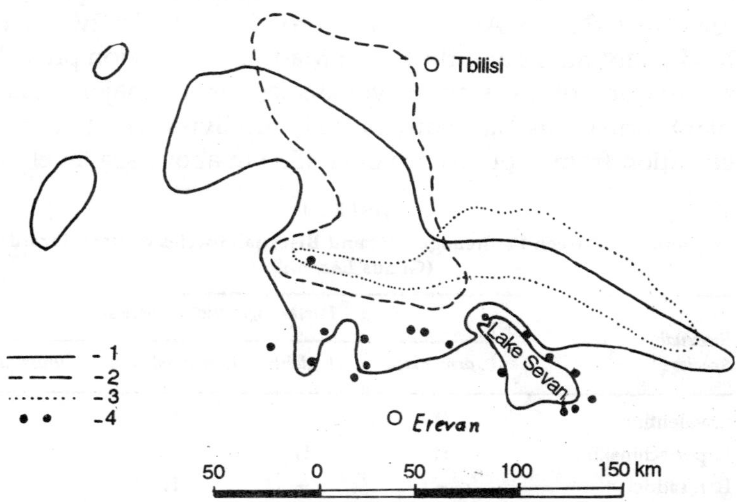

Fig. 6.11. Coincidence of the ranges of the four parthenogenetic species of Lacerta. (1) *L. armeniaca;* (2) *L. dahli;* (3) L. rostombekovi; (4) main localities of *L. unisexualis.*

Parthenogenetic species of *Lacerta* often have more or less distantly isolated populations with ranges that differ in size. Such gaps in the ranges appear to be the result of both tectonic activity and other postglacial environmental changes; they are associated, in particular, with a great reduction in the area that was occupied by forests. Thus, some relic populations of *L. armeniaca* and *L. dahli* and now confined to patches of forest vegetation that remain in parts of central Armenia and southern Georgia. Some isolates, however, may have separate hybrid origins.

The pronounced sympatry of parthenogenetic species of rock lizards is also characteristic, although to a lesser extent; of whiptail lizards of the genus *Cnemidophorus*. The parthenogenetic species of this genus can be arranged in three groups according to the amount of sympatry. The first group is composed of nine more or less sympatric species. They occupy one of the most complex physiographical and ecological regions in North America, the broad ecotone between the Rocky Mountains and the Sierra Madre. This region provides a multitude of areas and diverse sites for the major western plant formations and many of their subdivisions. It varies in elevation from about 540 to over 4500 m above sea level.

TABLE 6.1

Contacts between Parthenogenetic and Bisexual Species of Rock Lizards (Genus *Lacerta*)[a]

Bisexual Species	*Parthenogenetic Species*			
	L. armeniaca	*L. dahli*	*L. rostombekovi*	*L. unisexualis*
L. valentini	H		H	
L. portschinskii	H	H	+	
L. r. raddei	+	+	H	+
L. r. nairensis	H	+	+	
L. mixta	+			
L. rudis macromaculata	H			
L. r. obscuru	H	+		

[a]+, used if contacts between species take place.
H, used if triploid hybrids are also known.

The second group of parthenogenetic whiptails is geographically far removed from the first; it includes *Cnemidophorus cozumela* and *c. rodecki.* Their ranges are confined to the Yucatan Peninsula and adjacent small offshore islands in Mexico and bordering regions of Guatemala.

The last group includes one or several South American species close to *Cnemidophorus lemniscatus;* the distribution of this group is poorly known.

Distinct sympatry, generally at the periphery of the ranges, exists between two and more unisexual species of the first group: *Cnemidophorus exsanguis, C. flagellicaudus, C. neomexicanus, C. tesselatus, C. opatae, C. sonorae,* the triploid derivative of *C. tesselatus, C. uniparens,* and *C. velox* some populations include three distinct all-female taxa, for example, *C. exsanguis, C. flagellicaudus,* and *C. sonorae* in southwestern Arizona. In contrast, the three unisexual forms of the group from the Yucatan Peninsula are completely allopatric.

Sympatric contacts are quite common between parthenogenetic and biparentally reproducing species of whiptail lizards. According to *Wright* and *Lowe* (1965), *Cnemidophorus uniparens* occurs sympatrically, within its extensive range, with five all-female species and with three bisexuals. Mixed populations of three and more taxa occur. The Yucatecan unisexual species *C. rodecki* and *C. cozumela* have overlapping geographic ranges with the bisexual species *C. angusticeps* and *C. deppii.* The well-known sympatric contacts between all-female and bisexual species of *Cnemidophorus.* The parthenogenetic gecko *Hemidactylus garnotii* is widely sympatric with the related bisexual species *H. frenatus,* a pronounced antagonism being observed between the two. This apparently can be explained by the intrusion of the rapidly dispersing unisexual species into the range of its bisexual relative.

Many all-female lizards are confined to specific biotopes; in a number of cases these biotopes differ from those occupied by their bisexual relatives.

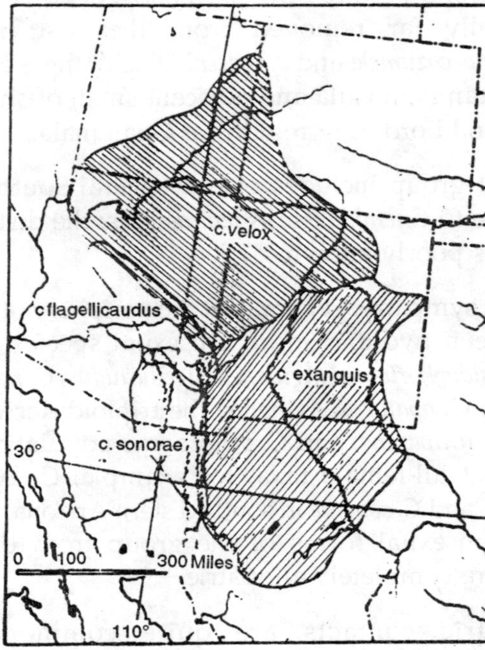

Fig. 6.12. Approximate geographical ranges of the four parthenogenetic species of *Cnemidophorus* centered ecologically in woodland habitats.

Parthenogenetic *Lacerta typically* inhabit mountains and are closely associated in their distributions with various kinds of rock surface and matrix exposures. A few of them settle secondarily in trees and strips of mountain meadow adjacent to the rocks. *Darevsky* (1967) distinguished three types of habitats of rock lizards, of which the following are typical for parthenogenetic species:

1. Dry rocks (usually sedimentary) and areas at their base covered with shrubs and exrophilous grassy vegetation. Such habitats are characteristic of the most xerophilous species of all parthenogenetic *Lacerta* — *L. rostombekovi.*

2. Moderately dry bedrocks and matrices, as well as stony channels of dried-up rivers in the montane forest zone. These biotopes are inhabited by *Lacerta dahli* and *L. armeniaca;* sometimes these two species occur on trunks of trees.

3. Bedrocks of mostly volcanic origin, stony screes, conglomerations of rocks, and large fragments of lava in the montane steppe zone, sometimes in sites of now vanished montane forests; rocks on the shores of mountain lakes. Such habitats are characteristic of *Lacerta unisexualis* and occasionally of *L. armeniaca*.

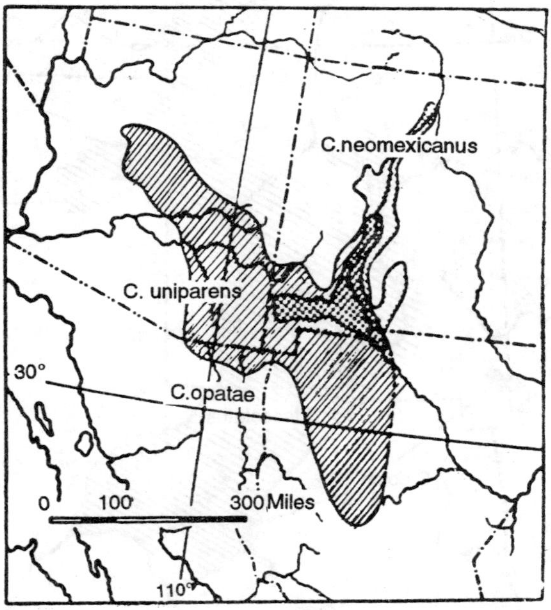

Fig. 6.13. Approximate geographical ranges of three parthenogenetic species of *Cnemidophorus* in western North America.

The occurrence on vertical rocks and secondarily on various buildings and stone walls is so specific to this group of lizards that it is a reliable ecological indicator for a specific biotope in nature. Nevertheless, the presence of suitable biotopes such as stones and rocks is by itself insufficient for these lizards to do well. Other favourable factors are necessary, the most important being temperature and humidity regimes that provide specific microclimates. The adaptation of these lizards to life on rock surfaces is also manifested in their ability to use various narrow crevices and cracks. Often such shelters are so narrow that the lizards obviously deform their heads as they squeeze into them.

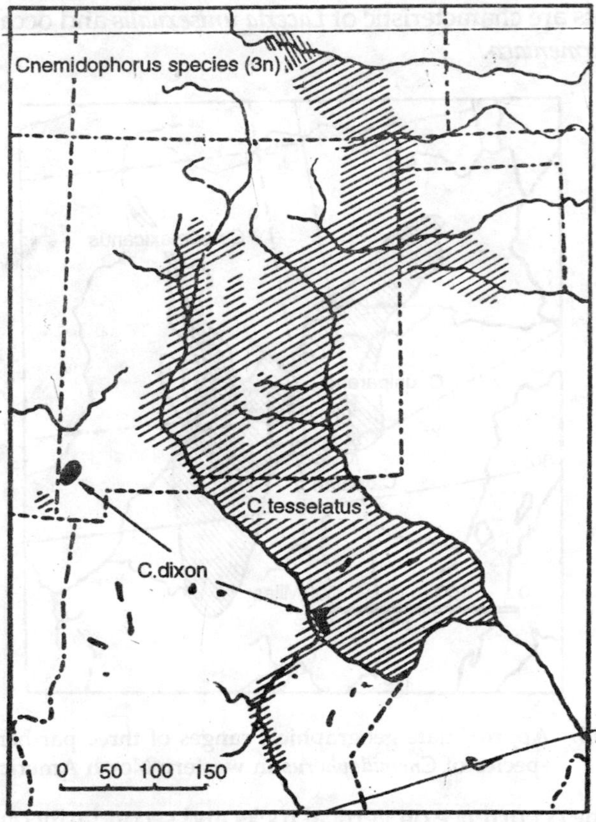

Fig. 6.14. **Approximate geographical ranges of three parthenogenetic species of *Cnemidophorus*.**

Rock lizards throughout the Caucasus are stenobionts

and practically free from competition for habitats with other species of lizards, most of which have not adapted themselves to life on the surface of rocks. Sympatric contacts of rock lizards and *Stellio caucasica* are observed in some places; even in this case, however, there is no noticeable competition between the species because large *Agama* prefer a different type of shelter and quite different food, being generally herbivorous.

Although two and even three parthenogenetic species may occupy a common biotope and be confined to the same habitat, they apparently posses some differences in their ecological requirements. Otherwise, as White observed, the best adapted species will in due course inevitably occupy the habitats of the others, unless it does not.

On the whole, the habitats of parthenogenetic Lacerta differ little form those of closely related bisexual forms. As a group, however, they tend to live in colder, dryer, or more changeable climates than their bisexual relatives. Perhaps this reflects the individual parentage of the parthenogenetic forms. Because of this similarity in habitats, in zones of sympatry between parthenogenetic and bisexual species, the latter are usually inhibited and cannot, evidently, stand competition with their unisexual relatives. This also seems to explain why parthenogenetic lizards usually occur along the periphery of the ranges of bisexual species and vice versa.

The ecological distribution of parthenogenetic *Cnemidophorus* has been analyzed in detail four major habitats for North American all-female species of this genus have been distinguished, namely woodland, grassland, desert-scrub, and thornscrub.

1. Woodland habitats are characteristic of the parthenogenetic species *Cnemidophorus exsanguis, C. flagellicaudus, C. sonorae,* and *C. velox,* the ecological distributions of which center in southwestern evergreen pine—fir and oak woodlands. The geographic distribution of *C. velox* is largely in pinon-juniper woodland, *C. exsanguis* in Chihuahuan Madrean

woodland, *C. sonorae* in Sonoran Madrean, and *C. flagellicaudus* in sub-Mogollon woodland. Populations of these species often penetrate into lower desert-grassland zones, usually along drainage patterns with their attendant river bank communities.

2. Unlike the four forest unisexual species of whiptail lizards, *Cnemidophorus neomexicanus* and *C. uniparens* occur in grassland and in desert-grassland habitats, the former usually being found at the edges of playas in closed basins, in other open sandy habitats (often along sandy arroyos), and in washes of varying types and sizes. For the last century, the range of both species has undergone considerable change due to sand intrusion into grasslands. Thus, grassland areas, inhabited only 50 to 100 years ago by bisexual *C. inornatus,* are now occupied by unisexual *C. uniparens.*

3. Thorn scrub-desert scrub habitats are typical for the parthenogenetic species *Cnemidophorus opatae,* which has a limited area of distribution. According to *Wright* and *Lower* (1968), this habitat is derived from thorn scrub by elimination of the subtropical species less tolerant to the more continental climate of the valley.

The all-female species *Cnemidophorus cozumela, C. rodecki* and *C. maslini* occupy primarily sandy beach strand thorn scrub ecotones on the coast and small off-shore islands of the Yucatan Peninsula.

4. Rockiness is the common denominator among habitats occupied by *Cnemidophorus tesselatus* and its triploid derivative. The rocks occur in bluffs, local outcrops, or foothill rock slopes; they may be scattered about in the coarse soil of alluvial fans and bajadas. Considerable ecological breadth is typical for the diploid, which has an extensive geographic range, as well as for the triploid, which has a lesser range; both occupy habitats in desert scrub and pinon — juniper woodland. Both species have an elevational range of 540 to 2400 m above sea level.

More than other whiptail lizards, *Cnemidophorus tesselatus* tends to occur as isolated local populations and small groups of individuals, even where the apparently suitable habitat is continuous, especially in areas adjacent to those in which bisexual forms are abundant.

The ecological distribution of North American parthenogenetic *Cnemidophorus* have been described in detail. The South American parthenogenetic lizard called *C. lemniscatus* is said to occur almost exclusively in and around towns and sizable settlements and is now extending its range in the valley of the Amazon. Populations of this parthenogenetic taxon also occur in Surinam.

Some authors have shown that the habitats of some sympatric all-female and bisexual whiptail lizards differ noticeably from each other thus providing a certain ecological isolation of the sympatriants. *Cnemidophorus velox*, for example, occupies principally oak-mountain mahogany habitats and pinon-juniper associations, whereas the bisexual *C. inornatus*, which occupies the same area, occurs primarily in grassland associations. Essential differences exist in the habitats of five species in Socorro County, New Mexico; three of them are parthenogenetic and two are bisexual.

The habitats of *Cnemidophorus neomexicanus* and *C. inornatus* differ in the valley of the Rio Grande and in the Albuquerque area of New Mexico. Milstead who established differences in habitat for a number of sympatric species of *Cnemidophorus* in the Sierra Vieja mountains of western Texas, concluded that this results from interspecific competition, which is an important factor in producing an ecological isolation of sympatriants. We agree with Zweifel however, that this conclusion may require review, because Milstead was dealing with six species rather than four as he thought. According to *Zweifel* (1965), the parthenogenetic *C. tesselatus* and bisexual *C. tigris* often share the same habitats, and *Christiansen* (1969) reported a common hibernation site of all-female *C. neomexicanus* and bisexual *C. inornatus.*

The ratio between individuals of various species in mixed sympatric populations of all-female species varies. Mixed populations of *Lacerta armeniaca* and *L. dahli* and *L. armeniaca* and *L. unisexualis* occur at many localities in the Caucasus; the ratio between the species is approximately 1:1.

On the northeastern shore of Lake Sevan, Armenian SSR, *Lacerta unisexualis* and *L. rostombekovi* occur together, and here the former accounts for 70% of individuals. In the Debet ravine in northern Armenia, however, *L. unisexualis* does not exceed 10% of the population, the rest being the smaller *L. rostombekovi.*

In a mixed population of three unisexual species in the Agstev ravine near Dilizhan, Armenian SSR, *Lacerta armeniaca* accounts for 60%, *L. dahli* for 30%, and *L. rostombekovi* for only 10% of the total population.

In sympatric populations of parthenogenetic and bisexual species of rock lizards the former, as a rule, dominate. For example, *Lacerta portschinskii,* in mixed populations with *L. armeniaca* and with *L. dahli,* very seldom exceeds 10 to 15%. On the other hand, in a mixed alpine population on the Gegam ridge in Armenia, *L. valentini,* which is adapted to life in high mountains, out numbers the parthenogenetic *L. armeniaca,* which is penetrating this region from below. Such contacts appear to be secondary.

In mixed populations of several all female and bisexual species of whiptail lizards in Sierra Vieja (western Texas), bisexual *Cnemidophorus inornatus* ("perplexus:) were more abundant in plains habitats whereas the all-female *C. exsanguis (= sackii)* and *C. tesselatus* accounted for 15 and 55 specimens, respectively. At the same time in the common creosote bush-catclaw habitat within the plains, *Cnemidophorus inornatus* and *C. tesselatus* were approximately equal in number (42 and 34 specimens). In roughland habitats *C. inornatus* were completely absent and *C. exsanguis* were considerably more abundant than *C. tesselatus*.

Within the city limits of Albuquerque, New Mexico, the bisexual *Cnemidophorus inornatus* is more numerous than the sympatrically occurring parthenogenetic *C. neomexicanus*. At one locality, 69% of the lizards were *C. inornatus* and 31% were *C. neomexicanus*; at a second locality, the percentage are 80 and 20, respectively.

The above data show no strict regularity in mixed populations of *Lacerta* and *Cnemidophorus*; the extent to which these data reflect temporal and spatial variations of patterns is not known.

SOME BIOLOGICAL PECULIARITIES OF PARTHENOGENETIC LIZARDS

Although parthenogenetic species of reptiles are often quite common, they are not free of problems that apparently are associated with their being parthenogenetic. We discuss below the failure of eggs to hatch, the presence of occasional males and hermaphrodites in parthenogenetic species, and the occurrence and results of mating between parthenogenetic reptiles and related biparentally reproducing taxa.

Interruption of Embryogenesis and Embryonic Death

Imperfections in parthenogenetic reproduction in some species of unisexual *Lacerta* are revealed by lethal abnormalities that arise at different stages of embryogenesis. Such monsters were usually discovered after their siblings had hatched, but dissection of prematurely spoiled eggs also indicated deaths at earlier stages of development. Thus, the majority of unhatched eggs were found to contain living monsters that could not break the egg shell; if extracted, these abnormal individuals survived from one hour to several days, depending on the type of deformity.

Abnormal embryos may be grouped on the basis of type and extent of anomaly. The most typical abnormalities in parthenogenetic species of rock lizards are illustrated.

More detailed characteristics of the monstrosities were given by *Danielyan* (1970). While examining abnormal embryos with X-rays, he found that they are teratopagous or teratodymous, according to *Nakamura's* (1938) classification. The former are twins with more or less joined axial skeletons; the latter are embryos with duplications in the anterior part of the trunk and head.

In most cases, monsters die in the final stage of incubation; at that time their sex is easily established by the presence or absence of hemipenes. Because normal one-year-old parthenogenetic individuals are females, one would expect the abnormal ones hatched from unfertilized eggs to be female also. Unexpectedly, nearly all abnormal individuals. Determining sex by the presence or absence of male genitalia only becomes possible at an advanced stages in the development of the embryo. The hemipenes of males and the cloacal sacs of females in *L. saxicola*, as well as in the other lizards, begin to develop externally as two processes projecting from the cloaca. After 25 to 30 days of incubation, these processes invert, but those of males are easily everted by pressing the base of the tail.

It is characteristic of male monsters of *Lacerta dahli* and *L. armeniaca* that their hemipenes do not invert, but remain on the outside throughout the rest of the incubation period. This, as well as other peculiarities such as open body cavity and lack of pigmentation, indicates that for some reason the development and growth of the male embryos stops somewhere in the middle of incubation. They remain alive in the egg, but do not develop further. Usually an abnormal male embryo develops only in one egg of a given clutch and is discovered alive after all its normal sisters have hatched. The pause in their growth usually sets in not earlier than the 20th and not later than the 45th day of incubation.

Abnormal embryos are much more common in parthenogenetic rock lizards than in bisexual ones. Furthermore, the deformations in bisexual forms are less marked. These data suggest an association between parthenogenesis and an interruption of embryogenesis that produces abnormalities. A similar association has been noticed in silkworms in Pacific herring.

Abnormal individuals may be aneuploids or chromosomal mosaics or polyploids arising from the confluence of more than two meiotic products. It is possible that at least some of the males result from recombination between the parental genomes. Probably genetic regulatory difficulties are involved in the formation of monsters; these may be expressed more strongly in males. The repeated recurrence of some specific monstrosities, for instance, agnathostomatosis, indicates a consistency in the genetic mechanism producing monstrosities, but the causal mechanism remains unclear. The number and frequency of abnormal embryos and monstrosities in parthenogenetic and bisexual forms are given.

Occasional Males and Hermaphrodites in Parthenogenetic Populations

Although males have been reported to occur sporadically in parthenogenetic species, they usually come from mixed populations of unisexual and bisexual species and belong to the bisexual species. Males do occur, although rarely, in some parthenogenetic species of lizards and male hybrids between parthenogenetic and bisexual species have been described; these will be discussed below.

Parthenogenetic males were compared with females of the same species of whiptail lizard. *Cnemidophorus exsanguis* and *C. tesselatus* were found to show little sexual dimorphism, although males of these species differ somewhat from females in scutellation and colour pattern. Both known males of C. *tesselatus* Class E have more femoral pores and more granules around the midbody than do females. The distinct spotting of the only known male C. velox, however, is in striking contrast to the unspotted condition of females of this species. A male C. *tesselatus* from near Presidio. Texas, did not appear to differ from the females in that area (*Saxon et al.*, 1967). No external sexual dimorphism was exhibited by the two known males of *Lacerta dahli* and two males of *L. armeniaca* their outward appearance in no way differs from that of female. In the closely related bisexual species *L. saxicola*, *L. rudis*, and *L. caucasica*, the males differ from the females in having a larger head, fewer transverse rows of abdominal and vertebral scales and blue, dark blue, or violet spots on the lateral abdominal fields.

TABLE 6.2

Frequency and Number of Monstrosities in Parthenogenetic and Bisexual *Lacerta*

Species	*Number of Embryos*	*Monsters*			
Parthenogenetic species		*Number*	*%*	oo	oo
L. armeniaca	777	35	4.6	1	34
L. dahli	182	12	7.9	8	4
L. rostombekovi	98	5	5.1	5	—
L. unisexualis	293	9	3.0	—	9
Bisexual species					
L. raddii nairensis	412	6	1.4	1	5
L. valentini	110	2	1.8	1	1

Although morphological criteria suggest that the males of some unisexual whiptail lizards belong to the corresponding parthenogenetic species a hybrid origin cannot be completely discounted because in each case they came from localities containing at least one of the related bisexual species.

Cnemidophorus exsanguis occurs with *C. inornatus* and with a subspecies of *C. tigris; C. tesselatus* is apparently sympatric with *C. tigris marmoratus;* and *C. velox* with *C. tigris septentrionalis.* Lowe et al. (1970a) assumed that male whiptail lizards are hybrids; they described two tetraploid hybrids between the triploid unisexual species *C. sonorae* and the bisexual *C. tigris.* The same arguments can be applied to the males of parthenogenetic species of rock lizards; these were found at localities where *L. portschinskii, L. raddei,* or *L. valentini* occur as well.

In two cases of apparent geographic parthenogenesis (geographical variations within species between sexual and clonal modes of reproduction), additional study has revealed the presence of a distinct parthenogenetic species. A study of sex ratios in populations of *Lacerta raddei nairensis* revealed the presence of bisexual and unisexual races that suggested geographical parthenogenesis this unisexual race was later recognized as *L. unisexualis Maslin* (1962) and *Duellman* and *Zweifel* (1962) suggested that the western population of the bisexual whiptail *C. inornatus* consists only of females; later it was recognized by *Wright* and *Lowe* (1965) as a distinct all-female species, *C. uniparens.* According to *Nussbaum* (1980), reports of males, in addition to taxonomic and distributional considerations, suggest that *Ramphotyphlops braminus* from the Seychelles, Lesser Sundas, and Philippines may be a complex of unisexual and bisexual species.

It has been suggested that in some populations of the South American whiptail *Cnemidophorus lemniscatus,* males are gradually disappearing. According to preliminary data, the shift from bisexuality to unisexuality seems still to be in rapid progress at Obidos on the Rio Amazonas. In Venezuela, 51% of adult *C. lemniscatus* and 77% of the young individuals are male. *Vanzolini* (1976) concluded that *Gymnophthalmus underwoodi* and *C. lemniscatus* are very probably going through a similar process, that is, individual populations of these species are independently evolving toward complete loss of males; as

TABLE 6.3

Records of Males in Various Unisexual Species of Lizards [a]

Species	*Location*	*Number of Males*	*Authors*
Lepidodactylus lugubris	Oahu, Hawaii	1	Cuellar and Kluge, 1972
	Solomon Islands	3	Cuellar and Kluge, 1972
	Esmeraldas, Ecuador	1	Schauenberg, 1968
Lacerta dahli	Kirovakan, north Armenia	1	Dobrovolskaya, 1964
	Stepanavan, north Armenia	1	Darevsky, 1966
L. armeniaca	Ltchashen, central Armenia	1	Darevsky, 1966
	Zurnabad, northwest Azerbaijan	1	Daresky et al., 1978
	Zigana pass, northeast Turkey	2	Darevsky, 1972
	Stepanavan, north Armenia	1	Darevsky and Kupriyanova, 1982
Cnemidophorus species 2	Pueblo County, Colorado	3	Maslin, 196, 1971b
C. velox	?	2	Maslin, 1962
	Mesa County, Colorado	1	Taylor, et al., 1967
C. exsanguis	Dona Ana County, New Mexico	1	Taylor et al., 1967
C. tesselatus class E	Sierra County, New Mexico	2	Taylor et al., 1967
	Presidio, Texas	10	Saxon et al., 1967; Saxon, 1968; Saxon (in Maslin, 1971b)
C. neomexicanus	Socorro County, New Mexico	1	Lowe and Zweifel, 1952
Gymnophthalmus underwoodi	Surinam, Brazil	4	Vanzolini, 1976

[a] Males of possible hybrid origin were excluded.

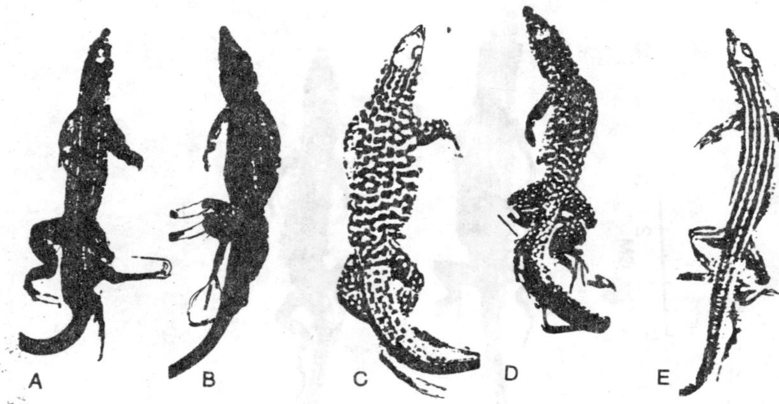

Fig. 6.15. Dorsal pattern of males of four normally parthenogenetic species of whiptail lizards. (A) Cnemidophorus exsanguis (S-V 71 mm); (B) triploid derivative of C. tesselatus (S-V 70 mm); (C) *C. tesselatus* (S-V79 mm); (D) *C. tesselatus* (S-V 79 mm); (E) *C. velox* (S-V 71 mm).

a result, the species show a mosaic of unisexual and bisexual populations. These distributions of parthenogenetic individuals could be equally well explained by rapid spread of parthenogenetic taxa.

There are no data concerning the fertility of males found among all-female species of lizards. The only male of the gecko *Lepidodactylus lugubris* discovered by *Cuellar* and *Kluge* (1972) among a great number of females from the island of Oahu, Hawaii, possessed well-developed testes and sperm ducts. At the same time, the testes of the three males from the Solomon Islands examined by these authors had some anomalies, reminiscent of the holotype of the Indonesian *Lepidodactylus intermedius* according to *Kluge* (1968), *L. intermedius* is identical to *L. lugubris*. Abnormal ovaries and oviducts were discovered by *Cuellar* and *Kluge* (1972) in a number of females of *Lepidodactylus lugubris* from the Solomon Islands. The microscopic structure of the testes of a male L. dahli did not differ from that of testes of bisexual *Lacerta*. Males of *L. armeniaca* had normal testes and fully developed copulatory

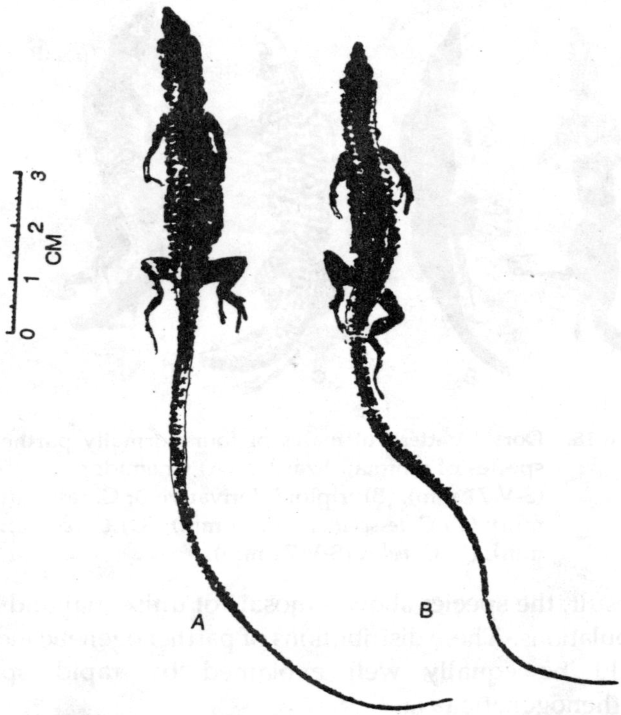

Fig. 6.16. Dorsal pattern of males of two normally parthenogenetic species of rock lizards. (A) *Lacerta dahli* (S-V 52 mm); (B) *L. armeniaca* S-V 50 mm).

organs. The genitalia of males described by *Taylor et al.* apparently were not examined. A male *C. tesselatus* from near Presidio, Tenas, had normal sexual behaviour manifested in a number of copulation attempts; motile sperm were found in the cloaca of the male after two copulations, although there were none in the female. Up to June, 1965, eight males of *C. tesselatus* had been collected from this area. Such a high frequency of males in one of the populations of all-female species may indicate that they may play a role in the reproduction of this clone (Maslin, personal communication). Even if rare males of some all-female species are fertile, however, their role in the maintenance of parthenogenetic populations is insignificant. Furthermore, the majority of known males of whiptail lizards belong to triploid species and therefore their reproductive capacity is also doubtful.

Parthenogenetic *Lacerta* have not lost typical responses to courting males; for instance, they continue to show the rapid movement of upraised forelegs and bobbing of the head. It is of interest to note that, just after emergence, females of the parthenogenetic butterfly *Solenobia triquetrella* strike a peculiar nuptial posture awaiting non-existent males. Such residual behaviour suggests that the parthenogenetic mode of reproduction is comparatively young, although the possibility that such behaviours have multiple functions cannot presently be excluded.

Pseudocopulation occurs in several all-female species of *Cnemidophorus, including C. uniparens* such pseudocopulation may be necessary for successful reproduction in this species. Such behaviour has been observed sporadically in C. *uniparens* and other species in the laboratory for 15 years, but copulation bites were rare among nearly 200 parthenogenetic individuals collected in the wild.

Among males of the bisexual *Cnemidophorus tigris,* "copulatory" behaviour is so common that subordinates become emaciated and may die from the perpetual harassment, unless the "sexual" offender is isolated. The implications are similar to those proposed by Crews and Fitzgerald, but it would be premature to propose that such abnormal courtship behaviour is essential for successful reproduction in C. *tigris*.

Similar "homosexual" behaviour, assumed to be a kind of social behaviour, was described in *Lepidodactylus lugubris* on Oahu Island. Such behaviour has never been observed in *Lacerta,* although some males caught in nature may have marks of bites on the belly, which they probably received during fights.

Besides normal males, rare specimens with hermaphroditic features (fully developed testes and underdeveloped oviducts) are sometimes found in populations of parthenogenetic rock lizards. Such individuals were first observed by *Lantz* (1923), who described two hermaphroditic *Lacerta dahli,* one each from northern Armenia and northeastern Turkey (Kars), respectively. The first specimen had normal testes and copulatory organs, and quite distinct oviducts as well. Poorly developed oviducts were also present in a male *L. dahli* and in a male *L. armeniaca* such specimens should be treated as males possessing certain secondary female characters and not called hermaphrodites. The reproductive potential of such individuals as well as of males from parthenogenetic populations is unclear.

NATURAL HYBRIDIZATION BETWEEN CLOSELY RELATED PARTHENOGENETIC AND BISEXUAL SPECIES

Principle

Fertilization of parthenogenetic females, as a result of mating with males of related bisexual species, usually results in adding an entire chromosome set, if the zygotes develop. If triploids are fertile and if their eggs are fertilized, a possibility exists that tetraploidy and reversion to sexuality may occur.

Hybrids in Lacerta

Because there is some sympatry between all-female rock lizards and closely related bisexual species of this group, natural hybridization is to be expected. Opportunities for hybridization occur in river canyons along which bisexual species penetrate the mountains and enter the ranges of parthenogenetic forms. Natural crossing in such areas results in hybrids that sometimes comprise 5 to 8% of the mixed population.

Female hybrids between *Lacerta armeniaca* and *L. valentini* as well as between the former and *L. portschinskii,* between *L. unisexualis* and *L. raddii nairensis,* and between *L. rostombekovi and L. portschinskii* are always sterile triploids.

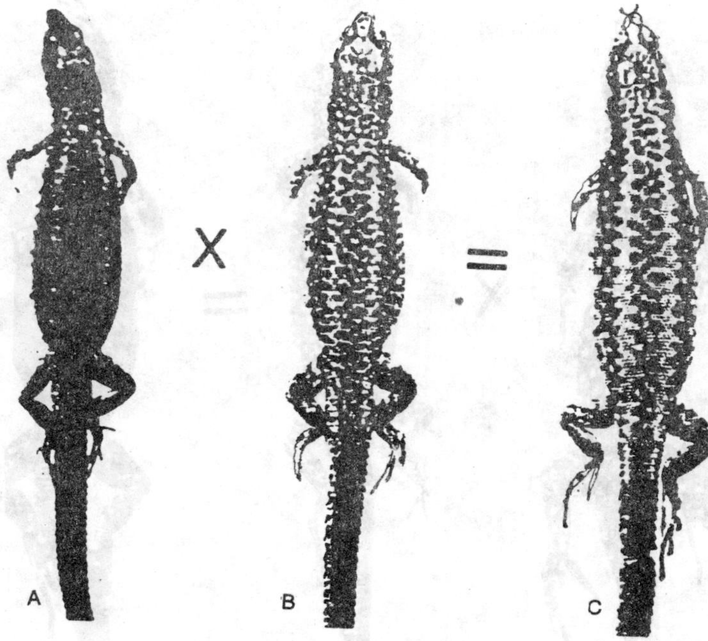

Fig. 6.17. Natural hybridization between a parthenogenetic female and a male of a bisexual species. (A) Diploid parthenogenetic female *Lacerta armeniaca* (2n = 38); (B) diploid male Lacerta valentini (2n = 38); (C) triploid hybrid female (3n = 57).

Hybrids between parthenogenetic females of *Lacerta armeniaca* and males of *L. valentini* are rather common on the southwestern shore of Lake Sevan in Armenia and have been studied intensively. In all, 62 adult and young hybrids were caught; 22 of these were used for cytological studies. Hybrid one-year-old specimens were reared in the laboratory from fertilized parthenogenetic females caught in July, one to two weeks before egg laying, or from specimens mated in the laboratory. Many hybrids were obtained from an artificial hybrid zone created by introducing males of *L. valentini* into an isolated population of *L. armeniaca*. Marks left by the jaws of males on the bellies of the females during copulation serve

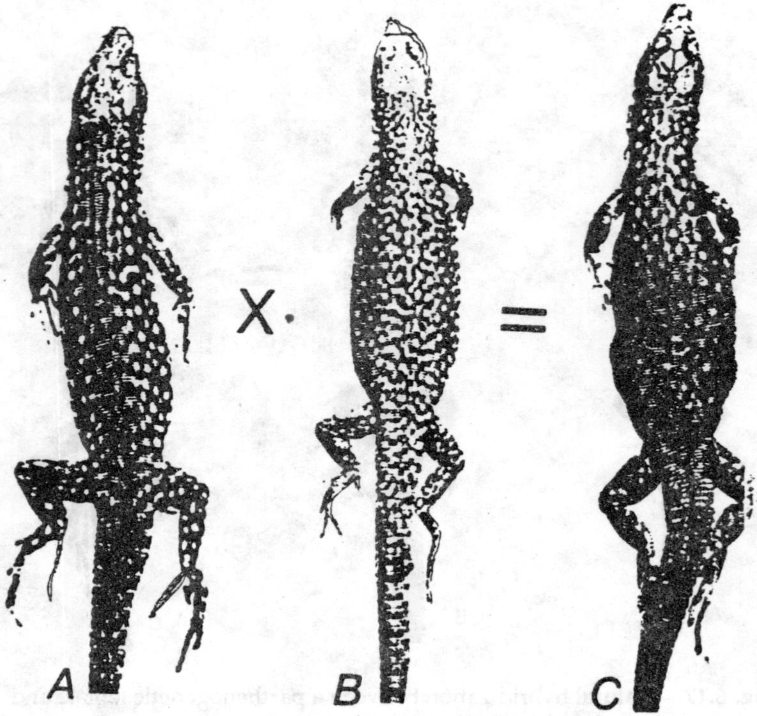

Fig. 6.18. Natural hybridization between a parthenogenetic female and a male of a bisexual species. (A) Diploid parthenogenetic female *Lacerta unisexualis* ($2n = 38$); (B) diploid male Lacerta raddei nairensis ($2n = 38$); (C) triploid hybrid female ($3n = 57$).

as evidence that mating occurred. Numerous marks on both sides of the body suggested that some of these females mated more than once, but other females had only one or no marks at all. During mating, the male rock lizards react most strongly to those females with colouration characteristic of the male's own species. The colouration of the lower part of the body seems to be of particular importance. In Caucasian rock lizards, it varies from dull white to pink, orange-red, yellow, or green. In *L. armeniaca* the belly is usually yolk yellow, and in *L. valentini,* it is greenish-yellow or pale lemon yellow. In some

L. armeniaca, however, the coloration is as bright as that of *L. valentini.* In open air cages, more brightly coloured females tend to mate more frequently.

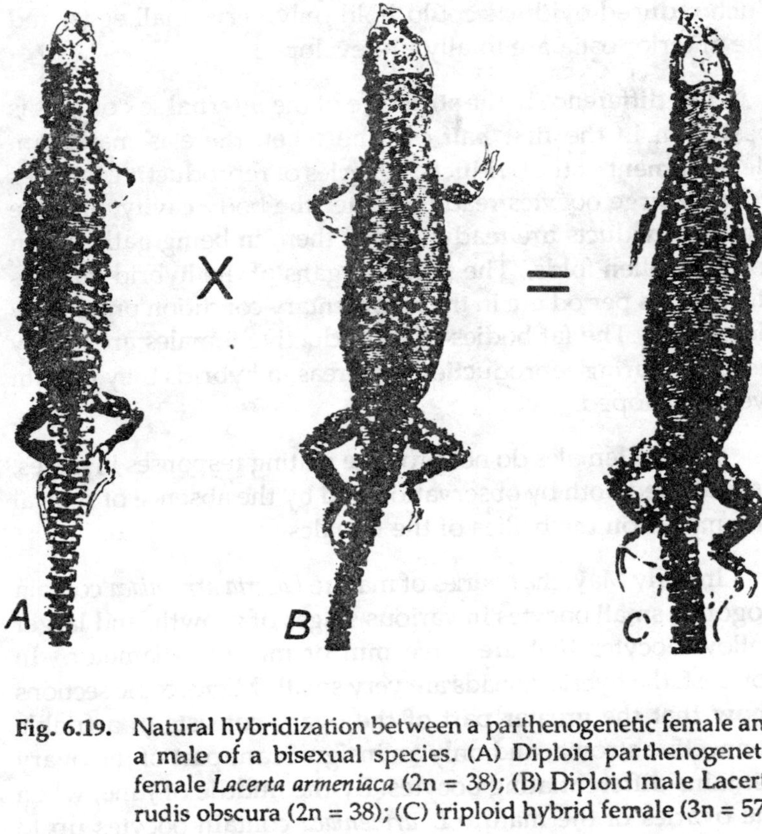

Fig. 6.19. Natural hybridization between a parthenogenetic female and a male of a bisexual species. (A) Diploid parthenogenetic female *Lacerta armeniaca* (2n = 38); (B) Diploid male Lacerta rudis obscura (2n = 38); (C) triploid hybrid female (3n = 57).

That the hybrids are females can be ascertained by the absence of typical male genitalia and the presence of poorly differentiated oviducts. Each ovary of the hybrid females is replaced by a small oval swelling in the middle of the mesovarium. The swellings are not clusters like normal ovaries, but resemble undeveloped testes. The size of these swellings varies only slightly during the reproductive cycle, indicating that maturation of oocytes and their subsequent entrance into the oviducts does not occur in hybrids. This is also evidenced by a considerable reduction of oviducts, which even in the largest hybrid individuals lack typical folds and look like two fused straight tubes with almost no clear space between them. Such reduced oviducts could hold only very small eggs; and the anterior ostia are totally undeveloped.

The difference in the structure of the internal sex organs is best seen in the first half of June when there is maximum development of the oviducts. Ovaries of reproductive females contain large oocytes ready to enter the body cavity, and the twisted oviducts are ready to take them in being gathered in loose swollen folds. The sexual organs of the hybrid females during this period are in the rudimentary condition previously described. The fat bodies of reproductive females are nearly used up during reproduction, whereas in hybrids they remain well-developed.

Hybrid females do not provoke mating responses in males, as indicated both by observation and by the absence of typical jaw marks on the bellies of the females.

In early May, the ovaries of mature *Lacerta armeniaca* contain oogonia, small oocytes in various stages of growth, and larger yellow oocytes that are three mm or more in diameter. In contrast, the hybrid gonads are very small. Microscopic sections show that the greater part of the ovary consists of a friable connective stroma, and only an insignificant part of the ovary contains differentiating oocytes. By the middle of June, when the ovaries of the mature *L. armeniaca* contain oocytes up to eight mm in diameter that are ready to enter the oviducts, the ovaries of hybrids have not increased in size. Some of the sex cells in the hybrids are at the early pachytene stage, but no typical pachytene stage with a complete chromosome conjugation has been found. On the other hand, sex cells in the ovaries of *L. armeniaca* frequently show a complete chromosome conjugation at this time. By the end of June, when the eggs of the parthenogenetic females have already entered the oviducts, the size of the ovaries has not changed in hybrids. Some hybrids have "false follicles," which are peculiar structures on the external surface of the ovary. At the end of July, empty spaces appear in the ovaries of some hybrids, and the uneven margins of sections gives the impression that false follicles have fallen off, leaving only traces at their former position.

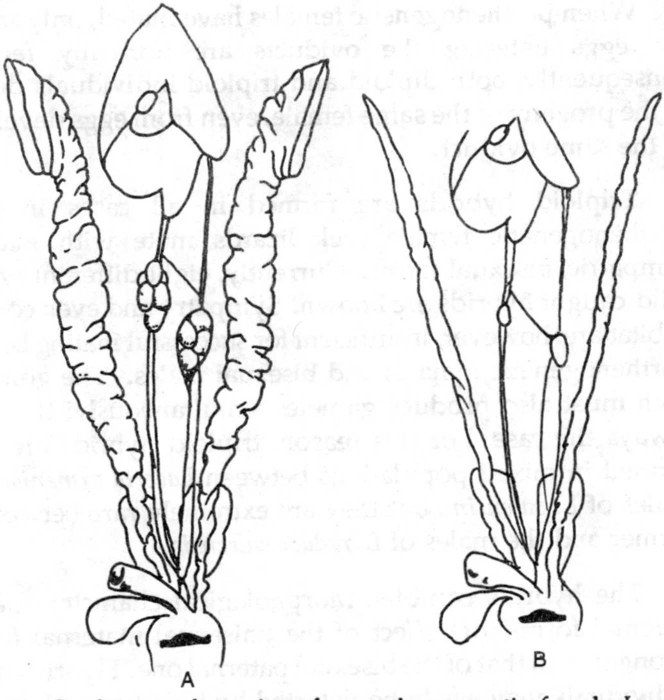

Fig. 6.20. Oviducts and ovaries of a parthenogenetic female *Lacerta armeniaca* (A) and of a sterile triploid hybrid female (B) during the breeding period in June.

Fertilization of the diploid egg of parthenogenetic females takes place in the oviducts, and the diploid female and haploid male pronuclei fuse, forming a triploid zygote. Development and hatching of the hybrid embryo proceed in a normal manner.

Differences appear only at some stage of the development of sexual organs; upon maturity the gonads of triploids are greatly reduced and are distinguishable only under the microscope. The sterility of hybrids and the absence of 3n oocytes can be explained by the absence of normal conjugation; meiosis is prevented from proceeding further than the early pachytene stage because the paternal chromosomes have no homologues.

When parthenogenetic females have mated, only some of the eggs entering the oviducts are normally fertilized consequently, both diploid and triploid individuals develop in the progeny of the same female, even from eggs developing in the same oviduct.

Triploid hybrids are formed in all cases in which parthenogenetic female rock lizards mate with males of sympatric, bisexual forms. Currently, eight different types of wild-caught hybrids are known. Sympatry and even common habitat are, however, insufficient for successful mating between parthenogenetic females and bisexual males. The gonads in both must also produce gametes simultaneously; this is not always the case. For this reason, triploid hybrids are easily formed in mixed populations between *Lacerta armeniaca* and males of *L. valentini*, but they are extremely rare between the former and the males of *L. raddei nairensis*.

The hybrids combine morphological characters of both parental forms; the effect of the unisexual maternal form is stronger than that of the bisexual paternal one. Hybrid triploid individuals may easily be detected by blood smears because the triploid erythrocytes average one-third larger than those of the diploid parental forms.

Variability of the pattern of scutellation, size of the cranium and cranial indices have been analyzed for hybrids between *Lacerta armeniaca* and males of *L. valentini* and their parental forms. The analysis of the mode of inheritance of parental characters by hybrids was based on evaluation of the null-hypothesis probability of similarity or difference between hybrids, and parental species for each character. Hybrids share five maternal and three paternal scutellation characters of 22 analyzed. These results support of the suggestion that maternal hereditary information prevails in triploid hybrids, which receive two sets of chromosomes from the mother, one from the father. The predominantly maternal inheritance in hybrids is supported also by hybrid indices calculated by *Shimansky*, which evaluate the hybrid characters as a percentage of the difference between parental characters.

The mean hybrid index for scale characters of the above hybrids is 15% compared to the maternal species, 72% compared to the paternal species. Hybrids considerably surpass both parents in cranial size. On the whole, the cranial size of hybrid individuals differs more from that of the maternal than of the parental species, because the head of male rock lizards is much larger. In general, the index of variability of all characters of hybrids is higher than that of the maternal *Lacerta armeniaca* and lower than that of the paternal *L. valentini.*

Hybrids resulting from mating of parthenogenetic females with males of bisexual species both in nature and under laboratory conditions are nearly always sterile triploid females. Consequently, a triploid male hybrid between *Lacerta rostombekovi* and *L. raddei raddei* from a sympatric area of these species in northern Armenia is of special interest. It had fully developed testes containing spermatogonia, spermatocytes I and II, and spermatids, but it lacked mature spermatozoids such as occurred in the testes of a simultaneously caught male of the parental species *L. r. raddei.* Besides testes and male genitalia, the hybrid also possessed the much reduced oviducts normally observed in sterile triploid females. Male hybrids between *L. valentini* and the parthenogenetic *L. armeniaca* have also been observed.

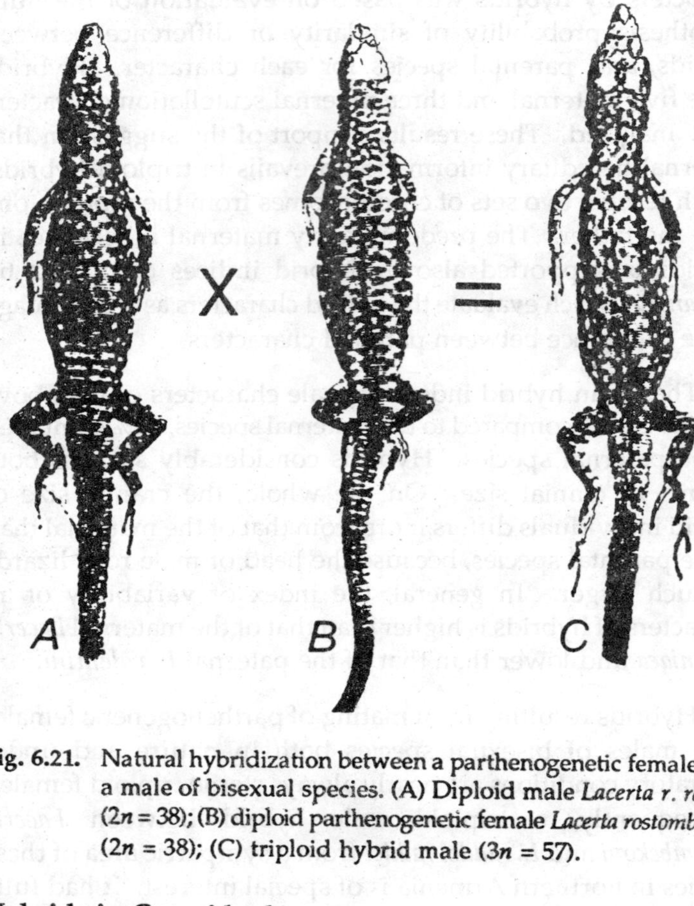

Fig. 6.21. Natural hybridization between a parthenogenetic female and a male of bisexual species. (A) Diploid male *Lacerta r. raddei* ($2n = 38$); (B) diploid parthenogenetic female *Lacerta rostombekovi* ($2n = 38$); (C) triploid hybrid male ($3n = 57$).

Hybrids in Cnemidophorus

Studies of all-female whiptail lizards by American workers soon resulted in the discovery of natural hybridization between some uniparental and biparental species of *Cnemidophorus*. In this respect, it is interesting to remember the history of C. "perplexus", a hybrid described as a distinct species in 1852 by Baird and Girard from a series of specimens collected primarily in the valley of the Rio San Pedro of the Rio Grande del Norte in western Texas. *Cope* (1893) stated that the type specimen is the largest obtained and *Burt* (1931) designated this specimen as lectotype.

The problem of the type locality and status of *Cnemidophorus perplexus* has been discussed repeatedly. Its taxonomic status remains uncertain; the name "perplexus" has at difference times been applied to at least 13 species of *Cnemidophorus*.

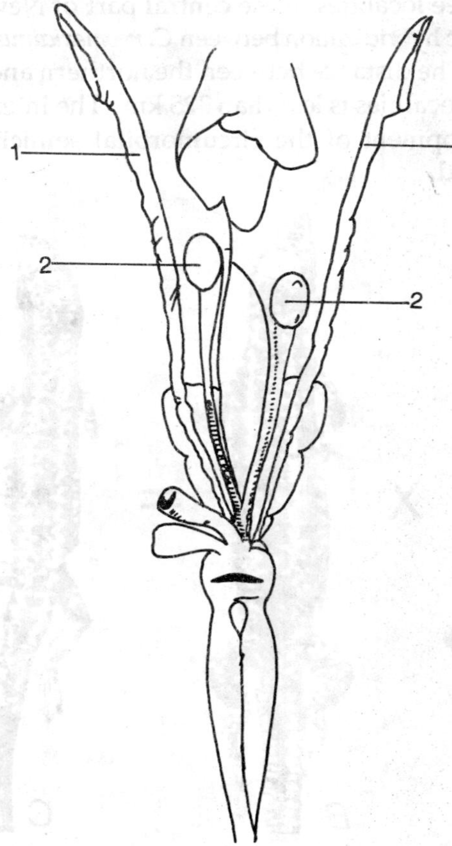

Fig. 6.22. Rudimentary oviducts and testes of a triploid hybrid male *L. rostombekovi* x *L. r. raddei*.

The lectotype designated by Burt is a natural hybrid between sympatric unisexual *(Cnemidophorus neomexicanus)* and bisexual species. The name *"perplexus"* was associated most often with these two species. In 1962 and 1963, six additional specimens of C. *"perplexus"* were collected in New Mexico in areas of sympatry of the parental forms. The newly collected specimens of *"perplexus"* also proved to be hybrids. All the hybrids, including the lectotype (three males, four females) were reported from three localities in the central part of New Mexico where sympatric hybridization between C. *neomexicanus* and C. *inornatus* occurs. The distance between the northern and southern most of these localities is less than 325 km. The intermediate degree of development of the circumorbital semicircle of scales is illustrated.

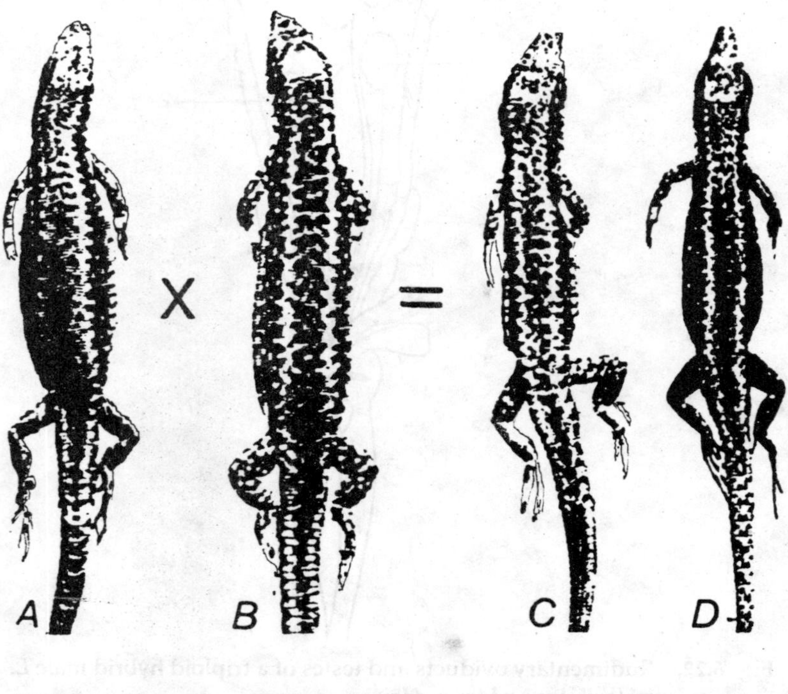

Fig. 6.23. Dorsal pattern of a diploid parthenogenetic female *Lacerta armeniaca* (A); diploid male L. valentini (B) and male hybrids between these two species (C,D).

Fig. 6.24. Left lectotype of *Cnemidophorus perplexus* (female, snout-vent length, 80.4 mm); right hybrid female (*C. neomexicanus x C. inornatus*). from 14 km west of Hatch, Dona Ana County, New Mexico (snout-vent length, 79 mm).

The seven hybrids are closer to the paternal species *(Cnemidophorus inornatus)* in scales around the midbody, in scales between paravertebral stripes, in scales of the circumorbital semicircle, in ventral body colour and tail colour, but they resemble the maternal form *(C. neomexicanus)* in the number of femoral pores, in toe lamellar scales and interlabial scales, and in spotting and femoral colour pattern. The large body size of female hybrids is apparently the result of heterosis that may

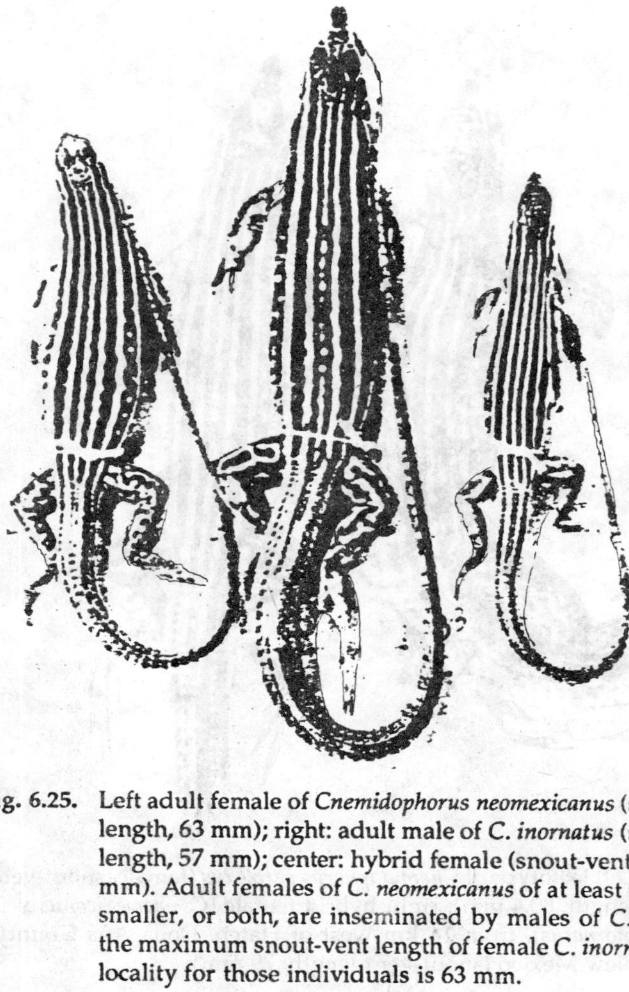

Fig. 6.25. Left adult female of *Cnemidophorus neomexicanus* (snout-vent length, 63 mm); right: adult male of *C. inornatus* (snout-vent length, 57 mm); center: hybrid female (snout-vent length, 78 mm). Adult females of *C. neomexicanus* of at least this size or smaller, or both, are inseminated by males of *C. inornatus*; the maximum snout-vent length of female *C. inornatus* at the locality for those individuals is 63 mm.

be attributed to their triploid nature. The value of the hybrid index for several characters of the seven individuals studied is variable. The grand mean hybrid index for the characters analyzed (calculated relative to the parental species) is 50.4%, indicating intermediacy of the seven hybrids. The hybrids differ in this respect from triploid hybrids of *Lacerta*, the hybrid index being 15% of the maternal and 72% of the parental species. These differences probably result from differing percentages of dominant traits in pairs of species.

In addition to the seven hybrids between *Cnemidophorus neomexicanus* and *C. inornatus* studied by *Wright* and *Lowe* (1967a), two other hybrid males were reported from Mesilla, Dona Ana County, New Mexico and two more, as well as one female, were reported by *Cuellar* and *McKinney* (1976). Two possible hybrids from Valencia County, New Mexico, were reported by *Axtell* (1966). In all, 14 hybrid males and six hybrid females are known from the sympatric area between these species in different regions of New Mexico. In patterns of scutellation, the two hybrids studied by *Taylor* and *Medica* (1966) differed somewhat from those described by *Wright* and *Lowe* (1967a), being close to the maternal unisexual species *C. neomexicanus* in circumorbital scales, orbital scales, and the number of femoral pores, but resembling the paternal *C. inornatus* in the number of scales around the mid-body.

TABLE 6.4

Grand Means for Taxonomic Characters Used in an Analysis of *Cnemidophorus velox* and *C. inornatus*

Character	*Number*	*Range*		*A*[a]		*B*[b]
FP	177	28-38	velox	33.3 ± 0.4	velox	33.6
	150	28-40	inornatus	32.8 ± 0.4	inornatus	33.0
GAB	181	64-80	velox	71.1 ± 0.6	velox	71.1
	155	57-76	inornatus	65.6 ± 0.9	inornatus	66.1
OR	171	148-200	velox	178.6 ± 1.3	velox	177.7
	132	143-190	inornatus	165.0 ± 2.1	inornatus	165.6
PV	180	0-10	velox	6.2 ± 0.3	velox	6.1
	161	3-13	inornatus	8.8 ± 0.4	inornatus	8.1
PV x 100	180	0.0-13.7	velox	8.4 ± 0.3	velox	8.4
GAB	155	4.7-21.0	inornatus	13.5 ± 0.5	inornatus	12.4[a]

A, Each locality treated as a separate entity; B, all individuals treated as a single sample; FP femoral pores; GAB, granules around midbody; OR, granules from occiput to rump; PV, granules separating paravertebral stripes at midbody.
[b] Paravertebral stripes fused at midbody.

The functional histology of testes and epididymides of the hybrids were first studied by *Christiansen* and *Ladman* (1968). The gonads were found to show various stages of spermatogenesis similar to those of bisexual *Cnemidophorus hyperythrus beldingi* from southern California and some other species of lizards. The cycle is characterized by an increase in spermatogenesis and testicular length, which attains a peak in April, May, or June. This is followed by a decrease in testicular size and the number of semniferous tubules and by cessation of spermatogenesis. It is not known whether or not the sperm produced by the hybrids are viable. Because the males are triploid, spermatogenesis may be abnormal, and although an endomitosis prior to meiosis would overcome this, the reduced amounts of sperm in the gonads of hybrids relative to those in the testes of males of parental *C. inornatus* caught at the same time suggests that meiotic problems occur. Studies of artificial hybridization between hybrids and the parental forms are needed because no data on the sterility or fertility of triploid hybrid females are available. Indirect evidence for sterility or of non-stability of the hybrids is provided by the fact that 125 years have passed since such hybrids were described as *C. "perplexus"* without these hybrids producing a population of any importance.

Among other known cases of natural hybridization involving all-female species of whiptail lizards, two probably allotriploid hybrid males of *Cnemidophorus maslini x C. angusticeps* from 26 km southwest of Champoton, Campeche, on the Yucatan Peninsula have been studied. Their detailed comparison with individuals from populations of both parental species discloses a number of intermediate morphological features that convincingly indicate their hybrid origin. Both possess features lacking in female *C. cozumela*, although normally characteristic of female *C. angusticeps*. The tests of one of the hybrids showed various stages of spermatogenesis; tails of spermatids were clustered in the human of seminiferous tubules, but the lumina

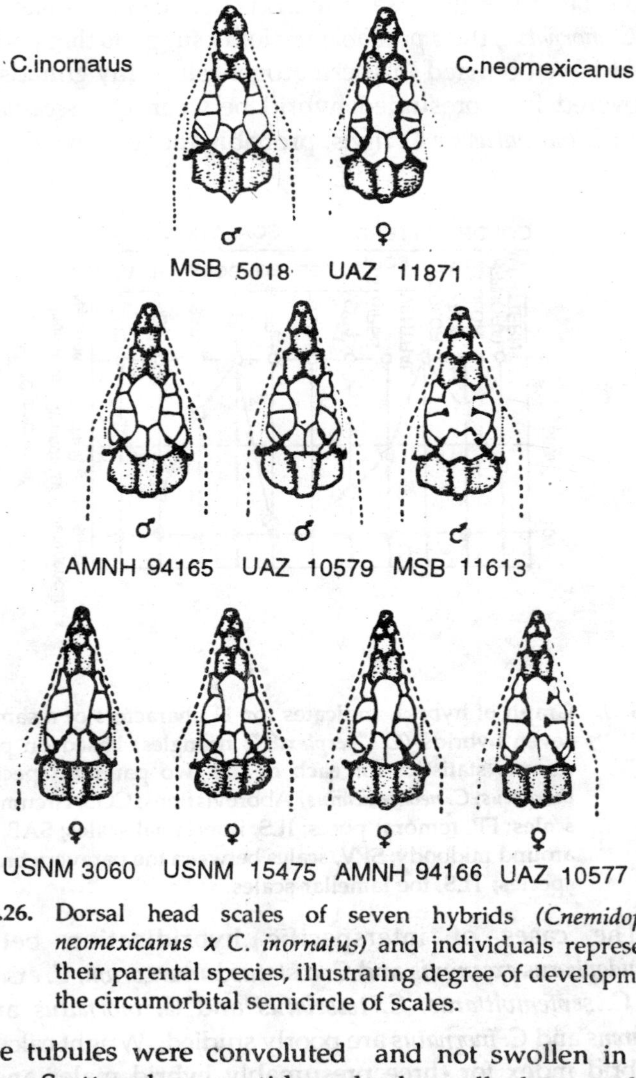

Fig. 6.26. Dorsal head scales of seven hybrids *(Cnemidophorus neomexicanus x* C. *inornatus)* and individuals representing their parental species, illustrating degree of development of the circumorbital semicircle of scales.

of the tubules were convoluted and not swollen in tight circles. Scattered spermatids and other cells, but no clearly mature sperm, were present in the vas deferens. The other hybrid apparently had a normal right testis; the vas deferens passed anterior to the level of the kidney and then abruptly formed a coil on the ventral surface of the kidney, ending as a blind tube. As in the case of hybrids between *C. neomexicanus* and *C. inornatus,* their probable triploidy suggests that problems would have occurred in spermatogenesis. Tiny gonads were discovered in a presumed hybrid between *C. tesselatus* and either *C. inornatus* or *C. tigris,* probably the former.

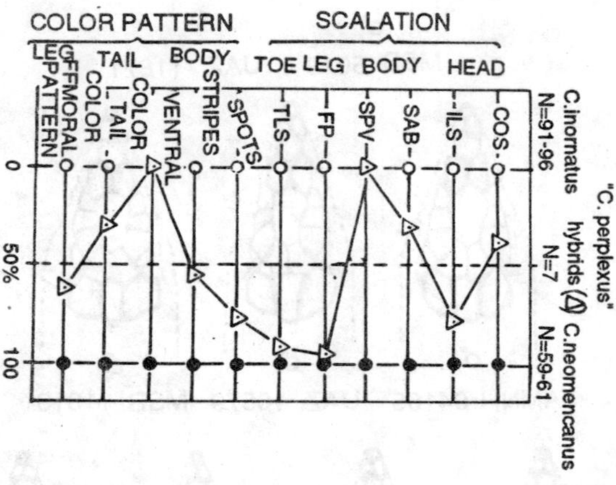

Fig. 6.27. Graph of hybrid indicates for 11 characters of a sample of seven hybrids *(C. "perplexus";* triangles) based on pooled-sample statistics for each of the two parental species *(C. inornatus, C. neomexicanus)* Abbreviations; COS, circumorbital scales; FP, femoral pores; ILS, interlabial scales; SAB, scales around midbody; SPV, scales between the paravertebral light species; TLS, toe lamellar scales.

The cases of interspecific hybridization between *Cnemidophorus exsanguis* and *C. costatus barrancorum, C. exsanguis* and *C. septemvittatus, C. tesselatus* and *C. inornatus* and *C. uniparens* and *C. inornatus* are poorly studied. Wright calculated a hybrid index for three presumably hybrid males and one female. The single feature indicating the hybrid nature of the specimens proved to be the number of scales around the midbody; other characters indicate their closeness to the presumed maternal species, *C. uniparens.* The hybrid nature of the above specimens is undemonstrated, and it is possible that three of them are males of the maternal species. That all the specimens were caught in a region of sympatric contact between the presumed parental species also speaks in favour of their hybrid origin. Hybridization between C. *inornatus* and C. *uniparens* has recently been documented by Cuellar and McKinney. The existence of localities of sympatry between other all-female and bisexual species of the genus *Cnemidophorus* together with the instance of hybridization already known, suggests that hybridization will be found between additional sexual and parthenogenetic species pairs, an in *Lacerta*.

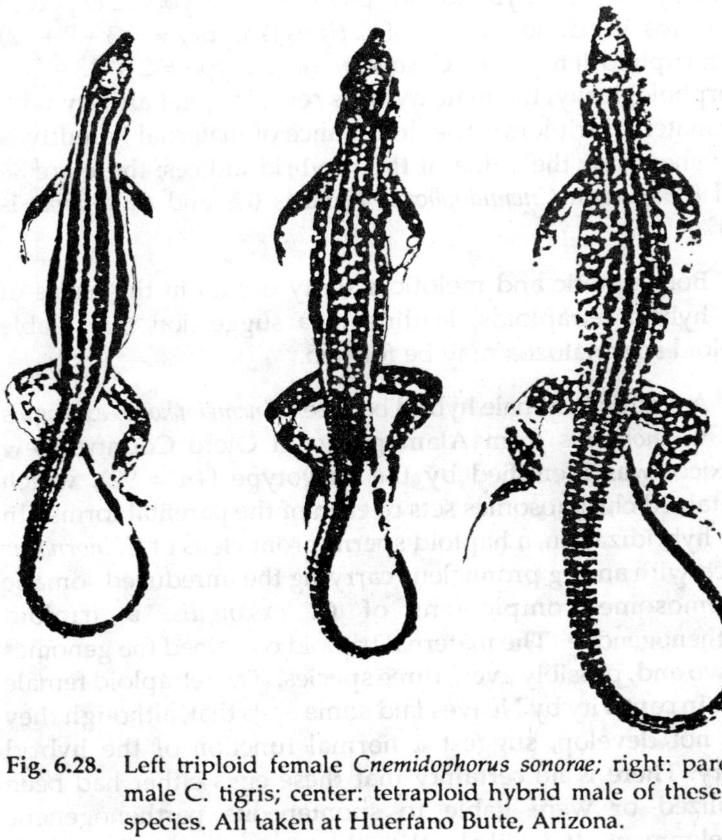

Fig. 6.28. Left triploid female *Cnemidophorus sonorae;* right: parental male C. tigris; center: tetraploid hybrid male of these two species. All taken at Huerfano Butte, Arizona.

The wide sympatry among some unisexual triploid and bisexual diploid species of whiptail lizards should lead to hybridization resulting in tetraploids, and a few presumably tetraploid specimens of both sexes that resulted from the crossing of triploid females of *Cnemidophorus uniparens* with males of *C. inornatus* have been reported. *Lowe et al.* (1970a) described two hybrid tetraploid (*4n* = 93) males and *Cole.* (1979) a female that originated through crossing of triploid (3n = 70) females of *C. sonorae* with diploid males of the bisexual *C. tigris*. Correspondingly, there are five chromosomes of S_I 46 chromosomes of S_{II}, and 42 chromosomes of S_{III} in the karyotype of the hybrids; the hybrids thus possess the karyotype expected from crossing diploid males of *C. tigris* (karyotype = 3 + 8 + 12) with triploid females of *C. sonorae* (karyotype = 2 + 38 + 30;. Morphologically, the male hybrids reveal a great affinity with the maternal triploid. The dominance of maternal heredity is also shown by the value of their hybrid indices; these are 86 and 81% when *Cnemidophorus tigris* is 0% and *C. sonorae* is 100%.

Both mitotic and meiotic activity occurs in the testes of the hybrid tetraploids, leading to a suggestion that viable diploid spermatozoa may be formed.

A tetraploid female hybrid between *Cnemidophorus exsanguis and C. inornatus* from Alamogordo in Otero County, New Mexico, was identified by the karyotype (*4n* = 92), which contained chromosomes sets of each of the parental forms. In this hybridization, a haploid sperm pronucleus of *C. inornatus* fused with an egg pronucleus carrying the unreduced somatic chromosome complement of *C. exsanguis*, a triploid parthenogenone. The maternal triploid contained the genomes of two and, possibly even, three species. The tetraploid female kept in captivity by Neaves laid some eggs that, although they did not develop, suggest a normal function of the hybrid ovary. There is no certainty that these eggs either had been fertilized or were liable to spontaneous parthenogenetic development. It is likely that the eggs were fertilized by a

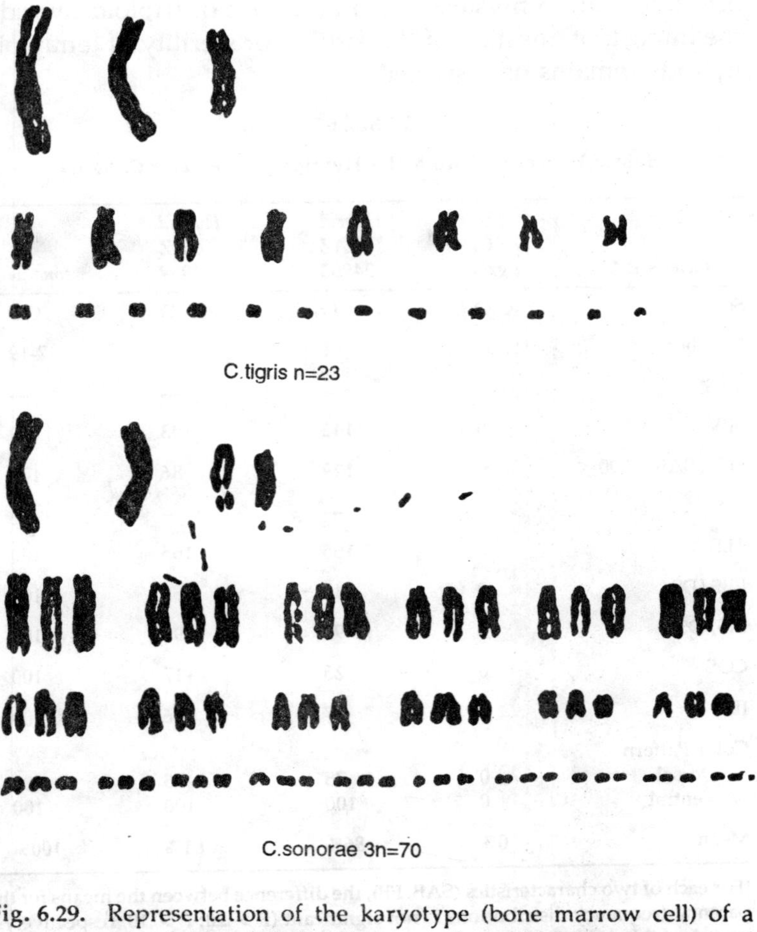

Fig. 6.29. Representation of the karyotype (bone marrow cell) of a tetraploid male hybrid between Cnemidophorus sonorae and *C. tigris*. This is a $4n = 93$ karyotype composed of 5 + 46 + 42 chromosome. A microchromosome that was lost in the preparation is represented by an inked in X at the lower right. This cell was selected since the chromosomes so clearly exhibit their characteristic morphology.

male of the parental species *C. inornatus*. A second aberrant individual with rudimentary gonads was caught by Neaves simultaneously with the above tetraploid hybrid female; its origin remains a mystery. As in the case of triploid hybrids, the important question of the fertility or sterility of tetraploid hybrids remains unanswered.

TABLE 6.5

Hybrid Indices for Two Males Hybrids (*C. sonorae* x *C. tigris*)

Characteristic [a]	C. *tigris*	*Hybrid UAZ 24953*	*Hybrid UAZ 24954*	C. *sonorae*
Sex	O,O	O	O	OO
Number	20-51	1	1	17-19
SAB[c]	—	—	—	—
SPV	0	142	103	100
SPV/SAB x 100	0	129	86	100
FP[c]	—	—	—	—
TLS	0	165	165	100
PBS/DS	0	43	58	100
MS/DS	0	90	90	100
COS	0	-25	+17	100
ILS	0	57	36	100
Color Pattern				
Dorsal	0	75	75	100
Ventral	0	100	100	100
Mean	0%	86%	81%	100%

[a]For each of two characteristics (SAB, FP), the difference between the means for the parental frequency distributions is not significant (P > .20; P > .10, respectively a meaningful hybrid index cannot therefore be established for these characteristics, and such values are not given.

It is possible that the sympatry and consequent hybridization of various species of whiptail lizards began comparatively recently and resulted from the destruction of primary habitats by man and from the disappearance of natural ecological barriers. The destruction of primary habitats, even in urban areas, usually results in only limited interspecific contacts because of the pronounced attachment of many species to their natural microhabitats. On the other hand, in captivity, the males of the bisexual species *Cnemidophorus inornatus* willingly mate with females of the parthenospecies *C. exsanguis* and *C. tesselatus*. Under laboratory conditions, male *C. tigris* also copulated with female *C. tesselatus* and *C. velox*.

MEIOTIC MECHANISMS AND VARIABILITY

Concept

Although the exact mechanisms by which most parthenogenetic reptiles produce ova containing the somatic chromosome number are not clear and in one case in dispute; the patterns of variability so far seen are consistent with clonal reproduction.

Meiotic Mechanisms

The discovery of a considerable number of species of fishes, amphibians, and lizards reproducing by means of hybridogenesis, gynogenesis, or parthenogenesis made the study of cytologic peculiarities of oogenesis in these vertebrates indispensable. As far as is known in unisexual species of vertebrates (with the possible exception of *Poecilia formosa)*, meiosis always occurs. The necessary restitution of the somatic chromosome number can be brought about in different ways. *Cognetti* (1961) distinguished three main restitutional types: premeiotic, intrameiotic and postmeiotic. *Uzzell* (1970) subdivided these into six types, some of which have not been observed.

A premeiotic restitution occurs in a number of invertebrates, for instance, in the earthworm *Lumbricus* and in the grasshopper *Warramaba virgo.* Prior to meiosis, a chromosomal endoduplication occurs, resulting in the formation of pseudobivalents. The meiotic prophase is thus entered at twice the somatic ploidy level, and after two maturation divisions, an ovum with the somatic ploidy level is formed. Among vertebrates, this way of restoring the somatic number is known to occur in triploid salamanders of the genus

Ambystoma in the triploid whiptail lizard *Cnemidophorus uniparens* and in triploid species of *Poeciliopsis*. An earlier suggestion that it also occurs in the triploid derivative of *Poecilia formosa* is apparently incorrect.

In the parthenogenetic triploid whiptail lizard *Cnemidophorus uniparens*, meiosis occurs in oogenesis. At this time, two maturation divisions take place, and both polar bodies are formed, but the count of bivalents at diakinesis and metaphase I, as well as of chromosomes in the first and second polar bodies, shows that their number approaches the triploid level, that is, about 69. Therefore, the chromosome number of this species doubles by a premeiotic endoduplication; the final restitution of the triploid level is achieved through two maturation divisions.

Intrameiotic restitution occurs after the start of meiosis. In the crustacern *Artemia salina*, the somatic number is restored either through fusion of two sets of chromosomes that are at the anaphase stage or through fusion of the pronucleus with the first polar body. Fusion of the pronucleus with one of the progeny of the first polar body is observed in oogenesis of the parthenogenetic species *Drosophila mangabeirai*. The possibility of this kind of diploidization in other species was discussed by *Narbel-Hofstetter* (1964). A typical meiosis takes place in the oogenesis of another crustacean, *Daphnia pulex*, in that all meiotic prophase stages are present; the division of bivalents does not, however, result in the formation of a spindle or the breakdown of the nuclear membrane.

According to *Cherfas* (1966a, b) an intrameiotic restitution occurs in the triploid unisexual form of the silver fish; the formation and isolation of the first polar body either does not occur at all or is accompanied by fusion of both chromosome groups. Meiosis in *Lacerta armeniaca* was originally thought to be intrameiotic but a premeiotic restitution has also been suggested; alternatively, restitution may be postmeiotic.

In postmeiotic restitution, the somatic number is restored either at or after the end of the maturation process. In parthenogenetic eggs of the scale insect *Lecanium hemisphaericum*, the second polar body fuses with the pronucleus. Restitution can occur through fusion of embryonic nuclear cleavage products as in *Gueriniella serratulae*. No cases of postmeiotic restitution have yet been suggested for vertebrates.

There has been some controversy concerning meiotic restitution in *Lacerta armeniaca*. According to *Darevsky* and *Kulikova* (1961), meiosis follows its normal course through the first, reductional, division. The first polar body and the metaphase plate of the second maturation division forms, but the second polar body does not. Although the chromosomes are very small and arranged very compactly, the number of bivalents at metaphase I, as well as the chromosome number at metaphase II, does not exceed the haploid number, that is, it is close to 19. On this basis, Darevsky and Kulikova concluded that the diploid number of chromosomes can be restored either by fusion of the second division nuclei or by suppressing the second meiotic division, with anaphase or the second maturation division not proceeding to the end.

Either of these diploidization mechanisms inevitably results in a high degree of homozygosity. Preliminary data obtained using starch gel electrophoresis of enzymes revealed a high degree of heterozygosity in these forms; this suggests that suppression of cytokinesis in the last premeiotic mitosis is a more probable mechanism of restitution of the diploid chromosome number in *Lacerta*. The presence of the heteromorphic pair of chromosomes in the karyotype of the parthenospecies *L. rostombekovi* is also at variance with the two mechanisms of meiosis suggested for *L. armeniaca*. Either mechanism would result in karyotypes with two acrocentric or two submetacentric chromosomes. A premeiotic endoduplication provides for the preservation of heterozygosis provided that formation of "pseudobivalents" occurs. This suggestion does not agree with the presence of a haploid number of bivalents and chromosomes at metaphases I and II.

One more mechanism of diploidization in the parthenogenetic *Lacerta* may be considered: the fusion of the pronucleus with one of the cleavage products of the first polar body. This restitution pathway is consistent with the count of bivalents and chromosomes at metaphases I and II and with a high degree of heterozygosis in these forms, as well as with the presence of a heteromorphic pair of chromosomes in all individuals of *L. rostombekovi*. Because triploid hybrids of *Cnemidophorus* are fertile, whereas those of Lacerta are sterile, oogenetic mechanisms may differ in whiptail and rock lizards, although the fertility differences could also depend on other factors. It should be pointed out, however, that if synapsis between homeologues occurs and results in bivalents rather than pseudobivalents, then suppression or elimination of crossovers between homeologues would be a necessary feature of meiosis in this species. Otherwise, genic heterozygosity and chromosomal heteromorphism would be lost unless retained by strong selection pressure.

No data on the meiotic mechanisms in other all-female species of lizards are available in the literature.

VARIATION IN PARTHENOGENETIC LIZARDS

Parthenogenesis implies that daughter are genetically identical to their mothers; phenotypic variability within parthenogenetic clones should therefore be restricted. The meiotic mechanisms observed in parthenogenetic reptiles are consistent with such reduced clonal variation, although if parthenogenesis arises as a result of hybridization, some alternative meiotic mechanisms would permit genetic variation within clones while heterozygosity is being lost. Recombinants and mutations also provide genetic diversity among progeny of individual females. On the other hand, multiplicity of clones, arising perhaps through mutation, recombination, or most probably, through numerous initial hybridizations, would increase the variability within a parthenogenetic taxon far beyond that expected within clonses; variability might even approach the levels seen within bisexually reproducing forms. Morphological, biochemical, and karyotype variation all occur in parthenogenetic lizards.

Morphological Variability

The coefficient of variation (CV%) is a convenient index for comparative study of phenotypic variability. Despite its defects, this index gives a general picture of the extent of variability in characters within a given population. Parthenogenetic *Lacerta armeniaca, L. dahli,* and *L. rostombekovi* exhibit less variation in scale counts than do the related bisexual *L. raddei nuirensis* and *L. valentini.* A detailed analysis of the pattern of scutellation in young lizards reared from eggs in the laboratory (i.e., within individual clones) and in wild-caught parthenogenetic specimens of the same age also demonstrates that the coefficient of variation is much lower in the former than in the latter.

Geographical and individual variability of parthenogenetic and biparentally reproducing Caucasian rock lizards has also been studied. The variability of single clones of parthenogenetic taxa is not very great, but the clones themselves differ considerably from each other, so that the degree of variability of parthenogenetic species as a whole resembles that of allied bisexual forms.

General similarity of variances of parthenogenetic and bisexual species of rock lizards has also been documented by more precise mathematical analysis. The degree of variability in populations of rock lizards was assessed by examining composite profiles of the deviation of each of their character means from the overall arithmetic mean of the given character (ME) expressed as a proportion of the overall standard deviation. Such composite profiles show that populations of *Lacerta dahli* and of *L. rostombekovi* differ substantially from each other in scutellation *L. rostombekovi* is typical in this respect. Some populations of this species differ considerably in the number of granules around midbody, femoral pores, preanal shields and in the number of granules around midtibia. Various populations also show marked variability, whereas in *L. dahli*, interpopulation variability is not great. Variation in the body length of these lizards depends on the elevation at which they live.

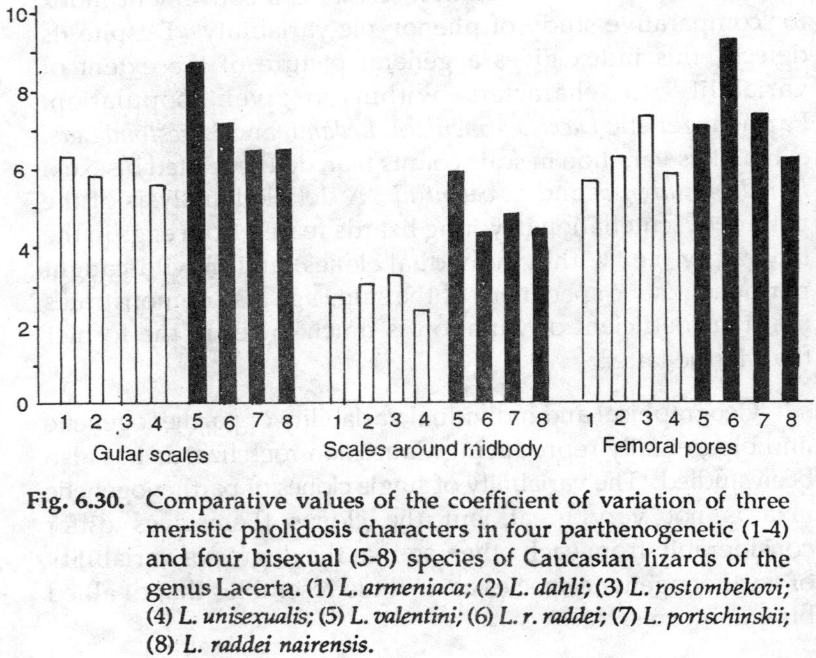

Fig. 6.30. Comparative values of the coefficient of variation of three meristic pholidosis characters in four parthenogenetic (1-4) and four bisexual (5-8) species of Caucasian lizards of the genus Lacerta. (1) *L. armeniaca;* (2) *L. dahli;* (3) *L. rostombekovi;* (4) *L. unisexualis;* (5) *L. valentini;* (6) *L. r. raddei;* (7) *L. portschinskii;* (8) *L. raddei nairensis.*

Investigations of morphological variability in selected populations of *Cnemidophorus velox* and *C. inornatus* have stimulated several studies of the relative variability of all-female and bisexual species of whiptail lizards. Taylor analyzed the colour pattern and certain meristic characters for seven samples of the parthenogenetic species *C. velox* and five samples of the bisexual *C. inornatus.* The variability of the bisexual species was found to be higher on the whole than that of the unisexual species although in several populations of *C. velox* the variability of some characters exceeds that of *C. inornatus.* A comparison of sympatric and allopatric samples shows that the differences are particularly stable within the contact zones; they are manifested in the number of femoral pores and in the number of granules separating the paravertebral stripes. There is, however, a similarity in the number of granules around midbody.

The considerable morphological similarity revealed by an analysis of relative variability of *Cnemidophorus velox* and *C. inornatus* in New Mexico led to their recognition as sympatric sibling species *C. inornatus* is more variable than the unisexual *C. velox* in a number of characters. Although their clinal and non-clinal variability allows local samples of the two species to be easily distinguished by scutellation in any part of their range, the forms are not necessarily distinguished by scale counts alone when compared as individuals or as pooled samples drawn from over the entire geographic ranges of the species. *C. velox* and *C. inornatus* are also morphologically similar to the all-female *C. uniparens;* the three form a group of sibling species with distinct sympatric areas. A detailed analysis of the variability of scutellation and colour pattern both in allopatric and sympatric populations of these species has shown that they possess distinct stable characters; the variability of the unisexual species is lower than that of bisexual *C. inornatus.* Relative variability in the number of scales around the mid-body, and the number of scales between the paravertebral light stripes in one of the mixed populations of the species considered.

Wright's (1966, 1968) analysis demonstrated that despite the possibility of occasional sympatric hybridization, the three sibling species preserve their individuality; hence, little or no gene flow following the formation of hypothetical "intergrading clones" occurs in this case. Theoretically, the transfer of genetic material could occur as a result of matings by occasional males of unisexual species or by matings with normal males of bisexual species.

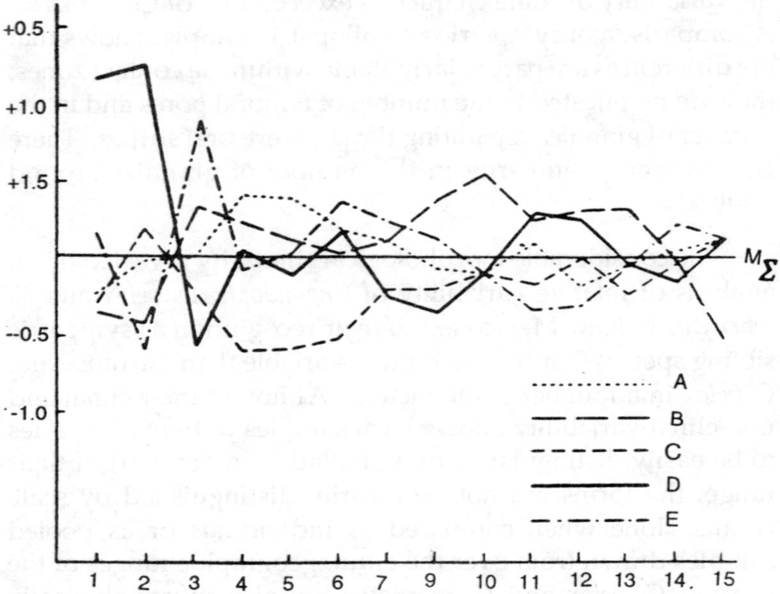

Fig. 6.31. Composite profile of variability in five samples of *Lacerta dahli* from different geographical populations: (A) Kirovakan (Armenia), (B) Stepanovan (Armenia); (C) Manglisi (Georgia); (D) Tsalka (Georgia); (E) an isolated population in the Tana river ravine in eastern Georgia. Ordinate: deviations of characters, expressed as the proportion of the overall standard deviation from the arithmetic mean of the given character (M2). Abscissa: characters; (1) snout-to-vent length; (2) tail length; (3) ratio of snout-to-vent length to tail length; (4) scales around midbody; (5) scales along middle of throat; (6) femoral pores; (7) number of granules between supraocular and supraciliary scutes; (9) number of ventral plates rows; (10) number of preanal scutes; (11) number of small scutes between central temporal and tympanic; (12) number of posttemporal scutes; (13) number of body scales on the border of one ventral scute; (14) scales around middle of shin; (15) rows of small scutes between femoral pores and extremeal thigh scutes.

Work on the widely distributed unisexual species *Cnemidophorus tesselatus* and its triploid derivatives added much

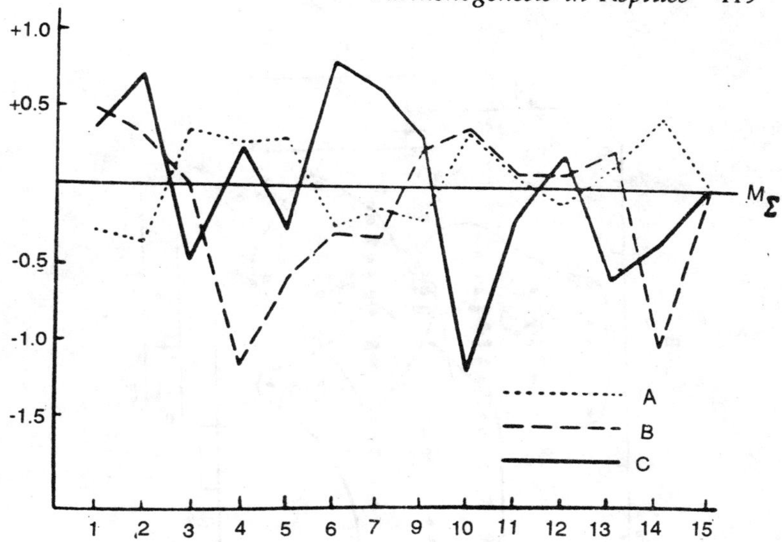

Fig. 6.32. Composite profile of variability in samples from three isolated geographical populations of *Lacerta rostombekovi.* (A) Spitak (Armenia); (B) Zagalu on the banks of Lake Sevan (Armenia); (C) Lake Gey-Gel'. (Azerbaijan, from Darevsky, 1967). Ordinate derivations of characters, expressed as the proporion of the overall standard deviation from the arithmetic mean of the given character (M). Abscissa: characters; (1) snout-to-vent legnth; (2) tail length; (3) ratio of snout-to-vent length to tail length; (4) scales around mid-body; (55) scales along middle of throat; (6) femoral pores; (7) number of granules between supraocular and supraciliary scutes; (9) number of ventral plates rows; (10) number of preanal scutes; (11) number of small scutes between central temporal and tympanic; (12) number of posttemporal scutes; (13) number of body scales on the border of one ventral scute; (14) scales around middle of shin; (15) rows of small scutes between femoral pores and external thigh scutes.

to our knowledge of variability of all-female species of whiptail lizards. Examination of the body pattern and scutellation of 545 specimen representation of the range of these two species led to the recognition of six distinct classes in *C. tesselatus*. Five of these (A, B, C, E and F) are allopatric, whereas populations

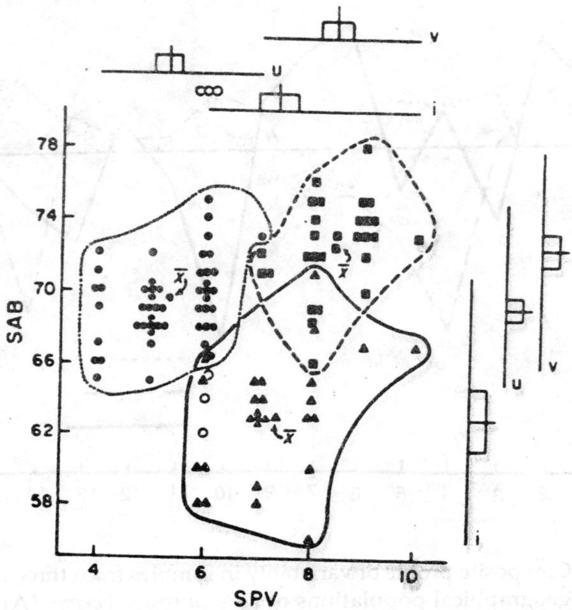

Fig. 6.33. Graph of the number of scales around midbody (SAB) versus the number of scales separating the paravertebral light stripes (SPV) for specimens comprising the samples of the three sympatric sibling species from the Socorro locality with corresponding Dice-Leraas Graphs (95% confidence limits) of the reduced data. Squares and v represent values for *C. velox;* closed circles and u are for *C. uniparens;* triangles and i are for *C. inornatus;* open circles are possible hybrids.

of class D occur in two widely separated areas — Colorado and New Mexico. A population with distinct colouration has been recognized in Texas and, together with *C. tesselatus* class F, has been named *C. dixoni* by *Scudday* (1973). In addition to pronounced differences in colour pattern, lizards of Zweifel's classes differ in scutellation. The sympatric populations show a consistent correlation between pattern differences and peculiarities of scutellation, indicating that they are genetically distinct clonal lines. On the whole, differences in scutellation and colour pattern among classes are as a great as those that distinguish subspecies in bisexual species of *Cnemidophorus.* Complications are introduced, however, by sympatric clones and by discordant variation Classes D and C, for instance, are sympatric. Furthermore, one of the populations within Class C has a pattern typical for the surrounding populations of the same class although its scutellation is closer to that of the neighbouring Class E. Because of these peculiarities of variation, Zweifel preferred to avoid using such categories as species or subspecies. The systematic status of some classes should be revised in light of the demonstration that classes A and B are formed by triploids whereas other classes are diploid. It is evident to us that the triploid classes (A and B) are not conspecific with *C. tesselatus.*

Morphological variability in *Cnemidophorus tigris and in C. tesselatus* has been compared using nine scale counts, two tooth counts, and three measurements of body dimensions. The measurements and tooth counts are size-dependent characters, but the nine scale counts basically are not. The size-related characters were comparably variable in both the biparentally and in the parthenogenetically reproducing species. This may be either because genotypes of the two species differ in the amounts of phenotypic response they show to environmental variability or because the size-related characters have such a large environmental component of phenotypic variation that genetic differences in variability are obscured. The reduced variability in six of the nine size independent characters in the parthenogenetic species suggests that the equal variability of the size related characters reflects environments influence rather than differences in genotypic response.

Despite the reduced variability within *Cnemidophorus tesselatus,* it is possible to detect clonal heterogeneity among sympatric diploid clones. Five of the seven electrophoretic-pattern clones occurring at Las Conchas show statistical heterogeneity for four of the nine scale counts; the two other clones, represented by single individuals, were excluded from the analysis. Clonal heterogeneity was also evident at other localities where only two clones co-occurred.

The classes recognized by Zweifel within *Cnemidophorus tesselatus* resemble in many ways the morphologically distinct populations of parthenogenetic species of rock lizards, in particular, of *Lacerta rostombekovi.* Variation of *C. tesselatus* has been compared with that in two subspecies of the bisexual *C. tigris.* The variation of *C. tesselatus* within a given area approaches that of bisexual forms, but samples of local populations of these species show much less variation in color pattern and scutellation in comparison to *C. tigris.* Thus variability of *C. tesselatus* is similar to that of other all-female species considered here.

Samples of three populations (51 specimens), of the parthenogenetic gecko *Hemidactylus garnotii* from Oahu in the Hawaiian Islands have been examined. The greatest distance between the populations is 27 km. Absence of bimodality in 21 characters and general homogeneity of character means and coefficients of variation indicated that there are no sympatric clones in this species.

A phenetic clustering technique shows that all members of the population from Puuola Point, excluding one specimen, are continuously linked to each other in the Prim network generated. In contrast, specimens from Waianae and Barbers Point do not show any tendency to be continuously linked. The related phenotypic dissimilarity of individuals observed within the samples from Waianae and Barbers Point was interpreted by Kluge and Eckhardt as probably the result of multiple colonization of the Hawaiian Islands by *H. garnotii* which may have introduced propagules from clones that resulted from multiple origins. Sympatric clones within unisexual populations, however, may result both from multiple origins and from perpetuation of mutations within an "ancestral" clone. Both phenomena have been reported in *Cnemidophorus tesselatus.*

The main meristic characters of *Hemidactylus garnotii* have a low coefficient of variation. The teiid lizard *Gymnophthalmus underwoodi* also has low variability of dorsal and ventral scales in comparison to bisexual species of that genus. On the other hand, the triploid parthenogenetic agamid *Leiolepts triplate* differs little in its range of variability from the related bisexual species.

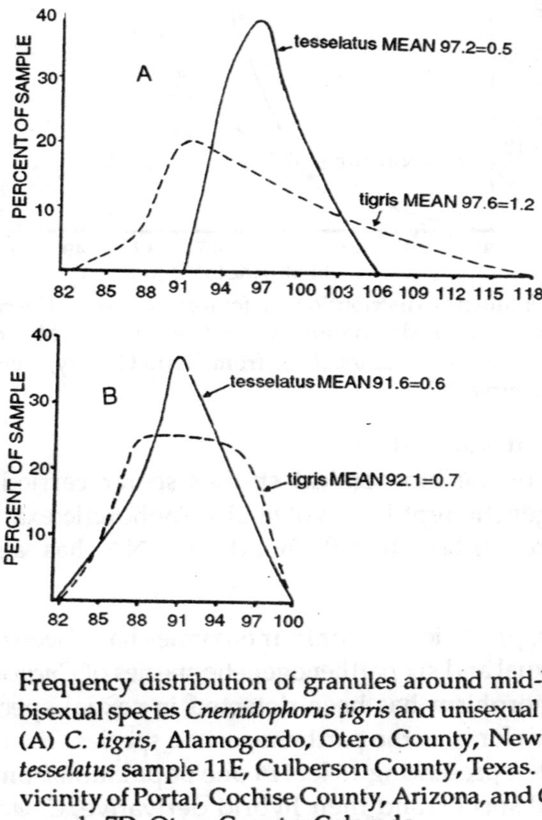

Fig. 6.34. Frequency distribution of granules around mid-body of the bisexual species *Cnemidophorus tigris* and unisexual *C. tesselatus*. (A) *C. tigris*, Alamogordo, Otero County, New Mexico; *C. tesselatus* sample 11E, Culberson County, Texas. (B) *C. tigris*, vicinity of Portal, Cochise County, Arizona, and *C. tesselatus*, sample 7D, Otero County, Colorado.

These results suggest that multiclonal unisexual species are as variable as bisexual species. Whenever, fewer clones exist, either in the species or in the unit being samples, the restrictive genetic mechanism results in low variability in the unisexuals, because in genetic variability, clones are equivalent to individuals or to monozygotic sibships. The variation within such clones will be solely environmental.

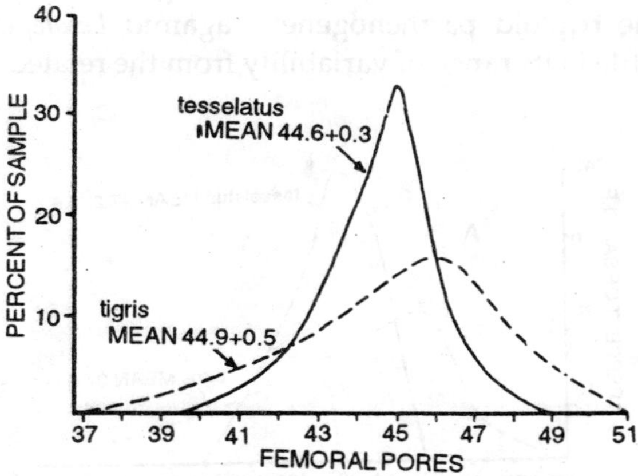

Fig. 6.35. Frequency distribution of femoral pores in *Cnemidophorus tigris* from Alamogordo, Otero County, New Mexico, and *C. tesselatus* and *C. tesselatus* from Sierra County, New Mexico, sample 13E.

Biochemical Variability

Most of the biochemical studies so far carried out on parthenogenetic reptiles involve electrophoretic examination of proteins, although mitochondrial DNA has also been analyzed.

Electrophoretic data for four enzymes have been obtained for five sexual and six parthenogenetic species of *Cnemidophorus*. No polymorphism has been detected in sexual species, and only single clones in the parthenogenetic species. Examination of proteins representing 15 loci in one population from each of two sexual species and their hybrid derivative, *C. laredoensis*, has shown no polymorphic loci either in *C. sexlineatus* or in *C. gularis;* all individuals of *C. laredoensis* belong to a single clone. Five electrophoretic markers revealed but a single clone for each of four parthenogenetic taxa of *Lacerta*. One of these loci was polymorphic in one of the sexual taxa.

A far more detailed electrophoretic study has been carried out on *Cnemidophorus tesselatus,* its triploid derivative, and on the three parental species. Of the 16 loci for which the three parental species were not fixed for the same allele, *C. tigris* was polymorphic for four, *C. septemvittatus* for seven, *C. sexlineatus* for three. The *C. tigris* samples came from nine localities, *C. septemvittatus* from three, and *C. sexlineatus* from two. Of the parthenogenetic lizards, *C. tesselatus* came from 25 localities, the triploid derivative from three.

All of the triploid individuals belong to a single clone, but the diploid populations of *C. tesselatus* contain 12. Use of six variable loci discloses as many as four clones at a single locality. Four of the 12 appear to be best accounted for as a result of multiple hybridizations. Three, however, may have arisen through mutation, because they are distinguished by an allele not seen in either parental species, and five may have been arisen by recombination, because they contain alleles seen only in one parent. The electrophoretically detected clones cut across the pattern classes described by *Zweifel* (1965).

Minor variability in mitochondrial DNA occurs in *Cnemidophorus tesselatus,* but it is not clear whether it arose before or after hybridization (Wright, Spolsky and Brown, personal communication).

Karyotypic Variability

The most frequent meiotic pathway known in clonal vertebrates, a premeiotic endoduplication of chromosomes resulting in formation of pseudobivalents, avoids any problems with synapsis and segregation of heterozygous chromosome arrangements. Alternative meiotic pathways, such as that proposed for *Lacerta armeniaca* do not avoid such problems. It is perhaps significant, therefore, that the heteromorphic pair of chromosomes observed in *L. rostombekovi* seems to be uniformly present.

The modal number of chromosomes in parthenogenetic

Lacerta is 38, but hypoploid figures occur, especially in *L. unisexualis;* their cause is unknown.

In contrast to parthenogenetic *Lacerta,* in which little karyotypic variation within a taxon has been reported, there are several accounts of chromosomal variation in parthenogenetic species of *Cnemidophorus,* at least one of which has premeiotic endoduplication of chromosomes as its method of restoring the somatic chromosome number despite a complete meiotic division. A small area within the range of *C. sonorae* contains three distinguishable karyotypes. It was noted that such karyotypic variation is a general rule specifically among the triploid parthenogenetic species of *Cnemidophorus*. Three karyotypes occur in the triploid *C. exsanguis*. Karyotypic variation also occurs in diploid parthenogenetic *Cnemidophorus;* three distinguishable karyotypes occur in the parthenogenetic species called *C. lemniscatus* and two occur in *C. cozumela maslini.*

In most of these cases, one karyotype, often that predicted on the basis of the karyotypes of the parental species, appears to be common and widespread, with the modified karyotypes rarer and more restricted. This suggests that the variant karyotypes have arisen by chromosomal mutation within parthenogenetic forms rather than representing multiple origins. The ubiquitous heteromorphism that is seen in *Lacerta rostombekovi,* however, appears to have arisen arisen early in the history of that taxon, because it does not represent the summation of the karyotypes of the two parental species. The widespread karyotypes in the parthenogenetic taxon called *Cnemidophorus lemniscatus* may also represent an early modification in that form.

Karyotypic polymorphism occurs in biparentally reproducing species of lizards, but it is rare. Meiosis in biparentally reproducing species probably tends to conserve karyotypes, whereas in clonal species, especially those with a premeiotic endoduplication, it may allow more rapid diversification.

EVOLUTIONARY BIOLOGY OF PARTHENOGENETIC REPTILES

Theoretical studies of parthenogenesis and studies that mainly gather data have often proceeded independently, so that research has lurched along like a chariot pulled by two unruly horses, giving knowledge a rather bumpy ride. We still know far too little to provide a satisfactory synthetic description of parthenogenetic reptiles, but we can make a start. We will try to organize the material to answer the following general questions: (1) How did parthenogenesis originate ? (2) How does it persist ? (3) What is its evolutionary potential ?

Parthenogenesis and triploidy are frequently associated in reptiles. Although it is not certain that the one precedes the other, the origin of parthenogenesis is discussed first. The persistence of parthenogenesis surely has at least two components: one is clearly molecular, genetic, and cytogenetic, the other is ecological. The latter has generally been emphasized to the neglect of the former.

THE ORIGIN OF PARTHENOGENETIC REPTILES

Hybridization and Parthenogenesis

Natural parthenogenesis in vertebrates is a relatively recent discovery. The earliest indication that we have found is *Taylor's* (1918) report that the entire type series of *Lepidodactylus divergens* (about 25 specimens) consisted of females; soon thereafter *Schmidt* (1919) reported that a series of 63 specimens of *Rhampholeon boulengeri* were all females. Other early observations on absence of males include those by *Smith* (1935) on *Hemidactylus garnotii* and by *Lantz* and *Cyren* (1936) on *Lacerta armeniaca*. Despite the clear demonstration of matroclinous inheritance in the fish *Poecilia formosa* and the similar demonstration for a salamander now called *Ambystoma tremblayi*, it was 20 years later that the idea that some lizards might be parthenogenetic was proposed for some taxa in the *Lacerta saxicola* complex.

Natural parthenogenesis soon became widely accepted, particularly after similar phenomena were discovered and studied in lizards of the genus *Cnemidophorus*, initially by *Minton* (1959), *Tinkle* (1959), *Duellman* and *Zweifel* (1962), Maslin (1962, 1966), and *Zweifel* (1965). A hybrid origin for these taxa was first suggested by the early studies of the chromosomes of several lizards. This is rather surprising, because the botanists *Winge* (1917) and Ernst (1918) had much earlier elaborated the theory that hybridization might result in apomixis. Ernst was particularly influenced by the striking similarity between cytological features of apomicts, such as the green alga *Chara crinita*, and of hybrids between distantly related species. Ernst concluded that parthenogenesis arises naturally and compensates hybrids between distantly related plants for their inability to reproduce sexually.

A hybrid origin of parthenogenesis once found widespread support among zoologists. On the basis of karyological observations, the hybrid origin of all-female species of the stick-insect *Carausius morosus* was suggested by Pehani (1925). *Harrison* and *Peacock* (1926) showed that hybrid butterflies, which result from the crossing of closely related species of the genus *Tefrosia*, can lay unfertilized eggs, some of which undergo a complete parthenogenetic development never observed in related forms. On this ground, a causal relationship was proposed between hybridization and the transition to parthenogenetic reproduction. Similar results were obtained by *Astaurov* (1940), who established that, in interbreed hybrids of the silkworm, rudimentary parthenogenesis manifests itself more clearly than in pure breeds; complete parthenogenesis often occurs. Nevertheless, the hybrid hypothesis, which treats hybridization as the main if not the only reason for the transition to parthenogenesis, did not gain permanent recognition in plants or in animals. It gradually lost support, especially after sharp criticism from *Winkler* (1920), who noted the existence in plants of sterile hybrids obviously incapable of parthenogenetic reproduction and the existence of parthenogenetic plant forms that certainly were non-hybrid in origin.

Although we believe that hybridity has been clearly established for 15 of the 32 confirmed or probable parthenogenetic species of reptiles, in no case has such as hybrid origin been duplicated in the laboratory, mainly because few hybridizations have been attempted. Maintaining a breeding colony in which mating between parental species can occur and in which resultant young can in turn be raised to maturity is the chief problem. Obvious candidates for synthesis are diploid species, such as several parthenogenetic *Cnemidophorus* and the parthenogenetic species of the *Lacerta sexicola* complex. Synthesis of clonal vertebrates has in general proved difficult, except for the hybridogenetic taxa *Rana esculenta* and *Poeciliopsis Cx*. Considerable effort has been expended to produce the gynogenetic diploid *Poecilia formosa*, for example by *Turner et al.* (1980), but with no success. On the other hand, *Drosopoulos* (1979) reported the laboratory synthesis of a triploid gynogenetic homopteran of the genus *Muellerianella;* the experimental crosses seem to occur easily and regularly. The production of parthenogenetic laboratory strains of *Bombyx* is also relevant.

The hybrid origin of clonal reproduction in vertebrates has remained controversial. We argued earlier that hybridity in diploid clonal taxa and in triploids that contain genomes from three different parental species must almost necessarily have preceded the establishment of clonal reproduction. Unfortunately, little is known about the hybridity of many of the probable parthenogenetic reptiles. Karyological and electrophoretic evidence indicates that at least 15 of the 32 are hybrids; nine are diploids; six, triploids. Of the 32 taxa, 15 are known to be diploid, and 11 are triploid.

The taxa that have been examined electrophoretically represent a random sample with respect to the probability of being hybrid. The *Cnemidophorus* were examined largely because they were taxonomically confusing and readily available; the

Lacerta were examined because of the possibility that they were not hybrids. To the extent that they represent a random sample, one can ask how common non-hybrids are likely to be among the parthenogenetic reptiles if we observe 16 that are hybrid out of 16 carefully examined. By this kind of analysis, it would appear that six or fewer of the 31 possible species are non-hybrid, at the 95% confidence level.

Although it is by no means clear that there is a random sample for these purposes, to whatever extent the analysis is correct, it suggests that reptiles are not parthenogenetic unless they are also hybrids. This in turn may only mean that hybridity is necessary for parthenogenetic taxa to persist, but we believe that hybridity is the cause of parthenogenesis.

The distribution of parthenogenetic reptiles is not random, however, in a conspicuous sense; this non-randomness holds for clonal vertebrate species as a group. Five different clonal taxa are known in *Lacerta;* 15 in *Cnemidophorus.* Certainly this is partly a result of additional searches for clonal taxa in these genera after the first were discovered. These are also genera in which there are many species and thus a wide variety of possible hybrids and possible parthenogenetic taxa.

There is, however, an additional feature that is probably even more significant in the clustering of these parthenogenetic taxa. With the possible exception of *Cnemidophorus lemniscatus,* all parthenogenetic species of *Cnemidophorus* that are known to be hybrids have a member of the *C. sexlineatus* group as at least one parental species. It is not impossible that the parthenogenetic taxon called *C. lemniscatus* is also a hybrid between *C. lemniscatus* and some member of the *C. sexlineatus group C. lemniscatus* ranges north in Central America, reaching as far as extreme southeastern Guatemala. *C. angusticeps* and *C. motaguae,* two members of the *C. sexlineatus* group, occur in Guatemala. We do not suggest that either of there is a parental species of the parthenogenetic taxon called *C. lemniscatus,* and in fact, *Cole et. al.* (1983) have suggested that *Ameiva ameina* may be a parental species. Nevertheless, contact between the two *Cnemidophorus* species groups is not impossible. It is quite striking that most populations of *C. lemniscatus* have 2n = 50 (2 + 24 + 24; *NF* = 52) chromosomes, whereas most individuals of the parthenogenetic taxon have 2*n* = 48 (1 + 25 + 22; *NF* = 49). One population has 2*n* = 48 (2 + 24 + 22; *NF* = 50), and another has 2n = 50 (1 + 25 + 24; *NF* = 51). The typical pattern for the *C. sexlineatus* group is 2*n* = 46 (2 + 24 + 20) so that intergroup hybrids might be expected to have 2n = 48 (2 + 24 + 22). The differences between the expected pattern and those usually seen in the parthenogenetic taxon called *C. lemniscatus* are of the kind often seen in modified patterns of parthenogenetic *Cnemidophorus* Sex ratios for *C. "lemniscatus"* from Central America have not been reported.

It is possible that the *Cnemidophorus sexlineatus* group contains two independent lineages, one of which has evolved a genetic factor or factors that determine clonal reproduction in hybrids. It is not impossible that all parthenogenetic *Cnemidophorus* have as one parental species a member of this subdivision of the *C. sexlineatus* group.

A similar situation may exist in the *Lacerta saxicola* complex. Electrophoretic data suggest that the parental species of the parthenogenetic taxa fall into an *L. r. raddei-L. r. nairensis-L. mixta* group and an *L. portschinskii-L. parvula-L. valentini* group. Each of the four parthenogenetic taxa for which the parentage has been determined has one parent from each group *L. uzzelli* probably derives from *L. r. nairensis* and *L. parvula*.

The widespread distribution of single genomes in clusters of clonal vertebrates (the monacha genome in the fish genus *Poeciliopsis*, the ridibunda genome in frogs of the genus *Rana*, a genome from the *Ambystoma jeffersonianum* complex in mole salamanders) suggests to us that there are genetic factors that control clonal reproduction in vertebrates and that these factors arose only one in each vertebrate genus that has clonal members.

These observations do not demonstrate that all or even most parthenogenetic reptiles are hybrid nor that hybridity causes the parthenogenetic reproduction. Exceptions have not yet been found, however, and we believe that the trend in results obtained since the mid-1960s tends to support the hypothesis of a hybrid origin. The data also provide a satisfactory answer to *Williams*' (1975) question, "Why has parthenogenesis not evolved more often in vertebrates?" It probably has never evolved in vertebrates and can only occur in those groups in which certain kinds of genetic differentiation have occurred, such that hybrids are automatically parthenogenetic. In this sense, the transition to parthenogenesis proposed by *Ernst* (1918), which supposedly compensates for the inability to reproduce sexually, may not occur, at least not in reptiles. We are aware that the difficulty in synthesizing *Poecilia formosa* may indicate that only certain specific genetic combinations produce clonal vertebrate taxa and that formation of triploid clonal taxa requires at least an additional step. This view does not exclude a gradual transition to parthenogenesis by drift or by selection, especially in the sense of "improving" parthenogenesis. It is more consistent, however, with an abrupt transition.

To the extent that parthenogenesis arose as a result of hybridization or that parthenogenetic taxa are hybrid, some degree of sympatry and syntopy between the parental species is required, but the parental species are often hot sympatric at present, which indicates that hybridization occurred at some time in the past.

Triploidy

In order to explain the connection between parthenogenesis and polyploidy in plants, some investigators advanced a polyploid hypothesis of apomixis. According to this view, autopolyploidy gives rise to vigorous but sterile polyploids and is in itself conducive to the transition to unisexual reproduction. Some experiments on hybridization of plants indicate that an increase in chromosome number stimulates the ability of egg cells to develop without fertilization. Many other facts, however, do not support this hypothesis.

Astaurov in a series of papers on artificial parthenogenesis in silkworms *(Bombyx mori)*, advanced and experimentally proved that odd-ploid parthenogenetic species, which are very frequent in nature, do not necessarily arise directly from an initial diploid bisexual ancestor. Instead they may originate indirectly as a result of the crossing of males of ancestral bisexual forms with females of diploid parthenogenetic races. Speciation and polyploidization in this case proceeds along the following sequence: (1) bisexual diploid species, (2) diploid parthenogenetic race, (3) polyploid parthenogenetic race, involving one or several with different levels of ploidy, and (4) bisexual polyploid species. The whole process, stage by stage, has been experimentally reproduced. Astaurov concluded that none of the known cases of polyploidy in animals, including triploidy in lizards, can be explained with certainty by direct polyploidization of a diploid bisexual ancestor; if often remains possible that they had an indirect origin.

A two-stage hypothesis explaining the hybrid origin of diploid and triploid parthenoforms has been proposed by *Lowe* and *Wright* (1966b) and Wright and Lowe. The first stage is the appearance of allodiploid parthenoforms as a result of hybridization of diploid bisexual species (2n = 46); the resultant parthenoforms are adapted to the immediate conditions of the environment. Although such forms evolve very slowly, because no recombination occurs, they may form diploid parthenogenetic species, some of which survive. The second stage represents a crossing of the diploid parthenoform with a parental or, more rarely, with some other bisexual species; this cross results in an allotriploid species.

The karyotypes of the triploid parthenospecies *Cnemidophorus uniparens* resulted from the hybridization of the bisexual species *C. inornatus* and, perhaps, *C. gularis*. An intermediate and now evidently extinct allodiploid parthenoform may have arisen first and given rise to the all-female *C. uniparens* by backcrossing with *C. inornatus*. The diploid parthenoform may have existed only for a single generation.

In a similar way, the allotriploid derivative of *Cnemidophorus tesselatus* may have arisen by hybridization of the allodiploid *C. tesselatus* with the bisexual species *C. sexlineatus*. The hybrid origin of the triploid is supported by the results of skin grafting experiments which show that the diploid taxon acquired two of its genomes from *C. tesselatus*, and the third one from some other source. Electrophoretic markers show that the bisexual species *C. sexlineatus* contributes the third genome.

Triploidy in reptiles could arise either from diploid parthenogenetic taxa, as suggested by Astaurov (1969a, b) but by Lowe and Wright (1966b), or it could arise subsequent to the origin of parthenogenesis within a formerly biparentally reproducing species. In the latter case, it could be associated with hybridization, or it could arise by matings of parthenogenetic members of a species with males of the same species, which results in autotriploids. *Peters* (1971) suggested that *Leiplepis triploida* is an example. In analogy with the triploid called *Cnemidophorus tesselatus*, the frequent evidence of genomes from at least two different parental species suggests that most triploid reptile species arose by mating of a diploid hybrid parthenogenetic form with one of its parental species or with yet a third biparentally reproducing species. Experimental production of triploid parthenogenetic species may help choose among these hypotheses.

THE PERSISTENCE OF PARTHENOGENESIS IN REPTILES

Molecular Aspects of Persistence

To the extent that they are hybrids, parthenogenetic reptiles are highly heterozygous, because their genomes derived from two or even three species. Such heterozygosity, which is both genic and chromosomal, may later be reduced by some mechanisms of meiotic restitution. Heterozygosity may also be decreased by unusual events, such as recombination, in the usual mechanism of meiosis (a premeiotic endoduplication without cytokinesis). Unless genetic changes within clonal taxa are highly deleterious, however, they will usually result in increased heterogeneity. Although increased karyotypic differentiation has been commonly reported in parthenogenetic reptiles, genic changes have been reported only by *Parker* and *Selander* (1976).

Virtually nothing has been reported on the molecular interactions of the differing genomes present in hybrid clonal taxa, although we can infer that the interactions exist, and we can see evidence of such interactions in both ecology and morphology.

One important feature in which molecular interactions are profoundly important is meiosis. Not only do most clonal vertebrates that are gynogenetic and parthenogenetic use a single mechanism of meiosis (a premeiotic endoduplication), but even some hybridogenetic taxa have such a meiotic mechanism. In *Rana esculenta*, for example, the lessonae chromosome set is lost or inactivated before meiosis is complete the ridibunda chromosome set then undergoes an endoduplication, and meiosis usually proceeds normally thereafter.

It seems probable that meiosis cannot occur in a haploid state. The molecular basis for this limitation is not known, but the frequency with which endoduplication occurs in clonal vertebrates suggests that the limitation is rather basic. For diploid and triploid parthenogenones, one or more chromosomes or genomes must be effectively in a haploid state prior to endoduplication. Certain meiotic patterns reported in clonal vertebrates are inconsistent with this generalization.

We argued earlier that there probably are genetic factors that control clonal reproduction in vertebrates and that these factors arose only once in each genus that has clonal members. Nothing is known, however, about the molecular aspects of these putative genetic factors.

One of the critical features of meiosis in parthenogenetic reptiles is that it generally produces progeny that are genetically identical to their mothers. This feature, which is the essence of clonal reproduction, is probably the most significant genetic feature in the persistence of hybrid parthenogenetic taxa. More commonly, viable interspecies hybrids that are able to complete meiosis show extensive recombination in the F_2 or B_1 generation; they produce a variety of genotypes, many of which result in disadvantaged phenotypes.

Ecological Aspects of Persistence

Much thought has been given to the adaptive significance of parthenogenetic reproduction. One possibility is that it permits polyploidy, which in turn is advantageous. A second is that parthenogenesis aids the colonization of islands, either real or ecological; it may also aid the utilization of transient habitats or the expansion of ranges. A third is that parthenogenesis saves time, for example, in finding mates, which is especially valuable with low population densities or short breeding seasons. A fourth is that, other things being equal, parthenogenesis doubles reproductive potential in any generation, because only female progeny are produced. These advantages are particularly cogent if parthenogenesis arises within sexually reproducing species. The advantages may also be applicable if parthenogenesis results from hybridization. In either case, however, physical and biotic factors will also be significant to the persistence of parthenogenetic taxa.

Suppositions about the ecological requirements for parthenogenetic reptiles span a considerable spectrum. On the one hand, the successful propagation of a clonal taxon supposedly requires a readily available habitat that is not used by the parental species; this keeps the newly arisen parthenogenetic taxon from being eliminated by competition with the parental species or by mating with parental males. Alternatively, because of their high heterozygosity, parthenogenetic taxa might in fact be bale to exclude the parental species from the habitats in which the parthenogenetic taxa arose and to which they are adapted. The present distribution of the parthenogenetic and biparentally reproducing taxa then represents a balance between the special adaptations that the parthenogenetic species received when it arose in the past and the ability of the biparentally reproducing taxon to evolve and adjust to slight but continuous changes in environmental conditions.

Drawing extensively on their own experience, *Wright* and *Lowe* (1968) showed that, although the habitats occupied by various all-female species are diverse, they share a common denominator. According to these authors, the habitats of the all-female species can be characterized by such descriptive terms a disclimax, ecotonal, transient, extreme, and perpetually disturbed. Such habitats, in the terminology of botanists, are "weed habitats," or habitats peculiar to weeds. Each of the parthenogenetic species of *Cnemidophorus* was regarded by Lowe and Wright as a weed, both geographically and ecologically.

The peculiarities of the habitats of all-female species generally make them poorly suited or unsuited for occupation by related bisexual species. Moreover, the present expansion of weedy habitat because of the activities of human beings leads to rapid distribution of all-female species at the expense of sympatric bisexual species. This, for instance, is observed within the city limits of Albuquerque, New Mexico, where *Cnemidophorus inornatus* is being ejected by the bigger and more aggressive parthenospecies *C. neomexicanus*.

The availability or absence of the "weed habitats" determines the final evolutionary success or failure of the newly arising parthenospecies of whiptail lizards. The South American all-female species called *Cnemidophorus lemniscatus*, which occurs in Amazonia mainly near settlements, has been considered a typical weed. The weed hypothesis may also fit the Central American all-female species of night lizards, *Lepidophyma obscurum*, which occurs on the southern border of the distributional range of the family Xantusiidae.

We feel that there is perhaps a more useful way to think about parthenogenetic species. Prior to trying to understand the patterns, if any, shown by parthenogenetic species as a group, it is necessary to consider the individual cases. For those taxa that are known to be hybrids, understanding of special features associated with parthenogenesis requires (1) comparative data for the parental species, and (2) some willingness to assume what ecological patterns such a hybrid would show. We assume that, in general, hybrids will be intermediate between their parents, although they may have an ecological spectrum that includes or even exceeds the ecologies of both.

Good comparative data for analysis are scarce. Many data have been accumulated by various authors on species of *Cnemidophorus* but these studies do not readily permit us to compare parthenogenetic taxa with their parental species. The conditions in parthenogenetic and closely related bisexual species of Lacerta have been analyzed. The pattern may be illustrated by considering the example of *L. armeniaca*, which is a hybrid derivative of *L. valentini* and *L. mixta*.

At present, *Lacerta valentini and L. mixta* are fully allopatric. *L. valentini* is confined to higher elevations and has a disjunct range. Known localities for *L. armeniaca* are largely allopatric to localities of each parental species although some overlap occurs along the edge of the range of each bisexual species.

Lacerta valentini occurs between 1900 and 3000 m above sea level. In contrast, *L. mixta* occupies lower elevations, between 800 and 2000 m above sea level. *L. armeniaca* is intermediate and occurs at elevations of 1400 to 2200 m above sea level. These elevational differences account for the general allopatry in this trio of taxa, because *L. armeniaca* and *L. valentini* are sympatric at the lower edge of the range of *L. valentini*, whereas *L. armeniaca* occurs with *L. mixta* near the upper limits of the range of the latter.

In the vegetational associations utilized, *Lacerta armeniaca* more resembles *L. mixta* than *L. valentini*. The latter inhabits montane steppes, whereas *L. mixta* and *L. armeniaca* occur in more wooded habitats. Both, however, persist in areas from which woods have been cleared, and it is in such an area that *L. mixta* and *L. armeniaca* are presently known to be sympatric.

In requirements of moisture; however, *Lacerta armeniaca* resembles neither of its parents, because it lives in much drier habitats than either. The tolerance of *L. armeniaca* for low humidity extends to its eggs, which survive better in low humidity situations than those of at least *L. valentini. (Danielyan,* 1971). In fact, the resistance of the eggs to drier conditions may be the critical aspect of the adaptation of *L. armeniaca*. This is especially true since the early, more sensitive, embryonic stages are shorter in the parthenogenetic species.

The pattern for *Lacerta armeniaca* is not atypical of the pattern of the other unisexual species of *Lacerta*. These species tend to be intermediate in elevational distribution, to be intermediate or resemble one parent in vegetational association, to live in relatively drier habitats, and generally to be allopatric to both parents. The exception is *L. dahli*, which inhabits relatively arid habitats, even though these habitats are slightly more humid than those occupied by *L. portschinskii*.

The unisexual species of *Lacerta* may be considered to be weed species. As a group, they clearly occupy extreme habitats, especially, with respect to aridity. This is true even for *L. dahli*. Indeed, many localities are occupied by two or more unisexual species, although all of the bisexual species of the *L. saxicola* complex are absent. In this sense, it seems possible that the occurrence of such extreme habitats at the time of origin of the unisexual species may have been a significant factor in their becoming successfully established.

Introduction of a historical component, however, permits us to suggest an alternative interpretation of these general habitat differences. Let us suppose that when the unisexual species arose all of Transcaucasia was relatively less humid than it is at present. Suppose also that the bisexual species had adjusted their ranges to remain as much as possible in the habitats they occupy today, but had also undergone changes in the allelic frequencies at many loci so that they could survive more arid conditions than they presently face. It would then be possible that hybridization resulted in highly heterozygous parthenogenetic species that were able to exclude the bisexual species from parts of what had been their ranges.

With later environmental changes, the bisexual species may have become adapted to slightly more humid conditions. Perhaps they could thus have maintained larger ranges as the environment shifted; they could also have escaped from competition with the parthenogenetic species. The parthenogenetic species, however, being relatively fixed genetically, would be expected to retain whatever adaptations they had when they arose—Such a view sees the extreme habitats of parthenogenetic species more as relics of the time of their origin than as weed habitats into which the parthenogenetic species moved to avoid competition with their parental species.

Another aspect of the biology of *Lacerta armeniaca* that reflects both its origin and its reproductive success is the number of eggs produced per clutch. This varies from three to eight, being generally five in *L. valentini*, from two to four in *L. mixta*, and generally from three to five in *L. armeniaca*. The data on clutch size in parthenogenetic *Lacerta* are, in our opinion, significant for a correct understanding of parthenogenesis in reptiles. When looked at as a group, parthenogenetic *Lacerta* appear to have larger clutch sizes than related biparentally reproducing species. Although this is correct, it also probably reflects the presence among the parthenogenetic taxa of two having *L. valentini* as a parental species; *L. valentini* produces the largest average clutch of the Transcaucasian taxa.

Similar data on clutch size have been provided for species of *Cnemidophorus.* In sympatric areas in the Santa Rita Mountains of Arizona, the clutch size of the unisexual *C. sonorae* is somewhat greater than that in the bisexual *C. tigris gracilis.* Most features of the reproductive cycle, including clutch size, are approximately the same for *C. inornatus* and for the sympatric parthenospecies *C. neomexicanus* in New Mexico. Two unisexuals, *C. sonorae* and *C. uniparens,* have mean clutch sizes of 4.1 and 3.2, respectively; the two sexual species, *C. tigris* and *C. inornatus,* have clutch sizes of 3.2 and 2.9, respectively. Do these data represent a special adaptation of parthenogenetic lizards, or are they simply consequences of hybrid origins ? We cannot answer the second part of the question, because the parental species of the parthenogenetic taxa are not known or only partly so *(C. uniparens, C. exanguis),* or because parental species have not been studied. Even if it is a consequence of hybridity, a larger clutch size is adaptive, other things being equal, both for the parthenogenetic *Lacerta* and for those species of *Cnemidophorus.* On the other hand, there are some indications of reduced fertility in the triploid agamid lizard, *Leiolepis triploida.* Many of the 30 females investigated had only one normally developed ovary, the second being completely or partly reduced the fertility of this triploid species requires further study. *Christiansen* (1971) speculated that the two-fold reproductive advantage per generation to *C. neomexicanus* might be leveled by analogously reduced viability of their eggs.

It is of interest to compare the ranges of the unisexual species with those of the parental species: Why do they not invade each other's habitats ? *Uzzell* and *Darevsky* (1975) suggested that, in addition to their high heterozygosity, special features of the adaptations of the unisexual *Lacerta* permit them to outcompete the bisexual taxa within the local range; at the same time, the special adaptations prevent them from moving into the range of the parental species. Another feature seems especially relevant in maintaining the allopatry of the two groups: whenever females of the parthenogenetic *Lacerta* occur in localities at which bisexual species also occur, the unisexual females occasionally become inseminated. In all cases so far studied, such matings produce sterile triploid females. It should thus be clear that in addition to competition and adaptation being limiting factors on the range of the unisexual species, the likelihood of leaving progeny capable of reproduction also decreases as the unisexuals expand into the ranges of the bisexuals. The bisexual species, in which the males clearly mate preferentially with conspecific females, will not be limited by matings of the males with unisexual females.

We have discussed the ecology of *Lacerta armeniaca* and the clutch size of parthenogenetic lizards to indicate what we see as a widespread problem in studies of natural parthenogenesis, namely, in establishing the nature of the proper comparison. Whether the data we discussed are of general significance or not, the problem is pervasive.

Although we feel that parthenogenetic species should generally be compared first with their parental species and only then with other parthenogenetic taxa, it is clearly important that they also be compared with related but nonparental sexual species. Because there are many zones in which parthenogenetic reptiles are sympatric with non-parental sexual species, the comparative ecologies of these forms also warrant investigation. Similar questions can be asked. What permits sympatry ? What maintains allopatry ?

In general, studies of the ecology of parthenogenetic reptiles and their sexual relatives have been descriptive. *Cuellar* (1979) removed adults of the parthenogenetic *Cnemidophorus uniparens* from a clearing to determine whether it would be invaded by

C. tigris. Such invasion would suggest that *C. tigris* had been excluded by competition, while repopulation by *C. uniparens* from adjacent areas would suggest that the observed allopatry at this locality reflected habitat-dependence. Initial results suggested that only *C. uniparens* reinvaded the depopulated area, but later results also seem to indicate parallel reinvasions by *C. tigris*. Consequently, it remains uncertain whether the allopatry between the two species at this particular site depends on competition, habitat, or on both.

EVOLUTIONARY POTENTIAL OF PARTHENOGENETIC REPTILES

Concept

The density of parthenogenetic *Lacerta* under natural conditions is very high, much higher than in closely related biparentally reproducing forms. A sampling showed 203 specimens of the unisexual *L. armeniaca* per kilometer of route sampled, 87 of *L. dahli*, 53 *of L. rostombekovi*, and 59 of *L. unisexualis;* in contrast there were 32 specimens of the bisexual *L. raddei nairensis* and 19 of *L. valentini*. Under particularly favorable conditions, the number of *L. armeniaca* can reach eight to ten individuals per square meter of rock surface. *Danielyan* (1970) calculated the number of individuals of the sympatric parthenogenetic species *L. armeniaca* and *L. unisexualis*, using a "random sampling" method that works well with small mammals and snakes. At different-times, *L. armeniaca* numbered 239 and 245 adult individuals per 0.12 km_2, whereas *L. unisexualis* numbered 149 and 189.

The present day ranges of parthenogenetic forms of rock lizards are much greater than was believed earlier. They far exceed the bounds of Armenia, extending to southern Georgia, northeastern Turkey, and northwestern Azerbaijan. In zones in which the distribution of parthenogenetic species falls within the range of biparentally reproducing forms, the former always outnumber the latter; in some of these zones, the parthenogenetic species tend to force out the sexual ones.

Such abundance and wide distribution of parthenogenetic species is usual, not only in reptiles but in other clonal vertebrates as well. Taxa that now appear to be scarce and restricted, such as *Leiolepis triploida*, *Lepidophyma obscurum*, and *Rhampholeon boulengeri*, may simply be less well known.

On the other hand, parthenogenetic reptiles are almost always terminal branches on evolutionary trees. Occasional diploids such as *Cnemidophorus tesselatus* have given rise to triploid derivatives, but no subgenera and genera are exclusively parthenogenetic. Certainly no subfamilies or families are so.

As judged by present abundance and distribution, parthenogenetic reptiles thus pose something of a paradox because their short-term success seems inconsistent with long-term success, as judged by phylogenetic restriction and limited genetic variability. This raises questions about the limitations and opportunities for evolution in parthenogenetic reptiles.

Limitations

Clonal reproduction imposes three kinds of limitations: (1) The lack of mechanisms for recombinations makes it difficult to avoid accumulating deleterious mutations in genomes, that is, clonal taxa are liable to be victims of *Muller's* (1964) ratchet. (2) The lack of a mechanism for recombination reduces clonal variation and keeps them from adapting to existing or changing physical and biological components of the environment. (3) When clonal taxa are sympatric with other congenerics, especially closely related ones or parental species, the males of these tend to mate with the clonal females.

Little is known concerning the accumulation of deleterious mutations in parthenogenetic reptiles. At present, it is impossible to produce an individual that is homozygous for any part of a genome of a parthenogenetic taxon, be it cistron, chromosome, or entire genome. Without such experimental manipulation, however, an assessment of the accumulation of deleterious mutations must remain speculative and indirect. This is in contrast to hybridogenetic taxa, in which it is readily possible to produce homozygotes for the hemiclonal genome. Indirect evidence for accumulation of deleterious mutants is also minimal. It is possible, however, that some of the inviability of ova of parthenogenetic *Lacerta* and *Cnemidophorus* and the partial sterility in *Leiolepis* represent not a phase in the selective improvement of parthenogenetic reproduction, but the beginnings of a mutational breakdown. The arguments given by *Muller* (1964) seem strong enough so that we would expect deleterious mutations to accumulate; they should be looked for.

Considerable thought has been given to the inability of parthenogenetic reptiles to track environmental change genetically, but relatively little information actually exists. Few reliable comparative physiological studies have used parthenogenetic reptiles, and none have been carried out in combination with studies of the appropriate biparentally reproducing species. For far too few of the taxa that have been shown to be hybrid have the parental species even been fully identified. Some studies, such as those by *Bulger* and *Schultz* (1979) on *Poeciliopsis*, indicate clearly that tolerance to environmental extremes reflects genomic composition.

In studying either of these genetic limitations, an important adjunct would be methods of dating the origins of clones. Various kinds of evidence can point toward greater or lesser relative age for a given clone as compared to others, but even good relative estimates are difficult.

In theory, the limitation imposed by mating with males of related biparentally reproducing taxa can be a significant feature of the biology of parthenogenetic reptiles. Such matings produce sterile triploid progeny in *Lacerta*. In *Cnemidophorus*, such matings also result in the addition of a genome; the reproductive potential of triploid and tetraploid hybrids between parthenogenetic and biparentally reproducing taxa in *Cnemidophorus* is not clear, but, in general, they have failed to form standing populations, even though such hybrids have been formed repeatedly. The triploid derivative of *C. tesselatus* is a conspicuous exception; it is probable that the other triploid *Cnemidophorus* species also arose this way, with their diploid antecedents having disappeared. Matings between parthenogenetic and biparentally reproducing *Lacerta* may be much restricted, however, even in areas of sympatry. In mixed populations of rock lizards, males always search for and mate with members of their own taxon. Thus, in the hybrid zone between *L. armeniaca* and *L. portschinskii* in northern Armenia (Stepanovan), females of *L. portschinskii* are rare, numbering approximately one per 50 females of *L. armeniaca*. Nevertheless, bisexual females had, on their bellies, numerous jaw marks (traces of mating), whereas such marks were seldom found on sympatric parthenogenetic females. Similarly, in a population of *L. dahli* and *L. portschinskii*, virtually every female of *L. portschinskii* had numerous sets of jaw marks made by males, whereas very few of the *L. dahli* had such marks, and those that were marked rarely had more than a single mark. Regardless of the frequency of such matings however, they probably impose a greater cost on individuals of the parthenogenetic taxon than on those of the biparentally reproducing one.

Given these real if largely undemonstrated limitations to the evolutionary success of parthenogenetic reptiles, there are, nevertheless, features of their biology that tend to mitigate these limitations, although their efficacy is as little explored as are their limitations.

The genome of each hybrid will contain approximately half the average load of delecterious mutations present in the population from which it was drawn. When the hybridization occurs, parthenogenetic taxa that are hybrid are about as out-crossed as it is possible to be; consequently the diploid or triploid parthenogenetic stocks should show initially some degree of heterosis. Thereafter, however, they will accumulate deleterious mutations and generally cannot shed these by recombination. In triploid parthenogenetic taxa, the accumulation of deleterious effects from recessive mutations should be slower, because there are generally three rather than two cistrons to be altered.

Each clone of a parthenogenetic taxon has a very limited ability to track environmental change by recombination and selection. This limitation may be offset in part by the high heterozygosity seen in those parthenogenetic taxa that are of hybrid origin or at least are hybrids and by the broader ecological spectrum that is available to them. Good comparative data from reptiles are, however, unavailable. We know that four species of parthenogenetic *Lacerta* usually live in more arid habitats than either to their parental species. When the ecologies of the two parental species differ markedly (as when *E. valentini* is one of the parental species), the parthenogenetic taxon usually is closer to one parental species. In many ways, this is reminiscent of the morphological data accumulated by *Parker* \(1979a, b) in that some features of their ecology are outside the range of both parental species, but usually closer to one than to the other, whereas other characters are intermediate, but still closer to one.

Opportunities

Several opportunities are available for species of parthenogenetic reptiles, although none appears to promise long-term survival.

(a) Parthenogenetic reptiles may revert to sexual reproduction : A general pattern for an evolutionary reversion to sexuality might involve a diploid clonal taxon to which a third and then a fourth genome is added through hybridization. The triploid taxon is also clonal, but the tetraploid reverts to sexuality, thereby forming a new tetraploid species. If such new tetraploid species were regularly formed, they would account for the general absence of parthenogenetic genera and subgenera. Tetraploid taxa of reptiles may not exist, however, and certainly they are not known in the genera in which parthenogenetic taxa are known.

A reversion to sexuality may come about in another way, but probably with less interesting evolutionary implications. Consider the tetraploid hybrid between *Cnemidophorus exsanguis* and *C. inornatus* described by *Neaves* (1971). It would presumably have two genomes from *C. inornatus*, one from *C. gularis*, and a fourth from a species not as yet identified. It is possible that in such a tetraploid, chromosomes from two inornatus genomes would synapse and undergo a normal meiosis, without participation of the remaining genomes, so that haploid *C. inornatus* ametes are formed. Such an occurrence seems unlikely because neither *C. uniparens nor C. velox*, each of which has two *C. inornatus* genomes, appears to produce haploid gametes. On the other hand, the haploid gametes of triploid *Rana esculenta* usually incorporate whichever chromosome set is present twice with normal recombination.

A third possible pattern of reversion to sexuality is suggested by *Cimino's* (1970) report of a diploid male fish with genomes from both *Poeciliopsis monacha* and *P. lucida* that was produced by a gynogenetic triploid with two monacha genomes and one from *P. lucida*.

Reversions to sexuality that involve a stable hybrid taxon probably are possible only at tetraploid or higher even-ploid levels. Productions of haploid gametes by triploids or tetraploids, which incorporate a complete genome from one species, is of evolutionary interest only to that species, and it may be significant only if that species is sympatric with the appropriate parthenogenetic one. There is no present evidence that reversion to sexuality offers much future to parthenogenetic reptiles.

(b) Parthenogenetic taxa may acquire additional genetic material : Production of haploid gametes that contain cistrons from more than one species seryes as a possible vehicle for introgression of alleles into a species. Such gametes may also help to form diploid trihybrid clonal species, as suggested by *Vrijenhoek* and *Schultz* (1974), although the recombination discussed by them may have taken place outside the context of a clonal taxon.

Acquisition of additional genomes by hybridization probably has been a significant pattern in parthenogenetic reptiles as well as in other clonal vertebrates, whether parthenogenesis preceded hybridity or followed it. Some parthenogenetic reptiles have been thought to be autotriploids, and autotriploidy has been occasionally reported in reptiles. Hall (personal communication), for example, observed one triploid out of some 400 *Sceloporus grammicus* examined and Witten (1978) observed one triploid male of *Amphibolurus nobbi*.

The clearest example of the acquisition of additional genomes is the triploid derivative of *Cnemidophorus tesselatus*, which has genomes from three species. Another triploid *Cnemidophorus* possibly also has genomes from three different species. Two triploids, *C. velox* and *C. uniparens*, have two genomes from *C. inornatus* plus one from some third species. It seems probable, although it is far from certain, that the second *C. inornatus* genome was acquired in a backross.

On the other hand, there are many instances in which additional chromosome sets have not led to the successful formation of a new parthenogenetic taxon. The triploid hybrids that result from fertilization of ova from parthenogenetic *Lacerta* are apparently all sterile females. The parthenogenetic *Cnemidophorus neomexicanus* apparently mates regularly, although infrequently, with *C. inornatus*. Nineteen male and six female hybrids have been reported. Their fertility has yet to be demonstrated, but these individuals appear to be sporadic hybrids; they are not members of a standing population. Whether just the proper sets of chromosomes will ultimately come together to form a new triploid parthenogenetic taxon is not clear, but seems unlikely.

Hybrids that should be or are known to be tetraploid have been reported for several species pairs of *Cnemidophorus*. Four probable hybrids between *C. inornatus* and *C. uniparens* were reported by *Wright* (1968) and two more by Cuellar and *McKinney* (1976); five of these six were males. Lowe et al.

(1970a) reported two male hybrids between *C. tigris* and *C. sonorae.* Other hybrids that should be tetraploid have been reported by Zweifel and by Axtell and Webb. One tetraploid that was female has been reported by Neaves this specimen laid eggs, but it is not known whether or not they were fertile or fertilized. The fertility of tetraploid *Cnemidophorus* has yet to be demonstrated.

Clearly, addition of genomes is a possibility for diploid parthenogens, and has the advantage of providing further genetic heterogeneity as well as additional buffering against deleterious mutations. Whether such addition is available to all diploid taxa, given the correct choice of genome, and whether there are attendant disadvantages in developmental complexity and gene regulation are unexplored questions.

New Clones of Parthenogenetic Taxa May Arise : Generation of new clones through gene mutation is probably a frequent occurrence in parthenogenetic reptiles. Many of these will be deleterious changes, but some may occasionally be advantageous. Several of the electrophoretic clones detected in *Cnemidophorus tesselatus* resulted from mutation or recombination. The restrictedness of these particular clones suggests that the events were either of recent origin, of no great adaptive advantage, or both. Many additional changes may have occurred that resulted in the loss of newly formed clones. Recombination, when it does occur, has the possibility on rare occasions of eliminating a deleterious mutation.

Apparently, a more frequent class of new clones is of karyological origin. Some of these also may be adaptive, but presumably many of them have little effect. Fritts (1969) believed that the modified karyotype that was widespread in his sample of *Cnemidophorus cozumela maslini* probably was adaptive; the alternative, that the single unmodified karyotype represents a recently arisen clone from renewed hybridization, cannot be excluded.

Without detailed studies, it will be difficult to determine whether most clones reflect multiple origins or changes after a single origin. *Parker* and *Selander's* (1976) study of clones of *Cnemidophorus tesselatus* is the best attempt to date to determine modes of clonal origins in reptiles. One can also imagine that sequences of electrophoretic and karyotypic modification among clones will provide information about clonal origins and the relative adaptive values of clones.

Parthenogenetic Reptiles May Change Their Ranges : While parthenogenetic reptiles may have little ability to track changing environments genetically, they clearly can often adjust their ranges to stay within the environments to which they were initially adapted. Probably the most general case is the gradual expansion of the range of parthenogenetic taxa from time to origin to present. Several particular cases are reasonably well-established. The best documented one concerns *Cnemidophorus uniparens,* which has apparently been expanding its range at the expense of *C. inornatus* for many years. A similar shift is occurring as *C. neomexicanus* expands, also at the expense of *C. inornatus.* These changes are associated with decrease in the area of grasslands; to what extent there is a competitive component in these changes and to what extent they depend solely on the changing habitat is unclear. From his work near Obidos in the Amazon Basin of Brazil, *Vanzolini* (1970) has reported a change from the sexual *C. lemniscatus* to the parthenogenetic taxon also referred to as *C. lemniscatus.*

Other, perhaps more intriguing, cases of changing ranges for parthenogenetic reptiles are confounded with possible range changes in related sexual species. Range changes of sexual species have surely occurred. For example, *Lacerta valentini* occupies disjunct relatively high regions of Transcaucasia that were uninhabitable during the Wurm glaciation. Clearly, its range has changed since that time. Furthermore, distribution records for two of the sets of parental species of parthenogenetic *Lacerta* in the Caucasus show them to be currently separated by considerable gaps. The hybridizations that gave rise to the clonal species must have occurred when the sexual species occupied ranges different from those occupied today.

In their study of *Cnemidophorus tesselatus* at a locality some 350 km northwest of the present range of *C. septemvittatus*. *Parker* and *Selander* (1976) observed an allele for glucosephosphate isomerase that apparently was derived from *C. septemvittatus*. This allele is absent from the populations of *C. tesselatus* that are sympatric with *C. septemvittatus*. Another *C. septemvittatus* glucosephosphate isomerase allele occurs in *C. tesselatus* some 400 km to the north of the range of *C. septemvittatus*. It is absent in populations of *C. tesselatus* sympatric with *C. septemvittatus* and in three intervening populations. We cannot tell whether these distributions of alleles reflect changes in the range (and thus allelic distribution) of *C. septemvittatus* or changes in the distribution of clones of *C. tesselatus*, or both.

Neaves (1969) was unable to account for the electrophoretic phenotypes that he observed in several parthenogenetic species of *Cnemidophorus*. He believed that certain alleles might have come from Mexican species. If so, either the parthenogenetic species or the parental species or both have changed their ranges.

Regardless of whether or not the environmental shifts result from human activities or from long-term climatic trends, the ability of parthenogenetic reptiles to change their ranges certainly provides the opportunity and time for new hybridizations and generation of new clones.

Selection May Act Differentially on Clones of a Parthenogenetic Species : Finally, there is the possibility of evolution through selection among clones. Unfortunately, little is known about the possibilities in reptiles. The most probable source of a multiplicity of clones is their origin from hybridization, which may be relatively frequent when it occurs at all. It seems certain that the greater the number of gene loci it is possible to examine, the greater will the clonal diversity in any parthenogenetic taxon appear until the actual finite number of clones begins to be detected. That some of these clones should be better adapted to specific environmental conditions seems highly probable, and there should, therefore, be a gradual sorting out of a clonal frequencies in space.

Even more interesting is the possibility that new clones will be generated within a taxon and that selection will occur among these. Such new clones could arise either through mutation or recombination. Those that arise through mutation, however, are likely to be less well adapted than their progenitor clones, because most mutations apparently are deleterious. It is, nevertheless, possible that occasional improvements will occur by chance. To what extent parthenogenetic reptiles can afford the generation of many deleterious clones through mutation in order to find the rare better adapted one is not known.

PARTHENOGENETIC REPTILES AS SPECIES

The usual definition of a biological species is based on the criterion of the actual or potential ability of individuals to cross with no reduction in progeny in the third generation. This definition implies the existence of two sexes and hence is inapplicable to all-female species. As *Mayr* (1963a) pointed out, even the word "population" can for this reason hardly be applied to the community of reproductively isolated females.

In view of this obvious limitation on the biological species concept, some authors have proposed that various kinds of species can be recognized, distinguishing the kind by mode of reproduction. *Cain* (1954), for instance, proposed the name "agamospecies" for forms lacking sexual reproduction. He noted, however, that the criteria for their specific independence must be the same as in the case of sexually reproducing organisms and that agamospecies must be placed in a taxonomic hierarchy beside their closest relatives. *Simpson* (1961) and *Mayr* (1963a) concurred.

Application to "agamospecies" of physiological criteria, such as the existence of genetic incompatibility in hybrids or the development of antihybridization mechanisms, poses problems, however, because the definitions of these criteria are closely connected with the existence of biparental reproduction. That is why *Darevsky* (1967) suggested that the taxonomic rank established for parthenogenetic forms should emphasize the fact that reproductive isolation exists, rather than the concrete mechanisms of its realization. This follows from the premise that there are various mechanisms of reproductive isolation on the genetic level, but that these mechanisms are of one and the same order on the evolutionary level because they insure the independent evolution of the species. This suggests that a systematist should in practice make no distinction between unisexual and bisexual species, but should evaluate them, as it were, with one morphological measure. This view is fully corroborated by the fact that many species of parthenogenetic lizards were described long before their parthenogenetic nature was discovered.

Subspecies in sexually reproducing taxa are most interesting biologically when it is uncertain whether the taxa are conspecific or not; such subspecies are frequently characterized by a high concordance in change of character state. In many other cases, descriptions of subspecies conceal variation or at least depend on our ignorance of it. The subspecies category in unisexual species is often relative, just as it usually is among sexually reproducing organisms. If they have cohesive ranges, subspecies of parthenogenetic taxa are reproductively isolated from each other. In this respect they do not differ from species, because they also lack the possibility of intraspecific hybridization. In this view, it would be best to treat *Cnemidophorus cozumela maslini Fritts*, 1969, as an independent species, or not to recognize it at all.

Peters (1971) suggested that all unisexual individuals that possess similar karyotypes and that originated from one known species should be regarded as forming one "agamospecies." An increase of ploidy or the fact that two unisexual forms with identical karyotypes originate from different species may also serve as the ground for distinguishing species. Correspondingly, subspecific status should be reserved for unisexual forms with similar karyotypes originating from different subspecies within one species. This suggestion is based on the idea that unisexual species appear spontaneously without hybridization. This approach could be useful in cases of the hybrid origin of unisexual species, if any parental species had distinctive subspecies, and more than one of the subspecies had been independently involved in the origin of a parthenogenetic taxon.

Peters (1971) stressed the idea that, unlike normal bisexual species, uniparental forms do not fir his criterion of "uniqueness in space and time." Nevertheless, he rightly considered that "agamospecies" represent as real a category as the bisexual ones do. Thus, it is expedient to apply binomial nomenclature when distinguishing them in taxonomy.

The International Code of Zoological Nomenclature gives no special recommendations about the taxonomic rank of parthenogenetic forms that arose by hybridization. Some systematists have proposed providing the Code with definitions to be applied to species of hybrid origin. Mayr, for instance, recommended supplementing Article 25b with the following wording "a name given to a hybrid species cannot be applied to either of the parental species." The amendment would be useful in the taxonomy of parental species of hybrid parthenogenetic taxa. The difficulty is that in addition to species that are of hybrid origin, there often occur in nature single hybrid individuals, and it may not always be possible to draw a distinction between the former and the latter. In the event of complete sterility of hybrids, as is the case with the triploid rock lizards, the question of their taxonomic rank has not been raised at all. The fertility of some hybrids of *Cnemidophorus*, as well as of some other groups of lizards may, however, present some difficulties.

RESEARCH PROBLEMS IN PARTHENOGENETIC REPTILES

Our review of published work on parthenogenetic reptiles discloses many problems, the most persistent one being the absence of basic data on numerous taxa. Relatively few have been studied extensively, and even for those genera, different kinds of data have often been collected. We have tried to present information on parthenogenetic reptiles and on related biparentally reproducing taxa. The review has also been hampered by the absence of reliable comparative data on the sexually reproducing taxa. The most immediate question, however, is probably whether all of the parthenogenetic taxa are hybrids or not. Identification of the parental species of those that are hybrids is an important subsidiary problem.

Many studies of parthenogenetic reptiles have emphasized their ecology. What kinds of distribution patterns do they have as a group ? What are their ecological interactions with their biparentally reproducing relatives ? An almost unexplored area concerns their molecular biology, not only for the parthenogenetic reptiles but for all clonal vertebrates. In molecular terms, what is the basis of the meiotic modifications shown by clonal taxa ? For those that contain genomes of more than a single species, how is development controlled ? To what extent and how are the differing genomes regulated ? How do allelic products from perhaps quite dissimilar cistrons interact ?

If parthenogenetic reptiles evolved by slow elimination of males from a population of a single biparentally reproducing species and if they were neither hybrid nor of hybrid origin, it would be appropriate to consider the various parthenogenetic forms as a group in order to look for commonalities in their evolution and biology. On the other hand, if they arise as a result of hybridization or at least have a hybrid genomic composition, the appropriate comparisons clearly ought to be with the parental species. It seems to us that the ecology of the parthenogenetic Caucasian rock lizards shows less evidence of their occupying weed habitats or ecotones than does the ecology of the parthenogenetic species of whiptails simply because comparable data are available on both the parthenogenetic forms and the parental species. Even most parthenogenetic rock lizards, however, occupy more arid habitats than either of their parental species. One of the features that permits them to do this is a greater tolerance of their eggs to aridity. A more rapid developmental rate in the early embryonic stages is also probably significant. The molecular basis of such differences in development is unknown. Is it simply a result of genomic heterozygosity ? Does it represent specially adapted sets of alleles within these genomes ? It seems to us that only comparative studies of parthenogenetic taxa and parental species can even permit the proper questions to be posed.

Although the hypothesis that parthenogenesis often arises as a consequence of hybridization between distantly related taxa was relatively well-accepted for a while, it later came into disfavour because many hybridizations did not result in clonal taxa and because, in plants, many clonal taxa were clearly not of hybrid origin. Nevertheless, all well-studied cases of parthenogenesis in reptiles have provided evidence that is best interpreted as indicating a hybrid origin. It has been difficult, however, to provide a cytogenetic argument as to how hybridization could produce clonal reproduction, such as seen in the reptiles. Perhaps the question is posed on the wrong organizational level. As far as we know, meiosis does not occur in a haploid state; meiosis in haploids is almost a contradiction in terms. If synapsis does not occur between homologues (and it probably does not in clonal vertebrates with the possible exceptions of *Lacerta armeniaca* and *Poecilia formosa),* a regulatory endodopulication must occur prior to meiosis. Why synapsis should not occur or why an endoduplication should, are molecular rather than cytogenetic questions.

This brings us back to the question of hybridity and hybrid origin. It is difficult but not impossible to imagine that sufficient molecular evolution would occur within a species so that no homeologues synapse. If a single dominant mutation ultimately producing that effect should occur, parthenogenesis would possibly arise immediately, because the necessary endoduplication might be achieved automatically by suppressing a cell division or by completing an additional mitotic division so that some biochemical threshold necessary to the onset of meiosis was attained. It is easier for us to imagine a show divergence of genetic control over meiosis in independent lineages. The consequence would be that, in a diploid hybrid, chromosomes of neither set would be able to synapse with their homeologues, although meiosis within either lineage might be completely normal. These questions must be investigated, however, at the molecular level.

So far, no parthenogenetic vertebrate has been synthesized in the laboratory, but few attempts have even been made. The success of future attempts may depend on careful selection of individuals to cross, or it might require selective "improvement" of clonal reproduction. Alternatively, it might be an almost automatic outcome of appropriate interspecies crosses.

Finally, in addition to the practical question of whether or not to recognize parthenogenetic lineages of reptiles as species, there is the matter of discerning proper names for them. It is obviously unsatisfactory to use the same name for biparentally reproducing and parthenogenetic lineages or for diploid and triploid populations. Often there are unanswered questions about the names for the parthenogenetic taxon itself. Thus, for example, *Mertens* (1966) used the name *Brookesia spectrum boulengeri* for the taxon we refer to as *Brookesia affinis,* but the three syntypes of *Rhampholeon boulengeri Steindachner,* 1911, include two males. These questions are probably compounded by the synonymies recognized for some clonal taxa. If it should turn out, for example, that *Hemidactylus garnotii* is of hybrid origin, what species are available as possible parental species ? The now lost holotype of *Doryura guadama* Theobald, 1868, was reported to have femoral pores. Was it merely another of the specimens of *H. garnotii* that has pores, or was it possibly a male of a parental species ?